MARINE POLLUTION

**PROCEEDINGS OF THE ARAB SCHOOL OF SCIENCE
AND TECHNOLOGY**

A. H. El-Abiad (editor)
Power Systems Analysis and Planning

G. Warfield (editor)
Solar Electric Systems

T. Kailath (editor)
Modern Signal Processing

R. Descout (editor)
**Applied Arabic Linguistics
and Information and Signal Processing**

E. E. Pickett (editor)
A. Habal and F. Abosamra (co-editors)
Atmospheric Pollution

P. D. T. O'Connor (editor)
Reliability Engineering

J. Albaigés (editor)
Marine Pollution

FORTHCOMING

R. Risebrough (editor)
Pollution and the Protection of Water Quality

P. Mackay (editor)
Computers and the Arabic Language

MARINE POLLUTION

Edited by

J. ALBAIGÉS
Centro de Investigación y Desarrollo
C.S.I.C., Barcelona, Spain

⊙HEMISPHERE PUBLISHING CORPORATION
A member of the Taylor & Francis Group

New York   Washington   Philadelphia   London

MARINE POLLUTION

Cover design by Sharon DePass.

1 2 3 4 5 6 7 8 9 0   B R B R   8 9 8 7 6 5 4 3 2 1 0 9

Library of Congress Cataloging-in-Publication Data

Marine pollution / edited by J. Albaigés.
      p.   cm. — (Proceedings of the Arab School on Science and Technology)
  Includes index.
  1. Marine pollution — Environmental aspects. I. Albaigés, J.
II. Series.
QH 545.W3M356   1989   89-1706
628.1'68 — dc19   CIP

ISBN 0-89116-862-1

Contents

Contributors   vii
Preface   ix
The Arab School of Science and Technology   xiii

MARINE POLLUTION: AN INTRODUCTION
J. Albaigés   1

**CHEMICAL CHANGES IN THE MARINE ENVIRONMENT:
ORIGINS AND CONCEPTS**
R. W. Risebrough   11

TRANSPORT OF MARINE POLLUTANTS
E. Özsoy and Ü. Ünlüata   35

**TRANSFORMATION OF POLLUTANTS IN THE MARINE
ENVIRONMENT**
J. Albaigés   97

THE BIOLOGICAL EFFECTS OF MARINE POLLUTANTS
B. L. Bayne   131

MONITORING AND SURVEILLANCE SYSTEMS
R. W. Risebrough   153

SOURCES AND TRANSPORT OF OIL POLLUTANTS
IN THE ARABIAN GULF
M. I. El Samra   177

TRACE METALS AND PETROLEUM HYDROCARBONS
IN SYRIAN COASTAL WATERS AND SELECTED
BIOTA SPECIES
F. Abosamra, R. Nahhas, N. Zabalawi, S. Baba,
G. Taljo, and F. Kassoumeh   185

INSTRUMENTAL TECHNIQUES FOR THE ANALYSIS
OF TRACE POLLUTANTS
J. O. Grimalt   201

SAMPLING, SAMPLE HANDLING, AND OPERATIONAL
METHODS FOR THE ANALYSIS OF TRACE POLLUTANTS
IN THE MARINE ENVIRONMENT
J. O. Grimalt   223

PRINCIPLES AND METHODS IN ENVIRONMENTAL
MANAGEMENT OF COASTAL MARINE WATERS
A. Randløv S. Ø. Dahl, and E. Poulsen   279

OCEAN WASTE MANAGEMENT
M. A. Champ and I. W. Duedall   305

INTERNATIONAL PROGRAMS FOR MARINE
POLLUTION STUDIES AND THE ROLE OF THE
OCEANS AS A WASTE DISPOSAL OPTION:
AN OVERVIEW
G. Kullenberg   347

Index   361

Contributors

F. Abosamra
Institute of Chemistry and Biology
Scientific Studies and Research
   Center
Syria

J. Albaigés
Centro de Investigación y Desarrolo
C.S.I.C.
Jorge Girona Salgado, 18-26
08034-Barcelona, Spain

B. L. Bayne
Institute for Marine Environmental
   Research
Prospect Place
The Hoe
Plymouth PL1 3DH, UK

M. A. Champ
Cross-Disciplinary Research
Division, Engineering
National Science Foundation
Washington, D.C. 20550, USA

M. I. El Samra
Marine Sciences Department
University of Qatar
Qatar

J. Grimalt
Centro de Investigación y Desarrollo
C.S.I.C.
Jorge Girona Salgado, 18-26
08034-Barcelona, Spain

G. Kullenberg
IOC, UNESCO
7, Place de Fontenoy
75700-Paris, France

R. Nahhas
National Oceanographic Committee
Supreme Council of Sciences
Syria

E. Özsoy
Institute of Marine Sciences
Middle East Technical University
P.O.B. 28
Erdemli, Icel, Turkey

A. Randlov
Cowiconsult
45, Teknikerbyen
DK-2830 Virum, Denmark

R. Risebrough
The Bodega Bay Institute
2711 Piedmont Avenue
Berkeley, CA 94705, USA

Preface

الآيـة ٢٠ مـن سـورة ٢١ الأنبـيـاء

''. . . and from water came all life.''
THE HOLY QUR'AN: Ayat 30, Sura xxi Anbiya

This book on Marine Pollution contains the lectures delivered at the Arab School of Science and Technology held in Zabadani, Syria, in August 1987. Past editions dealt with Pollution and the Protection of Water Quality (1982) and with Atmospheric Pollution (1985), so that the main topics in the field are now covered in this series of proceedings published by Hemiphere. This is consistent with the aim of the school of providing up-dated overviews of the latest advances in different fields of science and technology for postgraduate specialists of the Arab countries and incorporating them into the open literature for general knowledge.

Although Marine Pollution is a problem we have been aware of for a long time, the scientific basis for its assessment, prediction of effects, and management are subject to continuous development; there are still some gaps in our knowledge. Thus, from the pioneering manuals that focused the attention on the variety of pollutant sources for the marine environment, the attention has turned, recently, to the more dynamic aspects of the cycling of pollutants and to the assessment of their long-term effects. In fact, the Comprehensive Plan for the Global Investigation of Pollution in the Marine Environment (GIPME) encompasses a series of stages, some of them still not fully supported by research results, such as: mass-balance determinations for the different pollutants, that require information on boundary fluxes and oceanographic processes; pollution assessment, which depend on the possibility of evaluation of the effects of contaminants upon the different "hierarchical" levels of the marine ecosystem; and regulatory actions, that include the development of regulatory standards, the implementation of measures and the establishment of surveillance plans.

In this context, the selection of topics for the school program was a challenge, regarding the diversity of subjects and disciplines involved. The compromise was a wide coverage of topics that are fundamental for understanding the current problems on marine pollution, with a sufficient depth of treatment to satisfy the needs of both postgraduate students and the generally concerned or interested people.

In the first two chapters of the book, written respectively by Albaigés and Risebrough, general concept that have emerged from the modern approach to the problem are introduced. One of these concepts related to the biogeochemical cycle of pollutants, so that the physical, chemical, and biological processes affecting the distribution and fate of contaminants in the marine environment are dealt with in the following chapters, by Özsoy and Ünlüata and Albaigés. The first authors review the mathematical models for the simulation of pollutant transport processes whereas the latter gives a more experimentally based explanation of the transformation of pollutants in the marine environment.

One of the aspects exhibiting most development in recent years is that related with the assessment of the "environmental stress" of the different levels of the marine ecosystem. A comprehensive presentation of this topic is made by Bayne.

Two chapters, preceded by an overview on monitoring systems by Risebrough, exemplify field studies carried out in the Arabian Gulf and in the Syrian coast by, respectively, El Samra and Abosamra et al. As the design of monitoring networks require a search for common analytical approaches, an extensive review by Grimalt illustrates why the methodologies being employed for determining levels of contaminants are far from unique or universal. Here is a crucial task for the analytical chemists in selecting the most appropriate sampling, analytical methods, and techniques for each problem.

The last step in a marine pollution program is the environmental management strategies for marine water. An appropriate presentation of the subject requires a disctinction between the coastal zone and the zones that are truly marine. Two chapters by Randløv et al. and Champ and Duedall deal, respectively with these specific topics.

Finally, closing the book is the chapter by Kullenberg on International Programs for Marine Pollution studies, including the aforementioned CIPME Program. Oceans cover 71 percent of our planet and it is obvious that marine pollution represents a question not only of national but also of international concern.

At this stage it is worth expressing my sincere gratitude to all authors for their valuable contributions and understanding of the editor's duties. Drs. A. Habal, R. Nahhas, and F. Abosamra, among other scientists from Syrian Research Centers, contributed significantly with their ideas and friendship to the success of the school. Special thanks are due to Mr. A. Rifai, the General Secretary of the School and to his efficient staff who provided with extreme cordiality continuous support before, during, and after the sessoin. Finally, J. Estremera and N. Tur from our center, offered their time in the preparation of the material for publication.

It is our hope that this initiative will continue and contribute not only to the exchange of the scientific and technical knowledge but will also, on the basis of the international cooperation, promote nations to coexist in this watery planet whirling through space.

J. Albaigés

Arab School of Science and Technology

SPONSORS

Scientific Studies and Research Center, Syria
UNEP-Coordinating Unit for the Mediterranean Action Plan

GENERAL SUPERVISORY COMMITTEE

A. R. Kadoura
Assistant Director General, Sciences
UNESCO

A. W. Chahid
General Director
Scientific Studies & Research Center

A. Shihab Al-Din
Kuwait Institute for Scientific Research

SCHOOL GENERAL SECRETARY

M. A. Rifai
Arab School of Science & Technology

SCIENTIFIC COMMITTEE

A. Habal
Scientific Studies & Research Center, Syria

F. Abosamra
Scientific Studies & Research Center, Syria

H. Kharouf
Department of Zoology
Faculty of Science
University of Damascus, Syria

R. Nahhas
Supreme Council of Sciences
Scientific Studies & Research Center, Syria

ORGANIZING COMMITTEE

N. Nahas Armanazi
Session Coordinator
Arab School of Science & Technology

G. Akbik
Administrator, International Relations
Arab School of Science & Technology

A. Kouatly
Administrator, Local Arrangements
Arab School of Science & Technology

A. Bash
Public Relations Director
Scientific Studies & Research Center

OBJECTIVES

Through its program, the School aims at reaching the following objectives:

1. Bring Arab scientists up-to-date on the latest scientific developments, and state-of-the-art technologies in the field under study.
2. Facilitate direct contact among Arab scientists, and between Arab scientists and their international counterparts, to create a favorable atmosphere for scientific and technological cooperation.
3. Make Arab scientists working abroad aware of their home countries resources, and encourage them to contribute to their scientific and technological development.
4. Provide the international scientific community with an overview of the scientific activities in the Arab countries through the publication of each School session proceedings which cover the scientific and technological developments in the Arab world.

ABOUT THE ARAB SCHOOL

The Arab School of Science and Technology, a pan-Arab, non-profit organization headquartered in Damascus, Syria, was founded in 1978 by the initiative of the Kuwait Institute for Scientific Research (KISR) and the Scientific Studies and Research Center (SSRC) and the Supreme Council of Sciences (SCS) in Syria to provide a high level continuing education program to Arab scientists in fields that are judged crucial to the development of the Arab countries.

Since its establishment, the School has dealt with four major topics: Electronics, Energy, Environment and Informatics. The School has plans to expand into other areas of specialization, such as the transfer, adaptation, and development of technology in the Arabian World.

The School has attempted to create for each topic, a regular forum of scientific exchange in chosen areas of the Arab World. It has succeeded in establishing one for Electronics in Syria, and another one for Energy in Kuwait.

In addition, the Arab School of Science and Technology strives to increase and foster cooperation among Arab and non-Arab specialists by providing a framework for a close interaction in any given technological field.

Marine Pollution: An Introduction

J. ALBAIGÉS
Department of Environmental Chemistry
CID (CSIC)
Jorge Girona Salgado, 18-26
08034–Barcelona, Spain

The present book on Marine Pollution contains the lectures delivered at the Arab School of Science and Technology held in Zabadani, a resort area near Damascus, Syria, in August 1987. Previous editons of this Summer School in the field of the Environment dealt with Pollution and Protection of Water Quality and with Atmospheric Pollution.

Pollution is a concept that has been introduced in our social life to broadly indicate that some resource has become inappropriate for a certain purpose. The concern for the pollution of the oceans appeared in the late 50's when the studies of radioactive tracers of nuclear explosions, namely 90Sr and 137Cs, revealed a significant increase in seawater, marine biota and sediments, providing an early picture of the global dispersion of harmful substances through the atmosphere to the marine environment. Later, the reproductive failure of marine birds in the California coast was the first evidence of a biological effect derived from the world-wide dispersion of a contaminant, in this case the pesticide DDT. Although at present the direct impact of marine pollution on global fisheries and on human health has not yet been recognized, there are sufficient signs of serious damage to local resources and even back to man (e.g. the Minamata tragedy) as to destroy the myth that the oceans might be considered as one huge garbage dump for all man's wastes. Indeed, the oceans that hide the keys of our begining cannot be regarded as a sink of unlimited capacity for all the refuses produced by human progress.

The man's contribution to the oceans can be of two kinds. He can add more of something which is already present, for example nutrients into coastal waters, in which case the danger is of exceeding the ability of the oceans to cope with the material, or he can add substances totally extraneous to the system, in which case the problems are more difficult to anticipate.

The Intergovernmental Oceanographic Comission (IOC) described the marine pollution as the introduction by man, directly or indirectly, of substances or energy into the marine environment (including estuaries), resulting in such deleterious effects as: harm to living resources; hazards to human health; hindrance to marine activities including fishing; impairing the quality for use of seawater and reduction of amenities.

The assessment or determination of these damaging effects requires a precise knowledge of the functioning of the system. It is clear that the input of a new substance into the sea cannot be regarded as the pouring of a spoon of sugar into a cup of coffee. The marine environment is neither a static nor an homogeneous system, but a highly dynamic, both physically and chemically, and a well compartmentalized one. We also need to remind ourselves that the oceans

1

continually exchange chemical burdens with another dynamic system, the atmosphere.

THE MARINE ENVIRONMENT

The oceans can be divided into regions both vertically and horizontally. The vertical divisions respond to clearly marked natural gradients, whereas the limits of the horizontal ones are usually much less defined.

The parameter denoting the horizontal regions is the primary productivity. As shown in Fig. 1A we can identify enriched and impoverished zones. The first ones are usually closer to the coast and are determined by a higher phytoplankton growth, supported by a supply of nutrients brought into these waters by continental streams and coastal upwellings. Obviously, these are the most conflicting areas because they contain the major fisheries of the world and at the same time are those of higher interference by man.

If most of the coastal areas can be considered as the marine orchards the oceanic central gyres represent the marine deserts. These are regions where the surface waters show a relatively stable structure, with little exchange with others water masses.

The primary productivity upon which life in the whole marine system depends takes place in the upper layer of the ocean, the euphotic zone, where photosynthesis may occur. According to the differences in types and rates of the carbon turnover, the ocean is also divided into well defined vertical layers. From the air-sea interface downward, the first is the sea surface film. Although the composition of this film, of a few microns thick, is a matter of some dispute, there is a general consensus that it contains hydrophobic and surface-active materials from both natural and anthropogenic sources in much higher concentrations they could achieve in the water column. It plays a significant role in concentrating hydrophobic compounds carried by aerial transport.

The next natural division, the euphotic zone, extends from the surface film to the thermocline. Most of the reactions of interest to the environmental chemists, caused either primarily or secondarily by the penetration of light into the water, occur in this upper layer.

The waters below the thermocline have basically the properties carried down from the surface, modified by biological and chemical reactions occurring in situ, by the addition of some materials through advection and diffusion, and of others by sedimentation from the surface layers. Although, the organic matter in this zone is low and relatively uniform, there is a net downward flux of organic debris, which may also include anthropogenic substances, from the euphotic zone fo the sea floor. The resulting input to the sediments is governed by the oxygen gradient in the water column. In any case there is a clear coupling of all these vertical regions with the sediment as evidenced when comparing the organic carbon distributions in both the surface layer and the sediment (Figures 1A and B).

The greatest part of the chemical and biological activity in the deep waters occurs in the water-sediment interface. This is particularly relevant from the environmental point of view because reaction products in the interstitial waters may then escape into the water column. On the other hand, it is also well known, for example, the role that changes in redox potential has in mobilizing metals from sediments to the overlying waters. However, the fate of the pollutants in the marine environment deserves a separate consideration.

2

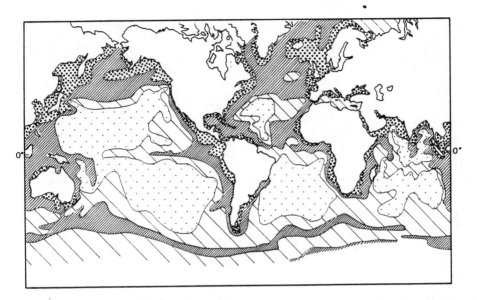

FIGURE 1A. Distribution of primary production in the world oceans
(adapted from Bolin et al., 1979). From   100 mg C/m day
to   250 mg C/m day

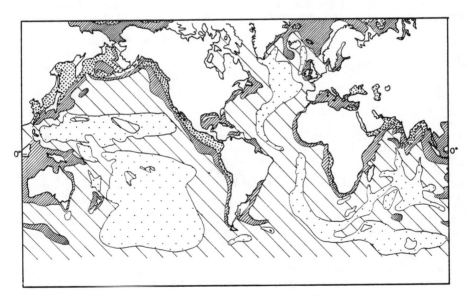

FIGURE 1B. Distribution of organic carbon in the surficial sediments
of the world oceans (adapted from Premuzic et al., 1982).
From < 0.25% O.C.   to > 1% O.C.

THE FATE OF MARINE POLLUTANTS

As soon as a pollutant is released into the sea it is subjected to a series of diverse processes that cause distribution of the product into the environment and, at the same time, produces changes in its physical and chemical characteristics. All these processes, that occur simultaneously and are interrelated define the biogeochemical cycle of the pollutant. The changes are continous and the reactions create a dynamic state with a great number of variables. Furthermore, the nature and the extent of the changes that occur depend upon the combined effects of the particular set of physical, chemical and biological environmental conditions existing at the moment or at the area of the input (temperature, winds, currents, nutrients, pE, etc...). We do not know all the reactions that occur, nor the factors that initiate or impinge upon such changes, nevertheless some general patterns can be drawn. As an illustration, diagramatic representations of the major processes involved in the fate of two representative pollutants, namely **petroleum** and **lead**, are given in Figures 2A and B.

Petroleum hydrocarbons, besides the aesthetic and ecological problems that may cause in the marine environment, e.g. in oil spills, can be a source of carbon for the low levels of marine life, thus becoming a part of the carbon cycle. On the contrary, lead being more conservative and not an essential trace element will accumulate in the marine reservoirs producing other detrimental effects.

Dispersal and destruction processes contribute to the cycling of these pollutants. The study of the routes of inputs and movement of chemicals through the system, the rates of these inputs and movements, the reaction pathways of chemicals and both the long and short-term reservoirs where compounds accumulate, that is the four R's (routes, rates, reactions and reservoirs) of the biogeochemical research, provides knowledge of where, for how long and in what form marine resources are exposed to a given pollutant.

As shown in Fig. 2A, following the bulk discharge of oil from any source into the sea, a surface film (slick) is usually formed. The hydrodynamics of the slick behaviour has been well studied so that its spreading and transport on the open ocean can be approximately predicted (Jeffery, 1973). **Evaporation** and **dispersion** by solution or emulsification are the most rapid processes to occur, decreasing in rate as time proceeds and the most easily volatile or soluble materials are removed from the spilled oil. These processes act during the first few weeks after the spill. Obviously, agitation by surface waves increases the rates as well as the formation of wind-produced sprays and bursting bubbles, that lead to a preferential transfer of surface active compounds into the atmosphere. As evaporation proceeds, it contributes to the formation of thick residues and, eventually, to tar balls that may become denser, increasing the possibility of dispersion into the water column and subsequent sedimentation.

Competing with dissolution and droplet or colloidal dispersions is the formation of **water-in-oil** emulsions, commonly referred to as "chocolate mousse", since the Torrey Canyon incident. They usually involve the more viscous and asphaltic fractions and are formed naturally on a dynamic sea surface, being stable for several months. Emulsification may also be promoted by naturally occuring or produced surface active materials.

Petroleum dispersed into the sea increases the chances of it being adsorbed on or mixed with particulate matter (sand, silt, clays, organic debris, etc...). **Adsorption**, then, may play an important role in the vertical transport and sedimentation, especially in shallow coastal waters. Formation and

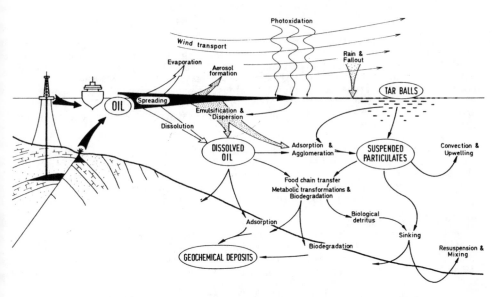

FIGURE 2A. Diagramatic representation of the fate of petroleum
spills at sea (from Albaigés 1980).

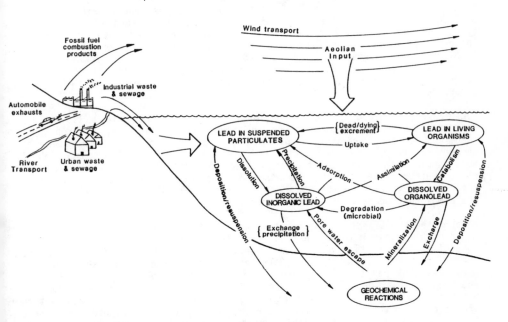

FIGURE 2B. Diagramatic representation of the cycling of lead in
the marine environment (adapted from Nriagu 1978).

recombination of petroleum particles can occur in several other ways, for example, by **uptake by organisms**. Zooplankton can take up dissolved or dispersed hydrocarbons that they excrete unmodified in the feces.

Degradative processes are less understood compared with physicochemical dispersion, and even qualitative predictions are difficult to make without good field data. They are mostly oxidative in nature; so, since much of the petroleum material floats on the surface, a major portion of the oxidation occurs there. Additional oxidation reactions probably occur in the upper part of the water column, where dissolved oxygen is available. The oxidation processes, which are generally long-term processes, are either chemical or biological in nature.

Auto-oxidation processes are brought about by the atmospheric oxygen and are enhanced by the photochemical action of sunlight and the catalytic action of photosensitizers present in the sea water or in petroleum. Many of the compounds formed are surface active and may enhance the formation of oil-water emulsions and the visual disappearence of the oil from the surface.

Microbial-oxidation of petroleum hydrocarbons has been demonstrated to be very effective either by bacteria, yeast and fungi. However, marine conditions are usually far from providing maximum biodegradation. It is generally assumed that biological processes begin to operate as the residence time of oil on the water increases from months to years, and gain in significance depending on a series of complex factors, among them temperature (that influences the type of bacterial populations), the availability of a carbon source, oxygen and nutrients for bacterial growth (Atlas, 1984).

As a result of the above reported wheathering processes, the remainder of the spilled oils appears in the sea mainly in two ways: as aggregates of different types and sizes (e.g., tar balls, hydrocarbons adsorbed onto organic and inorganic particles, etc.) and dissolved in the water. Whether the ultimate fate of petroleum residues is biodegradation or sedimentation remains to be determined, because despite the increased use of fossil fuels since the last 50 years the concentration of hydrocarbons in the oceans has not rised accordingly.

The cycling of lead has been extensively discussed by Nriagu (1978) and deserves several different aspects. As indicated in Fig. 2B, anthropogenic lead may enter the marine environment via run-off or streams and by atmospheric deposition, the latter being at present the major source (about 60% of the total input). At this respect it has been estimated that the concentration of lead in the atmosphere is now six times higher than it was in the pre-industrial time (Craig, 1980). In fact, only 4% of the present flux in the atmosphere comes from natural sources.

Whether the input is through the atmosphere or through continental streams the major form is particulate, so that most of the lead entering the sea from coastal sources settles on the continental shelves. Thus, sediments are the primary sink for this pollutant. Only the dissolved fraction and the very fine particles that can be transported to long distances, probably can reach the open sea and be made available to pelagic organisms. An approximate estimation of the present and past world average concentrations in surface waters indicate a tenfold increase. Consequently, Gerlach (1981) indicated that there is reason to suspect that lead concentrarion in organisms have increased 20-fold during the past few centuries. However, much of our knowledge in this area is tentative because of the inaccuracy of most analytical determinations, especially in sea water.

6

Within the water column lead partitions into four major pools: the dissolved, organic and inorganic, the suspended particulates, the biomass and the sediment. The rates and amounts of exchange among them are determined by the physiochemical conditions and the biological activity of the system. The extreme complexity of the processes involved, some of them not yet completely understood, difficults any quantitative assessment of this marine biogeochemical cycle.

Lead (II) oxide and carbonate seem to be the stable inorganic forms of undissolved lead in oxygenated sea waters, humic and fulvic acids playing a major role in their binding or complexation. One study suggested that lead in sea water was 66% combined with colloidal organic species, 24% bound to inorganic colloids and 10% in solution as free or complexed ions (Batley and Florence, 1976). In general, the unbound fractions are larger in oxidizing than in reducing conditions. These mainly occur in sediments with high inputs of organic matter and restricted water circulation. In anaerobic sediments most of the lead will be in the form of sulfide. Further, it may be methylated from where it may return to the biotic compartment (e.g. to benthic biota), to the overlying waters and even to the atmosphere. In aerobic conditions sulfide is oxidized to sulphate, limiting the lead concentration in solution.

Despite the complexity of these pictures, the present knowledge of the biogeochemistry of pollutants in the marine environment is sufficient to conclude that once we release a chemical to the ocean, there is the potential for a rapid distribution throughout many areas, but also oceanic processes may function in a manner which result in concentrating the pollutant in certain segments, with unforeseen consequences. Therefore, before we can make any statement concerning the effects of a possible pollutant, we must be able to predict their pathways of transport and regions of concentration and to know a good deal about the chemical and biological fate of the material.

THE MASS BALANCE ASSESSMENT OF MARINE POLLUTANTS

The mass balance approach provides confirmation of the significance of the principal biogeochemical processes that control the transport, distribution and fate of the pollutants in the marine environment (research objective) and guidelines for answer particular questions posed by the decision makers (management objective). This is why the approach has been incorporated into the General Plan for the Global Investigation of Pollution in the Marine Environment (GIPME) (IOC, 1984).

It includes the following basic concepts:

Input of pollutants	Standing stock of pollutants in an ocean basin or a regional sea	Output of pollutants
I_p $\longrightarrow$	$M_t = \Sigma_{\theta p} M_c$	$\longrightarrow$ E_p

where I_p represents the influx of the pollutant of concern (p) from all sources into the marine region in a given time (mass/time) and E_p the efflux from the system. M_t is the total mass of the pollutant within the system, calculated from the concentrations in each compartment (θ_p) and the corresponding mass of each zone (M_c).

7

When the pollutant p is distributed in a steady state, then

$$Ip = Ep \quad \text{and} \quad \frac{Mt}{Ip} = \frac{Mt}{Ep}$$

which is the residence time of the pollutant in the ecosystem (τ_p).

A formulation of these concepts has been exemplified by Li (1981), who proposed the following equation for the calculation of mass balances of elements in coastal zones, under the assumption of a steady state:

$$C_{rp}^j \cdot F_{rp} + C_{rd}^j \cdot F_{rd} = (F_{rp} + F_{rd} \sum_{\text{All } j} C_{rd}^j - F_{op}) C_{sh}^j + C_{op}^j \cdot F_{op}$$

where C_{rp}^j is the average concentration of element j in riverborne particles

C_{rd}^j is the average concentration of element j in river discharge water

C_{sn}^j is the average concentration of element j in coastal and shelf sediments

C_{op}^j is the average concentration of element j in pelagic clay sediments

F_{rp} is the aggregate rate of suspended particle discharge from rivers (kg/yr)

F_{rd} is the aggregate rate of water discharge from rivers (kg/yr)

F_{op} is the rate of pelagic clay sediment accumulation (kg/yr)

These calculations can be established for the sea on a global or regional basis. A number of mass-balances for individual trace metals in the ocean have been constructed. Examples of these are balances for copper (Boyle et al, 1977), nickel (Sclater et al, 1976), cadmium (Simpson, 1981) lead (Settle and Patterson, 1980) and mercury (Millward, 1982).

Some of these elements (i.e. nickel and copper) appeared to be in a reasonable balance, but for others (i.e. lead, cadmium and mercury) the oceanic influx and efflux are imbalanced. In this situation the use of non-steady state models is mandatory, as it is, for example, in most of the coastal or estuarine regions. In these areas the fluxes indicated in Figure 3 should be taken into account. However, it is not always easy to obtain adequate and reliable information for all these processes. The absence of reproducible sampling techniques for some compartments, the lack of data on solubilities, partition coefficients amd other physicochemical properties (e.g. metal speciation) in natural waters or the poor estimations of exchanges in the interfaces (air/water, water/sediment) are some of the aspects that difficult such calculations.

The situation is even more defavorable with respect to the organic pollutants. The analysis of these components in seawater is more complex. In many instances very large samples have to be taken in order to obtain measurable signals over blanks, which poses numerous methodological problems, as will be shown later. On the other hand, components with rather structural similarities behave differently in the marine environment (e.g. PCB isomers), being their biogeochemical cycle unknown. The atmospheric compartment and particularly the air/sea interface plays an important role in controlling the concentration of organics in the open ocean. Garrels et al. (1975) estimated that about 25% of the DDT added to land annually reaches the oceans by atmospheric deposition whereas only 0.5% is transported by streams. The widespread distribution of the chlorinated camphenes (toxaphene) in the oceans has been attributed also to the atmospheric transport and deposition in remote areas.

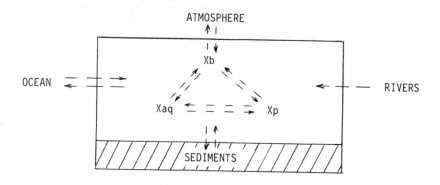

FIGURE 3. Simple compartment model of pollutant (X) with transfer between aqueous Xaq, particulate Xp and biogenic Xb forms and their exchanges in the water column.

Any effort to promote fundamental research for the understanding of these aspects should be encouraged, because the mass balance calculations are the mean to estimate whether human activities have contributed to the increase of concentrations of certain pollutants in the sea or, by the contrary, whether the pollutant load is safely assimilated by the marine environment. These quantitative assessments, then, permit to determine the assimilative capacity of the corresponding areas, providing a guide for management of waste disposal.

From what has been presented it can be concluded that marine pollution studies cannot be restricted to a simple exercise of determination and reporting of pollutant concentrations. Environmental problems must be tackled not only with measurements but with the development of concepts. In this sense, measurements of pollutants should be made within the context of the processes that affect their input, transport, degradation or final destination and monitoring programmes should be established and modified according to the answers provided by research activities. The development of sampling and analytical techniques to obtain reliable and intercomparable data in the different parts of the ocean as well as the application of bioassays for identifying chronic effects of pollution on marine organisms and ecosystems are necessary components of such studies. Good predictions are indeed very much dependent on our understanding of the processes going on in the ecosystem. Finally, we need environmental criteria scientifically established to develop appropriate management strategies for the marine environment. It is generally the case that international and national administrative regulations reflect more the popular media interests rather than the scientific priorities. Some views on these questions are presented in the following chapters.

REFERENCES

Albaigés J (1980). The fate and source identification of petroleum tars in the marine environment. Coll. Internat. CNRS No. 293, 233-247.
Atlas RM (1984). Petroleum microbiology. Mac Millan Pub. New York. 692 pp.
Batley GE and TM Florence (1976). Determination of dissolved cadmium, lead and cooper in sea water. Mar. Chem., 4, 347-363.
Bolin B, ET Degens, S Kempe and P Ketner (Eds.)(1979). The Global Carbon Cycle, J. Wiley, New York. 491 pp.
Boyle EA, F Sclater and JM Edmond (1977). The distribution of dissolved copper in the Pacific. Earth and Planetary Sci. Lett. 37, 38-54.
Craig PJ (1980). Metal cycles and biological methylation. The Handbook of Environmental Chemistry, (Ed. O. Hutzinger) Vol. 1A, pp 185-197. Springer-Verlag. Berlin.
Garrels RM, FT Mackenzie and C Hunt (1975). Chemical cycles and the global environment. William Kaufmann, Inc. Ca. 206 pp.
Gerlach SA (1981). Marine Pollution. Springer-Verlag. Berlin 218 pp.
IOC (1984). A framework for the implementation of the Comprehensive Plan for the global Investigation of Pollution in the Marine Environment. IOC. Techn. Ser. 25. UNESCO.
Jeffery PG (1973). Large-scale experiments on the spreading of oil at sea and its desappearence by natural factors. Proc. Joint Conf. on Prevention and Control of Oil Spills, pp. 469-474. A.P.I. Wash. D.C.
Li YH (1977). Geochemical cycles of elements and human perturbation. Geochim. Cosmochim. Acta, 45, 2073-2084.
Millward GE (1982). Nonsteady state simulations of the global mercury cycle. J. Geophys. Res. 87, 8891-8897.
Nriagu JR (Ed.)(1978). Biogeochemistry of lead in the environment. Elsevier. Amsterdam.
Premuzic ET, CM Benkovitz, JS Gaffney and JJ Walsh (1982). The nature and distribution of organic matter in the surface sediments of world oceans and seas. Organic Geochem., 4, 63-77 pp.
Sclater FR, EA Boyle and JM Edmond (1966). On the marine geochemistry of nickel. Earth and Planetary Sci. Lett. 31, 119-128 pp.
Settle DM and CC Patterson (1980). Lead in albacore: Guide to lead pollution in Americans. Science, 207, 1167-1176.
Simpson WR (1981). A critical review of cadmium in the marine environment. P   in Oceanography, 10, 1-70.

Chemical Changes in the Marine Environment: Origins and Concepts

ROBERT W. RISEBROUGH
Institute of Marine Sciences
University of California
Santa Cruz, California 95064, USA

Industrial growth over the past 50 years has produced many changes in the chemical environment of the world's ecosystems and these changes are proceeding at an accelerating rate. The marine environment is no exception to this global phenomenon, and the changes that are occurring in the coastal zones and in the oceans are the topic of this course of the Arab School of Science and Technology. In this chapter we focus on the origins of the elements, synthetic compounds, and classes of compounds, both synthetic and natural, that have contributed to the changes in the chemical compositions of the waters and organisms of the oceans, and also, of the marine atmosphere. Only, however, when we consider these changes within the framework of several broader concepts do consistent patterns emerge, and only then is it possible to make generalizations, conclusions, and predictions. Primarily, these are the concepts of biogeochemical cycles, residence times, mass balance, assimilative capacity, chemical markers, and of global dimensions. They have been immensely useful in assessing the significance of the chemical changes we usually refer to as pollution or contamination.

There is an important distinction between "pollution" and "contamination" which we shall consider below, after a more general consideration of the concept of chemical change and the effects such changes may have on organisms and ecosystems. Virtually any kind of chemical change above a critical magnitude has the potential to elicit a biological repsonse, either directly on organisms, or globally through a modification of climate.

With respect to the responses to chemical changes, we first address the question whether marine and coastal systems are qualitatively different from freshwater systems. As an example of the latter I consider a river that flows from the mountains to the sea. In earlier times, when the human population was small, we would have considered the water to be "pure"; we know, however, that it contained dissolved salts and minerals, a wide variety of organic compounds, as well as dissolved gases. Yet none of these additional components detracted from the uses to which the water was put by the first human inhabitants of the area; instead, on balance they enhanced the value of the water and frequently the river was the principal dietary source of essential trace minerals. With time not only was more water taken from the river,

11

but the river proved to be a convenient place to put waste chemicals, particularly human organic waste. In low amounts such material was quickly degraded by chemical and biological processes to elementary materials, thereby accomplishing the process of re-cycling. The river had, however, a finite capacity to degrade organic waste before exhausting the oxygen supply. Other kinds of wastes, from factories and agricultural activities, not only added to the organic burden but occasionally included toxic materials and materials that were resistant to chemical or biological degradation.

At a certain level and kind of chemical change the waters could no longer be used for drinking, for many industrial applications, or even for irrigation. A lack of sufficient oxygen or the presence of toxic chemicals frequently lead to the disappearance of fish and other aquatic organisms. The water had become "polluted". Its chemical composition had changed to a degree that some or all of the uses were no longer possible. Further changes, through treatment processes or reductions in the amount of waste put into the river, would be necessary to restore one or all of these uses.

This concept of pollution, involving the loss of one or more traditional uses is equivalent to the definition of marine pollution offered by Dr. Albaiges (this volume):

> The introduction by man, directly or indirectly, of substances or energy into the marine environment (including estuaries) resulting in such deleterious effects as harm to living resources, hazards to human health, hindrance to marine activities including fishing, impairment of quality for use of sea water and reduction of amenities.

This definition was compiled by a group of experts convened by the International Oceanographic Commission (GESAMP, 1980). It is widely accepted by both administrative agencies and working scientists. A number of similar definitions have been proposed (Champ, 1983). Evidently it applies equally to both freshwater and marine systems which differ only in the relative magnitudes of the receiving waters and the amounts of waste introduced into them.

Dr. Bayne (this volume) has used a slightly different approach, stressing the association of pollution with a biological response. Furthermore, Dr. Bayne has made the distinction between "pollution" and "contamination", which "refers only to the chemical consequences of environmental impact from industrial and other sources". Such a theoretical distinction between a "pollutant" and a "contaminant" has been adopted by many scientists working in tbe field who must attach labels to the many compounds they detect and measure in marine systems.

In practice, however, almost all of us are very sloppy in the use of these terms. Individual organisms may respond to the presence of a pollutant/contaminant by attempting to move to a cleaner area (Randlov et al., this volume), or by synthesizing enzymes which detoxify the pollutant/contaminant compound(s) (Bayne, this volume). Such a biological response frequently has no adverse effect on the individual organism, no effect at all upon populations or upon population structure, and has no association with a loss of a traditional use. Many compounds are frequently called "pollutants" even though they are not associated with any loss of a traditional use or an impact upon a biological system. This confusion apparently derives from the diversity of the scientific disciplines involved and from the legal and administrative mandates of public agencies: chemists detect a wide variety of synthetic compounds in environmental samples and document changes in the relative amounts of native elements such as mercury and lead; biologists and ecologists think in terms of changes in population structure and abundance; physiologists and biochemists think in terms of enzymatic induction and action; administrators and public agencies must meet the requirements of law and implement policy decisions to protect resources (Champ, 1983).

Virtually all of the substances we think of as pollutants in the marine environment, - radioactive isotopes, heavy metals, and all of the synthetic organics are not presently associated with any loss or damage to a traditional use, except in local coastal situations in the case of several of the synthetic organics. Even a major oil spill is associated only with a temporary loss of traditional uses. Furthermore, in almost all cases there is no detectable biological response at the population level of sensitive species.

It was, however, the inability of several species of birds to reproduce that sounded one of the earlier alarms about pollution/contamination of the environment and many of the examples came from coastal areas. The earliest example was the Bald Eagles, _Haliaetus leucocephalus_, of Florida (Broley, 1958). This species feeds to a large extent on fish. It disappeared from the southern California marine environment in the 1950s (Kiff, 1980). The coastal population in the north-eastern United States of another fish-eating species, the Osprey, _Pandion haliaetus_, declined rapidly in the 1960s (Ames and Mersereau, 1964; Ames, 1966). In both Europe and North America, populations of the Peregrine Falcon, _Falco peregrinus_, either had disappeared or had been greatly reduced by the mid-1960s (Hickey, 1969).

There is now convincing evidence for a pollutant effect, most likely of PCBs, in a mammalian population in a coastal environment. The reproduction of Harbor Seals, _Phoca vitulina_, in the western area of the Wadden Sea of the Netherlands, where the species has declined is significantly depressed (Reijnders, 1986). But the documented cases of a pollutant effect upon the population structure of a vertebrate species are exceedingly few.

Changes in the structure of marine benthic communities in the immediate vicinity of wastewater outfalls and other sites of waste discharges are well documented, but there appears to be no convincing evidence of population changes or changes in population structure induced by pollutants/contaminants at short distances away, even in coastal areas near Los Angeles and in San Francisco Bay which receive a high burden of contaminants/pollutants. The elegant studies of Dr. Bayne and his colleagues have shown that there are "biological effects of contamination that are proving appropriate in field studies and in monitoring of environmental

13

impact" (Bayne, this volume). It is now the task of invertebrate field biologists to demonstrate the biological changes on the population level in areas removed from the immediate sites of waste discharge that can be linked to chemical change.

At sites of waste discharge in the coastal zone, the occurrence of tumours, lesions and fin abnormalities in fish that are induced by pollutants/contaminants is well known (Malins et al., 1984, 1987; Murchelano, 1975; Sherwood, 1982). To date there apparently have been no documented effects on the populations of any of the species affected, but a reduced production of gametes by a sizeable fraction of the population would not necessarily decrease recruitment if the reproductive potential of the population as a whole is maintained.

A further difficulty arises in the application of the words "pollutant" or "contaminant" to elements such as lead and mercury whose fluxes into the marine environment have increased as a result of man's activities. Present levels are higher than those in the pre-industrial era. Here, rather than "pollution" or "contamination" we might think of changes in fluxes.

In the absence of documented effects, should marine pollution studies, or more accurately studies of the changing chemistry of the oceans, have a low priority and should this course have been devoted to a more important topic? Not at all. The potential exists for future loss of valuable uses of marine resources. As scientists we very much need to understand the rapid changes that are occurring in our global environment, in order to provide guidance and to predict, as far as is possible, the future consequences of present choices. As both a reservoir and sink, the oceans play a critical role in the global distribution of many contaminants. At a day-to-day level there are important decisions to be made about the coastal zones, which are inevitably used as sites for waste disposal. How much treatment, if any, is necessary for sewage wastes discharged into coastal waters? Can factory effluents be discharged untreated into coastal waters, and if not, what level and kind of treatment is necessary? And over the longer term, what changes are necessary in our chemical technologies such that persistent compounds like the PCBs will not be future problems? Our survey of the chemical changes that are occurring in the oceans and coastal zones will hopefully contribute to this kind of decision-making.

Our concept of chemical change in the oceans is not therefore fundamentally different from the chemical changes that occur in the stream that descends from the mountains. The capacity of the oceans to absorb waste is, however, much greater. A large quantity of human sewage could be discharged at some distance from shore, and within a short time would be broken down to the simple nutrients, carbon dioxide, and water. The components of greatest concern, the pathogenic bacteria and viruses, would not survive for long periods in this environment. Nutrients would be incorporated into the food webs and there would be no long-term deleterious or even detectable chemical change. In our stream, however, and in enclosed coastal waters, the oxygen supply would soon be depleted. Excess of nutrients might produce undesirable levels and kinds of phytoplankton growth. Two concepts are immediately important here, those of assimilative capacity and of natural cycles, but the others we shall consider, those of residence times, mass balance, and of global dimensions, are also relevant.   A further discussion of marine pollution or contamination requires that they be considered next.

CONCEPTS USED IN MARINE POLLUTANT/CONTAMINANT STUDIES

Biogeochemical Cycles.   Of the cycles in the natural world, the water cycle is surely the most familiar.   Water evaporates from the ocean surface, is brought by the winds to the continents, descends as rain, and enters the sea again through rivers and streams.   The other major cycles, referred to as biogeochemical cycles since they have both biological and geological components, that are of greatest importance to both terrestrial and marine living communities, those of carbon, sulfur, nitrogen, and phosphorous, have recently been the subject of intensive study by SCOPE, the Scientific Committee on Problems of the Environment (Svensson and Soderlund, 1976; Bolin et al., 1979).   SCOPE is a non-governmental organization that draws upon the world's scientific resources to produce reports on important environmental questions.   The report on the carbon cycle is particularly relevant to the marine environment, since the continuing increase in atmospheric carbon dioxide (Keeling et al., 1976a, 1976b; Freyer, 1979) is expected to cause a major increase in seawater levels throughout the world as the polar ice caps melt.   All of these cycles have been significantly changed by human activities.

One   cycle that has only an incidental biological component but which has been profoundly changed by man is that of lead (Nriagu, 1979).   In pre-historic times this cycle was relatively simple. With erosion of the land, lead and all other components of the rocks were dissolved or dispersed in particles, and were brought to the sea in the rivers.   A relatively small amount of lead entered the atmosphere in volcanic emissions, to descend on both land and sea.   With time the lead entered the sediments, to be incorporated again into rock.   The smelting of lead was one of the first industrial activities, and over the past two thousand years the input of lead into the atmosphere has continued to increase, with a greatly accelerated increase over the past century.   This history is recorded in the Greenland Icecap; analysis of the layers of snow and ice, which can be dated by radioactive tracers, has shown an increase in the amounts of lead in snowfall, beginning in the layers deposited about two thousand years ago (Murozumi et al., 1969).   There has therefore been a corresponding increase in the deposition of lead from the atmosphere to the sea, and because of the relatively low amount of lead introduced by natural pathways, most of the lead in seawater is anthropogenic in its origin (Tatsumoto and Patterson, 1963; Schaule and Patterson, 1981, 1983; Settle and Patterson, 1982; Flegal and Patterson, 1983; Flegal et al., 1984).

The synthetic organochlorines that have become ubiquitous global pollutants follow a very different cycle.   That of DDE (p,p'-DDE; 4,4'(para-chlorophenyl)dichloroethylene) is taken as an example because of the exceptional importance of this compound.   A derivative of the principal ingredient of the insecticide DDT, it

15

is usually the most abundant single compound among the organochlorines in an environmental sample, whether this be human tissue, a fish, or a marine invertebrate. Except for a minor contribution by other DDT compounds, it is responsible for the thinning of the eggshells of sensitive bird species, which has been widely documented among eggs laid after 1946, when DDT came into widespread use (Ratcliffe, 1967; Hickey and Anderson, 1968; Heath et al., 1969; Wiemeyer and Porter, 1970; Risebrough, 1986). The origin of DDE is principally p,p'-DDT, with a minor contribution from another DDT compound present in preparations of the miticide dicofol (Risebrough et al., 1986).

The first difference from the behavior of lead derives from its non-polar properties; it is poorly soluble in water but highly soluble in lipids. Unlike lead it therefore accumulates to a much greater extent in organisms, and moreover, accumulates in food webs, passing from lower levels, generally herbivores, to the carnivorous species at the higher levels. Although the vapor pressure is low, it readily volatilizes into the atmosphere to maintain the equilibrium, dictated by the laws of thermodynamics, of chemical potential of a compound in all the phases of a multi-phasic system (Mackay, 1979; Mackay and Leinonen, 1975; Mackay and Paterson, 1981; Mackay et al., 1983). From the atmosphere it readily passes into lipid cuticles of plants and lipid-rich microlayers of water bodies, or into any environmental component where it is present at levels below those dictated by the thermodynamic constraints. Thus the removal processes from the atmosphere may drive the entry processes in spite of the comparatively low volatility. Similarly, in spite of low solubility in water, there will be a net flux of DDE into water from the atmosphere or from aquatic organisms if the water concentration is too low. The surface waters of the oceans mix only very slowly with the deeper waters, but DDE and other organochlorines have been detected in benthic organisms of the deep oceans (Barber and Warlen, 1979). They reach the deeper waters through a biological process. The faecal pellets of many zooplankton, particularly copepods, are relatively heavy and sink comparatively rapidly through the water column. DDE and the other organochlorines ingested or absorbed by the zooplankton are thereby transported to the deeper waters (Elder and Fowler, 1977; Fowler and Knauer, 1986). Once reaching the sediments the DDE may again be incoporated into food webs, and final, permanent deposition in the sediments may occur only after many re-cyclings through the food webs.

Differences in vapor pressure and water solubilities may profoundly alter the patterns of the cycles. Compounds that have both higher vapor pressures and higher water solubilities than do DDE would be expected to more readily enter the atmosphere and partition less into organisms from aqueous solution. Their rate of descent into the deeper waters of the oceans is expected therefore to be slower, and the relative amount of time spent in the atmosphere to be longer. Such factors may explain the environmental distribution of the toxaphene mixture and the hexachlorocyclohexanes (BHC, including lindane). Together, they are the most abundant of the synthetic organochlorines in the medium molecular weight range in surface waters of the eastern Pacific (de Lappe et al., 1983; B.W. de Lappe and R.W. Risebrough, unpublished).

The cycles in which other contaminant compounds move, including the radioactive isotopes, lower-molecular-weight organics, etc., will also be a function of basic physical and chemical properties. It should be evident, however, that an understanding of the cycles through which compounds move is central to an understanding of their environmental significance.

Mass Balance. The concept of mass balance derives directly from that of biogeochemical cycles. How much is in each environmental component and how much is moving in and out? A calculation of the mass balance is therefore a calculation of the global budget of the substance.

A global mass balance for lead has been prepared by Nriagu (1978). The "steady state" amount of lead in the atmosphere at any one time is calculated to be in the order of 1.8×10^{10} grams. Yearly input from anthropogenic emissions is determined to be 44×10^{10} g; input from wind-blown dusts, originally derived principally from anthropogenic sources, is estimated to be 1.6×10^{10} g. Inputs from other sources, including volcanic emissions, organic particulates, and seasalt sprays, are relatively small. The inputs are balanced by deposition to the land and sea, estimated to be 32 and 14×10^{10} g/yr, respectively. The reservoir of dissolved lead in the oceans is estimated to be 27×10^{12} g, three orders of magnitude higher than the atmospheric reservoir. This reservoir is believed to be still increasing, such that deposition in the sediments does not yet balance input from the atmosphere.

It has not yet been possible to draw up global mass balances of the principal synthetic organics such as the PCBs, the DDT, chlordane, HCH and toxaphene compounds, although reasonably good estimates of the total amounts that have been synthesized and released into the environment are available. The principal difficulty has been obtaining reliable measurements of their concentrations in seawater, particularly throughout the water column, and of the fluxes from the sea back to the atmosphere.

In spite of uncertainties in the values for the concentrations of petroleum hydrocarbons in seawater, Burns and Saliot (1986) have constructed a mass balance of petroleum hydrocarbons in the Mediterranean Sea. Inputs from all sources were balanced by outputs through volatilization, the formation of tar, sedimentation, and degradation. The inventories of petroleum hydrocarbons in each of the components of the ecosystem were estimated.

Because of the increasing levels of atmospheric carbon dioxide, the carbon mass balance, particularly the fluxes between the atmosphere and the sea, are of special interest (Bolin et al., 1979).

Residence times. The residence time is the average length of time a contaminant or other environmental chemical spends in a reservoir such as the atmosphere or the water column of the sea. It can readily be calculated if there are good estimates of the total amount in the reservoir, if a steady-state has been achieved, and if there is a good estimate of either input or output. Thus for lead, dividing the estimate of total lead in the atmosphere by either the yearly input or the yearly output produces an estimate of the residence time of about 14 days.

Estimates of residence times in the sea of contaminant organic compounds are as yet preliminary, because of uncertainties in the measured concentrations in seawater and the relatively few determinations of vertical fluxes. Earlier determinations of residence times of the principal elements in seawater vary widely from element to element, from 100 years for aluminum to 300 million years for sodium (Goldberg and Arrhenius, 1958; Bewers and Yeats, 1977).

Assimilative capacity. We have seen that the stream descending from the mountains to the sea has a much lower capacity to absorb an added chemical burden than does the open ocean. The assimilative capacity is therefore a quantitative measure of the amount of a given compound or of materials that waters can receive so as to achieve degradation to elementary materials or such that concentrations of materials already present are not sufficiently altered to affect a beneficial use.

The assimilative capacity of smaller rivers, such as the majority of those in the Arab countries, is evidently rather low, much lower than that of coastal waters. Nevertheless there is obviously a limit to the capacity of coastal waters, particularly those in shallow areas with low current movement, to degrade organic sewage waste without exhausting available oxygen. In 1976, oxygen became depleted in bottom waters off New York City, killing many benthic organisms (Swanson and Sindermann, 1979). Although no clear-cut cause could be established, it has been associated with the dumping of sewage sludge from the New York City area, some 7 million wet tons in 1985, at a site 12 miles (19 km) from shore. The U.S. Environmental Protection Agency has recently designated a new dumping site 106 miles (171 km) from shore, over deeper waters of the Continental Shelf, for deposition of these wastes. The Agency and the New York City Department of Environmental Protection have begun an extensive monitoring program for levels of dissolved oxygen, phytoplankton abundance, nutrients, and other relevant parameters in this area (Suszkowski and Santoro, 1986); Levine, 1986).

The Federal Water Pollution Control Act of 1972 of the USA, known as the Clean Water Act required that all sewage wastes discharged into coastal waters receive secondary treatment by 1977. Such treatment would degrade organic waste into the basic nutrients, carbon dioxide, and water, thereby reducing the need for oxygen in receiving waters. The act was ammended in 1977 to permit waivers to this requirement, on the argument that in many coastal areas the assimilative capacity is large, and that the available oxygen is more than adequate to achieve degradation. Many coastal cities in the United States, like coastal cities elsewhere in the world, have therefore continued to discharge wastewaters that have received only primary treatment. The City of Los Angeles, however, has recently been refused permission to continue this practice, on the argument that the continued discharge of large quantities of organic matter would result in significant modifications of the benthic environment in the vicinity of the disharge site. Because of the very substantial cost of secondary treatment, this argument would be given lesser weight in other areas of the world which have fewer financial resources.

The concept of assimilative capacity hardly applies to highly persistent compounds that accumulate in food webs and that possess toxic properties. The dumping of wastes from a DDT factory in Los Angeles in offshore waters during the 1950s lead to the disappearance of the Bald Eagle and Peregrine Falcon from nearby islands (Risebrough, 1987; Kiff, 1980). Continued discharge of wastes from the DDT factory during the 1960s, in the order of 200-300 kg/day, resulted in reproductive failures of fish-eating birds such as the Brown Pelican, Pelecanus occidentalis (Carry and Redner, 1970; MacGregor, 1974; Risebrough et al., 1971; Risebrough, 1986); after the discharge was stopped, the reproductive success of fish-eating birds quickly improved (Anderson et al., 1975). Similarly, the concept does not apply to other persistent compounds such as the PCBs, which, moreover, like the DDT comopounds are largely dispersed through the atmosphere.

It has been said that the solution to pollution is dilution; rather, the solution is efficient re-cycling to simpler compounds, and phasing out use of toxic, environmentally persistent compounds.

Chemical markers. The concept of markers, compounds or combinations of compounds whose characteristics provide information on origins and about the recent past, has had a wide application in many human activities. Vocal accents yield information on the origins of the speaker. The presence of a chemical, or of combinations of chemicals, frequently yields information about processes, events, and origins (Chesler et al., 1978). Petroleum chemists frequently associate the relative amounts of compounds characteristic of petroleum, such as pentacyclic triterpanes, steranes, or isoprenoids with the petroleums of a given area, or even with petroleums of different strata of an oilfield (Seifert and Moldowan, 1978). This same information can be used to determine sources of petroleum contamination; Simoneit and Kaplan (1980) used the "markers" in petroleum isolated from southern California marine sediments to determine the origins of the petroleums. From the relative amounts of several pentacyclic triterpanes in extracts of mussels from the Mediterranean coast of Spain, it was possible to conclude that local petroleum contamination did not derive from offshore oilfields (Risebrough et al., 1983).

It is frequently important to determine the extent of contamination in areas of waste discharge. Compounds characteristic of this waste that are more refractory to degradation processes are therefore potentially useful as markers.

19

For sewage wastes they include several components or degradation products of detergents such as alkylbenzenes and nonylphenoxy acids (Eganhouse et al., 1983; Giger et al., 1984; Ahel et al., 1987). Coprostanol is being increasingly used as a "marker" compound (Hatcher and McGillivary, 1979; Wade et al., 1983). It is a fecal steroid present in human wastes (but also in the wastes of marine mammals), and is relatively persistent in coastal environments. Detection of these compounds in coastal sediments indicates therefore the presence of components of domestic sewage. Concentration gradients in sediments would provide information on the accumulation and movements of wastes in local coastal areas.

Global dimensions. Each pollution event has a component of dimensions. A lesser amount of pollutant, or a larger volume of receiving water, would reduce the impact, and perhaps even avoid it. A consideration of marine pollution within a global perspective requires therefore a determination of the dimensions of the system. The most meaningful appears to be the total amount of organic carbon synthesized by the phytoplankton in the marine environment; this parameter sets an upper limit to possible biological activity. The earliest estimates of total marine productivity, using radiocarbon techniques and expressed as grams carbon incorporated, were in the order of 1.5×10^{16} g/yr (Steemann Nielsen, 1953; 1960). With improvements of this technique, the estimates have been revised upward, in the order of 4.4×10^{16} g/yr (De Vooys, 1979). The estimate of terrestrial productivity is of the same order of magnitude, 6×10^{16} g/yr (Ajtay et al., 1979). Besides setting upper limits to the biological dimensions of each system, these numbers provide also a basis for comparison with anthropogenic inputs. Thus the yearly input of 10^{11} g of lead into the sea from the atmosphere is about 400,000 times less than the marine productivity. In the mid-1960s global yearly production of DDT was in the order of 10^{10} g, a million times lower, but DDT has biological effects at the part per million level. Biological effects of this compound on a global scale would therefore be anticipated.

The global consumption of fossil fuels, expressed as grams of carbon emitted into the atmosphere as carbon dioxide, is now about an order of magnitude less than the total biological productivity on land and in the sea (Bolin et al., 1979b). Man's activities are therefore approaching the level of nature's activities. Pollution, or chemical perturbations, would appear to be inevitable.

ANOTHER ASPECT OF THE GLOBAL DIMENSIONS

Thirty years ago there were no marine pollution problems and it was thought that the oceans were much too large to be affected by the activities of man. In the short intervening period our perspectives have vastly changed. The first impetus came from the global fallout of radioactive isotopes following the test explosions of atomic weapons undertaken by the Soviet Union and the United States in the 1950s. Events half-away around the world affected the local chemical environment in ways that had immediate relevance to human health. The "dirty" hydrogen bomb explosions produced the major effects, but even a low yield nuclear explosion in Nevada in the western United States yielded fission products that were detected in Paris, Cairo, and Japan (Miyake et al., 1956). No part of the world was any longer "remote".

In the mid 1960s the detection of DDT compounds in antarctic wildlife (Sladen et al., 1965; George and Frear, 1966) first aroused scepticism, but was quickly followed by reports of the global distribution not only of the DDT compounds, but also of the hexachlorocyclohexanes and the PCBs. These compounds readily entered the atmosphere to be carried long distances in the wind systems.

Two meetings were held in 1970 that called attention to problems of marine pollution. Both were chaired by Professor Edward Goldberg of the Scripps Institution of Oceanography. The first produced a report "Chlorinated Hydrocarbons in the Marine Environment" which predicted that 25% of the DDT produced would reach the oceans (Goldberg et al., 1971). All of the assumptions used at that time would no longer be considered valid, yet subsequent work has confirmed that a large fraction of compounds such as DDT that are released into the environment would eventually reach the oceans because of their mobility in the atmosphere. The second meeting was held in Rome under the auspices of the Food and Agriculture Organization of the United Nations. It expanded the international interest beyond the chlorinated hydrocarbons to include also mercury and petroleum. Marine pollution had become a topic to be addressed by the international community.

MARINE POLLUTION OR COASTAL POLLUTION?

Oceanographers have traditionally considered processes that occur in the "blue water" areas of the oceans separately from those that occur in the coastal zones. In addition to the estuaries and deltas, the coastal zones include much of the continental shelves. These are the areas where the rivers and wastewater outfalls disharge their contents into the sea, where the organic materials in these waters are degraded to elementary compounds, and where sedimentary processes transfer a large fraction of the pollutant/contaminant compounds and elements to the sediments. In part this consists of the sinking of inorganic particles with associated materials, and in part of the sinking of fecal pellets and organic detritus deriving from the food webs.

Yeats and Bewers (1983; Figure 1) estimate that 95% of the metals associated with particulate matter in fresh waters entering the coastal zones are transferred to the sediments; the amount of particulate material transferred from the coastal zones to the oceans is of the same order of magnitude, estimated to be about twice as much, as the amount transferred from the atmosphere to the sea. The oceans are therefore distinguished from the coastal zones by the relative amounts of particulate materials received from the atmosphere and from adjacent water masses.

A glance at the list of titles in any recent issue of the MARINE POLLUTION BULLETIN would confirm that most of the studies report

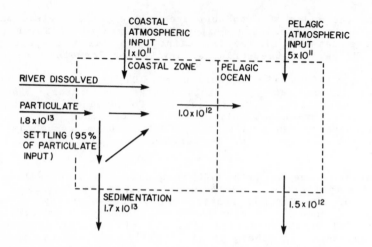

Figure 1. Generic coastal zone model illustrated with particulate
matter fluxes (kilograms per year) (Yeats and Bewers,
1983; reproduced with permission of Canadian Journal of
Fisheries and Aquatic Sciences.

on problems encountered in coastal zones rather than in areas that
are truly marine. A large fraction of the problems mentioned in
this course, probably the majority, are encountered in the coastal
zones rather than in the marine areas defined by oceanographers.
A discussion of the origins of the pollutants/ contaminants
requires therefore a distinction between the coastal zones and the
zones that are truly marine.

RADIOACTIVE ISOTOPES

The detection in fish from the open ocean of radioactive isotopes
deriving from the weapons testing had provided the first major
case of global contamination, prompting a major research effort on
marine radioactivity. With time the input from the testing has
declined; this can be demonstrated by the analysis of layers of
permanent snowfields, such as those in Antarctica, which show
peaks of activity in the snow deposited in the years of intensive
testing, declining to very low levels at the present (Cutter et
al., 1979). Those components of the fallout with longer
half-lives, however, particularly the plutonium isotopes,
americium-241, and cesium-137, have become components of the
marine environment, persisting in surface waters and participating
in biological processes (Bowen et al., 1971;1980; Noshkin and
Bowen, 1973; Fowler et al., 1983). The levels of radioactivity
emitted are low, a small fraction of the radioactivity emitted by
natural isotopes such as those of the uranium decay series.

The possibility of nuclear accidents, of renewed atomic explosions
in the atmosphere, or of atomic war, would be the factors that
would increase the levels of radioactivity in oceanic waters.

Current concern is for the coastal zones adjacent to the sites of nuclear reactors. Monitoring to detect any leaks into local environments and studies of the behavior of the isotopes that are released are the current priorities (Goldberg et al., 1980). The United States national "Mussel Watch" program measured the levels of plutonium isotopes and of americium-241 in mussels, Mytilus californianus and M. edulis, and in oysters, Crassostrea sp. from U.S. coastal sites over a period of three years, thereby establishing a baseline against which future changes might be compared (Goldberg et al., 1978; Goldberg et al., 1983). The focus is clearly on the coastal rather than on the marine environment; the consequence of leakage of radioactivity into rivers would be substantially greater.

HEAVY METALS AND TRACE ELEMENTS

We have noted an increase in the input of lead into the marine environment beyond the coastal zones; lead concentrations in seawater are many times above those in the pre-technological era. There has also been an increase in the global fluxes of mercury caused by an increased mobilization of this element by man. Beyond a documented increase of mercury levels in the food webs of the Baltic, measured by an increase in mercury levels in the feathers of birds, particularly the Sea Eagle, Haliaetus albicilla, (Berg et al., 1966), increases have not been detected or quantified beyond the coastal zone. Levels considered high of mercury in tuna from the Mediterranean or in swordfish have recently been attributed to natural sources of mercury.

Similarly, any changes in the fluxes of aluminum, iron, magnanese, cobalt, nickel, and copper into the marine environment beyond the coastal zone have not been found; current fluxes are considered to be the same as those over recent geological timescales. The available data on the levels of zinc and cadmium in river waters suggest that the fluxes of these elements, particularly of cadmium, into the sea have increased as a result of man's activities (Yeats and Bewers, 1983). Because of its incorporation into the biological food webs, cadmium is rapidly re-cycled in surface waters (Knauer and Martin, 1981), such that the refluxing of cadmium in the surface waters as a result of this activity is much higher than the fluxes from rivers into the coastal zones and into the oceans.

A biological response to the higher levels of these metals in seawater has not been documented. Any such response would surely be detected first in coastal areas. The mercury poisoning at Minimata occurred because high amounts of organic mercury were discharged into a relatively small estuary (Irukayama, 1966). Principal sources are wastewater outfalls that include industrial waste from metal proceessing facilities, particularly the metal plating industries, and the effluents from mining activities. Locally, unregulated atmospheric discharges would be expected to increase levels of contamination by selected metals in coastal ecosystems.

23

PETROLEUM

Contamination by petroleum extends over both the coastal and marine zones. The distribution of petroleum hydrocarbons in the marine environment has been discussed in two extensive reviews by the U.S. National Academy of Sciences (NAS, 1975; 1985). Yearly input from all sources is in the order of 4 million tonnes, or 4×10^{12} g. Losses and discharges during transportation, including spilled oil, deballasting and loading operations, bilge and tank washings are the major input source, accounting for 42% of the total input (NAS, 1985). In the Mediterranean, losses associated with transportation are relatively more important, amounting to 68% of the estimated total input (Burns and Saliot, 1986). Land-based industrial discharges and wastewater outfalls are the second and third major input sources. In their estimates of a mass balance for the Mediterranean of petroleum compounds, Burns and Saliot (1986) used figures published by Eganhouse and Kaplan (1981) for per capita discharges from urban and rural populations of 1 kg and 0.4 kg, respectively. Total input into the Mediterranean as waste from human activities was thereby estimated to be 100,000 and 50,000 tonnes from urban and rural areas, respectively (Burns and Saliot, 1986).

In the Arab World, the areas of concern are the Arabian Gulf, the Red Sea, the eastern Mediterranean, and selected areas of the north African coast, encompassing the majority of the region.

SYNTHETIC ORGANICS

The problems encountered with the synthetic organics have been principally associated with local discharges into the coastal zones of wastes from manufacturing or processing factories. These have included the incidental discharge of the highly toxic pesticides dieldrin and telodrin from a manufacturing plant into coastal waters of the Netherlands, resulting in the deaths of many birds and local declines in the populations of several species (Koeman et al., 1972). The populations recovered after the discharges were stopped.

Discharges of DDT wastes from a factory in Los Angeles, by dumping in offshore waters and into the wastewater system of Los Angeles County, have been disscussed above. They produced the highest levels of DDT contamination of a coastal environment in the world with many effects on local wildlife.

High levels of PCBs in the harbor of New Bedford in the U.S. state of Massachusetts were detected in the mid 1970s through the sampling of seawater and of mussels, Mytilus edulis (de Lappe et al., 1980). They derived from the wastes of factories manufacturing electrical capacitors; persisting high levels have resulted in the closing of local fisheries (Weaver, 1984).

Wastewaters have been shown to be significant sources of PCBs into local coastal environments. Globally, however, the distribution of these compounds derives from their atmospheric dispersal which quantitatively is much more important than the local discharges from wastewater outfalls. Compounds such as toxaphene and the

alpha isomer of hexachlorocyclohexane are the most abundant
synthetic organics in the medium molecular weight range in the
eastern Pacific; since the hexachlorocyclohexane mixture is no
longer used in the United States and Canada, and since the wind
circulation patterns indicate that origins must be in Asia or
beyond, the origins must be global. The global distribution of
these and other chlorinated hydrocarbons in the marine environment
is governed by the net environmental input in the world as a
whole.

In developing countries, problems associated with the synthetic
organics might therefore be anticipated in the immediate vicinity
of pesticide manufacturing plants, and of factories manufacturing
PCBs or using PCBs in the manufacture of other products.
Discharge of wastes from these factories into local coastal waters
would cause problems similar to those documented elsewhere.

PLASTICS, REMNANTS OF NYLON FISHING NETS, AND OTHER FLOATING
DEBRIS

This category is added to the traditional four categories of
marine pollutants/contaminants considered above because of the
increasing volume of these wastes that are visible at sea
(Carpenter and Smith, 1972; Carpenter et al., 1972; Colton et al.,
1974; Morris, 1980a;1980b; Pruter, 1987) and that are deposited on
beaches (Gregory, 1977). Documented harm to date consists of the
drowning of marine mammals and birds and of other marine species
through entanglement (Ogi and Tsujita, 1973; Tull et al., 1972;
Piatt and Nettleship, 1987; Laist, 1987). Although there is no
evidence at this time for reductions in populations, the potential
for major, serious long-term effects exists (Laist, 1987). Small
plastic particules accumulate in the digestive tracts of marine
birds of small plastic particles (Baltz and Morejohn, 1976; Bourne
and Imber, 1982; Connors and Smith, 1982; Furness, 1983). These
latter consist of both the particles which are the starting
materials in plastic manufacture and small particles produced by
abrasion of plastic containers discarded at sea. Continued growth
of the plastics industry and indiscriminate disharges into the sea
of plastic products pose therefore a major long-term threat to
marine resources.

Globally, there is also a massive affront to our beaches and
shorelines which poses a challenge to our aesthetic values and to
the values of our civilization.

Ahel, M., T. Conrad, and W. Giger. 1987. Persistent organic chemicals in sewage effluents. 3. Determinations of nonyl-phenoxy carboxylic acids by high-resolution gas chromatography/mass spectrometry and high-performance liquid chromatography. Environ. Sci. Technol. 21:697-703.

Ajtay, G.L., P. Ketner, and P. Duvigneaud. Terrestrial primary production and phytomass. Pages 129-181 in B. Bolin, E.T. Degens, S. Kempe, and P. Ketner, eds. The Global Carbon Cycle. SCOPE 13. John Wiley and Sons, Chichester and New York. 491 pp.

Ames, P.L. 1966. DDT residues in eggs of the Osprey in the northeastern United States and their relation to nesting success. J. Appl. Ecol. 3:87-97.

Ames, P.L., and G.S. Mersereau. 1964. Some factors in the decline of the Osprey in Connecticut. Auk 81:173-185

Anderson, D.W., J.R. Jehl, Jr., R.W. Risebrough, L.A. Woods, Jr., L.R. Deweese, and W.G. Edgecomb. 1975. Brown Pelicans: improved reproduction off the southern California coast. Science 190: 806-808.

Baltz, D.M., and G.V. Morejohn. 1976. Evidence from seabirds of plastic particle pollution of central California. West. Birds 7:111-112.

Barber, R.T., and S.M. Warlen. 1979. Organochlorine insecticide residues in deep see fish from 2500 m in the Atlantic Ocean. Environ. Sci. Technol. 13:1146-1148.

Bayne, B.L. The biological effects of marine pollutants. This volume.

Berg, W., A. Johnels, B. Sjostrand, and T. Westermark. 1966. Mercury content in feathers of Swedish birds from the past 100 years. Oikos 17:71-83.

Bewers, J.M., and P.A. Yeats. 1977. Oceanic residence times of trace metals. Nature (London) 268:595-598.

Bolin, B., E.T. Degens, S. Kempe, and P. Ketner, eds. 1979. The Global Carbon Cycle. SCOPE 13. John Wiley and Sons, Chichester and New York. 491 pp.

Bolin, B., E.T. Degens, P. Duvigneaud, and S. Kempe. 1979b. The global biogeochemical carbon cycle. Pages 1-56 in B. Bolin, E.T. Degens, S. Kempe, and P. Ketner, eds. The Global Carbon Cycle. SCOPE 13. John Wiley and Sons, Chichester and New York. 491 pp.

Bourne, W.R.P., and M.J. Imber. 1982. Plastic pellets collected by
    a prion on Gough Island, central South Atlantic Ocean. Mar.
    Pollut. Bull. 13:20-21.

Bowen, V.T., K.M. Wong, and V.E. Noshkin. 1971. Plutonium-239 in
    and over the Atlantic ocean. J. Mar. Res. 29:1-10.

Bowen, V.T., V.E. Noshkin, H.D. Livingston, and H.L. Volchok.
    1980. Fallout radionuclides in the Pacific Ocean: Vertical
    and horizontal distributions, largely from GEOSECS stations.
    Earth Planet. Sci. Lett. 49:411-434.

Broley, C.E. 1958. The plight of the American Bald Eagle. Audubon
    Magazine 162-163.

Burns, K.A., and A. Saliot. 1986. Petroleum hydrocarbons in
    the Mediterranean Sea| A mass balance. Mar. Chem.
    20:141-157.

Carpenter, E.J., and Jr. K.L. Smith. 1972. Plastics on the
    Sargasso Sea surface. Science 175:1240-1241.

Carpenter, E.J., S.J. Anderson, G.R. Harvey, H.P. Miklas,
    and B.B. Peck. 1972. Polystyrene spherules in coastal
    waters. Science 178:749-750.

Carry, C.W., and J.A. Redner. 1970. Pesticides and heavy metals:
    Progress Report, December, 1970. County Sanitation Districts
    of Los Angeles County. P.O. Box 4998, Whittier, CA 90607. 51
    pp.

Champ, M.A. 1983. Etymology and the use of the term
    "pollution". Can. J. Fish. Aquat. Sci. 40 (Suppl.2):5-8

Chesler, S.N., H.S. Hertz, W.E. May, S.A. Wise, and F.R.
    Guenther. 1978. Organic marker compounds for environmental
    analysis. Intern. J. Environ. Anal. Chem. 5:259-271.

Colton, J.B. Jr., F.D. Knapp, and B.R. Burns. 1974.
    Plastic particles in surface waters of the Northwestern
    Atlantic. Science 185:491-497.

Connors, P.G., and K.G. Smith. 1982. Organic plastic particle
    pollution: suspected effect on fat deposited in red
    phalaropes. Mar. Pollut. Bull. 13:18-20.

Cutter, G.A., K.W. Bruland, and R.W. Risebrough. 1979. Deposition
    and accumulation of plutonium isotopes in Antarctica.
    Nature (London) 279:628-629.

de Lappe, B.W., R.W. Risebrough, A.M. Springer, T.T. Schmidt,
    J.C. Shropshire, E.F. Letterman and J.R. Payne. 1980. The
    sampling and measurement of hydrocarbons in natural waters.
    Pages 29-68 in B.K. Afghan and D. Mackay, eds., Hydrocarbons
    and Halogenated Hydrocarbons in the Aquatic Environment.
    Plenum Press, New York. 588 pp.

de Lappe, B.W., R.W. Risebrough and W. Walker II. 1983. A large-volume sampling assembly for the determination of synthetic organic and petroleum compounds in the dissolved and particulate phases of seawater. Can. J. Fish. Aquat. Sci. 40(Suppl.2) 322-336.

De Vooys, C.G.N. 1979. Primary production in aquatic environments. Pages 259-292 in B. Bolin, E.T. Degens, S. Kempe, and P.j Ketner, eds. The Global Carbon Cycle. SCOPE 13. John Wiley and Sons, Chichester and New York. 491 pp.

Eganhouse, R.P., and I.R. Kaplan. 1981. Extractable organic matter in urban stormwater runoff. 1. Transport dynamics and mass emission rates. Environ. Sci. Technol. 15:310-315.

Eganhouse, R.P., D.L. Blumenfield, and I.R. Kaplan. 1983. Long-chain alkylbenzenes as molecular tracers of domestic wastes in the marine environment. Environ. Sci. Technol. 17:523-530.

Elder, D.L., and S.W. Fowler. 1977. Polychlorinated biphenyls: Penetration into the deep ocean by zooplankton fecal pellet transport. Science 197:459-461.

Flegal, A.R., and C.C. Patterson. 1983. Vertical concentration profiles of lead in the Central Pacific at 15° N and 20°S. Earth Planet. Sci. Lett. 64:19-32.

Flegal, A.R., B.K. Schaule, and C.C. Patterson. 1984. Stable isotopic ratios of lead in surface waters of the Central Pacific. Mar. Chem. 14:281-287.

Fowler, S.W., S. Ballestra, J. La Rosa, and R. Fukai. 1983. Vertical transport of particulate-associated plutonium and americium in the upper water column of the Northeast Pacific. Deep-Sea Research 30:1221-1233.

Fowler, S.W., and G.A. Knauer. 1986. Role of large particles in the transport of elements and organic compounds through the oceanic water column. Prog. Oceanog. 16:147-194.

Freyer, H.-D. 1979. Variations in the Atmospheric CO_2 Content. Pages 79-99 in B. Bolin, E.T. Degens, S. Kempe, and P. Ketner, eds. The Global Carbon Cycle. SCOPE 13. John Wiley and Sons, Chichester and New York. 491 pp.

Furness, B.L. 1983. Plastic particles in three procellariiform seabirds from the Benguela Current, South Africa. Mar. Pollut. Bull. 14:307-308

George, J.L., and D.E.H. Frear. 1966. Pesticides in the Antarctic. J. Appl. Ecol. 3(suppl.)155-167.

GESAMP (Joint Group Of Experts on the Scientific Aspects of Marine Pollution). IMCO/FAO/UNESCO/WMO. 1980. Report of the Eleventh Session, Dubrovnik, Yugoslavia, 25-29 Feb. 1980. Reports and Studies No. 10. 14 pp. + 1x Annexes

Giger, W., P.H. Brunner, and C. Schaffner. 1984. 4-Nonylphenol in sewage sludge: Accumulation of toxic metabolites from non-ionic surfactants. Science (Washington, D.C.) 225:623-625.

Goldberg, E.D., and G.O.S. Arrhenius. 1958. Chemistry of Pacific pelagic sediments. Geochim. Cosmochim. Acta 13:153-212.

Goldberg, E.D., P. Butler, P. Meier, D. Menzel, G. Paulik, R.W. Risebrough, and L.F. Stickel. 1971. Chlorinated hydrocarbons in the marine environment. National Academy of Sciences, Washington, D.C.

Goldberg, E.D., V.T. Bowen, J.W. Farrington, G. Harvey, J.H. Martin, P.L. Parker, R.W. Risebrough, W. Robertson, E. Schneider and E. Gamble. 1978. The mussel Watch. Environmental Conservation 5: 1-25.

Goldberg, E.D., S. Fowler, and R. Fukai. 1980. Radionuclides. Pages 143-162 in The International Mussel Watch National Academy of Sciences, Washington, D.C. 248 pp.

Goldberg, E.D., M. Koide, V. Hodge, A.R. Flegal, and J. Martin. 1983. U.S. Mussel Watch: 1977-1978. Results on trace metals and radionuclides. Est. Coast. Shelf Sci. 16:69-93.

Gregory, M.R. 1977. Plastic pellets on New Zealand beaches. Mar. Pollut. Bull. 8:82-84.

Hatcher, P.G., and P.A. McGillivary. 1979. Sewage contamination in the New York Bight. Coprostanol as an indicator. Environ. Sci. Technol. 13:1225-1229.

Heath, R.G., J.W. Spann, and J.F. Kreitzer. 1969. Marked DDE impairment of Mallard reproduction in controlled studies. Nature 224:47-48.

Hickey, J.J. (Ed.). 1969. Peregrine Falcon Populations: Their Biology and Decline. Madison, University of Wisconsin Press. 569 pp.

Hickey, J.J., and D.W. Anderson. 1968. Chlorinated hydrocarbons and eggshell changes in raptorial and fish-eating birds. Science 162:272-273.

Irukayama, K. 1966. The pollution of Minimata Bay and Minimata disease. Adv. Water Pollut. Res. 3:153-180.

Keeling, C.D., R.B. Bacastow, A.E. Bainbridge, C.A. Ekdahl, P.R. Guenther, L.S. Waterman, and J.F.S. Chin. 1976a. Atmospheric carbon dioxide variations at Mauna Loa Observatory, Hawaii. Tellus 28:538-551.

Keeling, C.D., J.A. Adams, C.A. Ekdahl, and P.R. Guenther. 1976b. Atmospheric carbon dioxide variations at the South Pole. Tellus 28:552-564.

Kiff, L.F. 1980. Historical changes in resident populations of California islands raptors. Pages 651-673 in D.M. Power, ed. The California Islands: Proceedings of a Multi-discicplinary Symposium. Santa Barbara Museum of Natural History. Santa Barbara.

Knauer, G.A., and J.H. Martin. 1981. Phosphorus-cadmium cycling in northeast Pacific waters. J. Mar. Res. 39:65-76.

Koeman, J.H., Th. Bothof, R. De Vries, H. Van Velzen-Blad, and J.G. Vos. 1972. The impact of persistent pollutants on piscivorous and molluscivorous birds. TNO-nieuws 1972:561-569.

Laist, D.W. 1987. Overview of the biological effects of lost and discarded plastic debris in the marine environment. 1987. Mar. Pollut. Bull. 18:319-326.

Levine, E.A. 1986. Program for monitoring sewage dump site in the New York Bight. Pages 760-763 in Oceans '86 Conference Record. IEEE Service Center, Piscataway, New Jersey, and Marine Technology Society, Washington, D.C.

MacGregor, J.S. 1974. Changes in the amount and porportions of DDT and its metabolites, DDE and DDD, in the marine environment off Southern California, 1949-72. Fish. Bull. 72:275-293.

Mackay, D. 1979. Finding fugacity feasible. Environ. Sci. Technol. 13: 1218-1223.

Mackay, D., and P.J. Leinonen. 1975. Rate of evaporation of low-solubility contaminants from water bodies to atmosphere. Environ. Sci. Technol. 9:1178-1180.

Mackay, D., and S. Paterson. 1981. Calculating fugacity. Environ. Sci. Technol. 15:1006-1014.

Mackay, D., W.Y. Shiu, and E. Chau. 1983. Calculation of diffusion resistances controlling volatilization rates of organic contaminants from water. Can. J. Fish. Aquat. Sci. 40 (Suppl.2):295-303

Malins, D.C., B.B. McCain, D.W. Brown, S.-L. Chan, M.S. Myers, J.T. Landahl, P.G. Prohaska, A.J. Friedman, L.D. Rhodes, D.G. Burrows, W.D. Gronlund, and H.O. Hodgins. 1984. Chemical pollutants in sediments and diseases of bottom-dwelling fish in Puget Sound, Washington. Environ. Sci. Technol. 18:705-713.

Malins, D.C., B.B. McCain, D.W. Brown, M.S. Myers, M.M. Krahn, S.-L. Chan. 1987. Toxic chemicals, including aromatic and chlorinated hydrocarbons and their derivatives, and liver lesions in white croaker (Genyonemus lineatus) from the vicinity of Los Angeles. Environ. Sci. Technol. 21:765-770.

Miyake, Y., Y. Sugiura, and Y. Katsuragi. 1956. Radioactive fallout at Asahikawa, Hokkaido in April, 1955. J. Met. Soc. Japan, ser. 2 34:226-230.

Morris, R.J. 1980a. Floating plastic debris in the Mediterranean. Mar. Pollut. Bull. 11:125.

Morris, R.J. 1980b. Plastic debris in the surface waters of the South Atlantic. Mar. Pollut. Bull. 11:164-166.

Murchelano, R.A. 1975. The histopathology of fin rot disease in winter flounder from the New York bight. J. Wildl. Manage. 11:262-267.

Murozumi, M., T.J. Chow, and C.C. Patterson. 1969. Chemical concentrations of pollutant lead aerosols, terrestrial dusts, and sea salts in Greenland and Antarctic snow strata. Geochim. Cosmochim. Acta. 33:1247-1294.

National Academy Of Sciences. 1975. Petroleum in the Marine Environment. National Academy of Sciences, Washington, D.C. 107 pp.

National Academy Of Sciences. 1985. Oil in the Sea: Inputs, Fates, and Effects. National Academy of Sciences, Washington D.C. 601 pp.

Noshkin, V.E., and V.T. Bowen. 1973. Concentrations and distributions of long-lived fallout radionuclides in open ocean sediments. Pages 671-686 in Radioactive Contamination of the Marine Environment. International Atomic Energy Agency.

Nriagu, J.O. 1978. Pages 137-184 in J.O. Nriagu (ed.) Biogeochemistry of Lead in the Environment. Elsevier. Amsterdam.

Nriagu, J.O. 1979. Global inventory of natural and anthropogenic emissions of trace metals to the atmosphere. Nature (London) 279:409-411.

Piatt, J.F., and D.N, Nettleship. 1987. Incidental catch of marine birds and mammals in fishing nets off Newfoundland, Canada. Mar. Pollut. Bull. 18:344-349.

Pruter, A.T. 1987. Sources, quantities and distribution of persistent plastics in the marine environment. Mar. Pollut. Bull. 18:305-310.

Randlov, A., S.O. Dahl, and E. Poulsen. Principles and methods in environmental management of coastal marine waters. This volume.

Ratcliffe, D.A. 1967. Decrease in eggshell weight in certain birds of prey. Nature 215:208-210.

Reijnders, P.J.H.   1986.   Reproductive failure in common seals feeding on fish from polluted coastal waters. Nature (London) 324:456-457,418.

Risebrough, R.W., F.C. Sibley, and M.N. Kirven. 1971. Reproductive failure of the brown pelican on Anacapa Island in 1969. American Birds 25(1): 8-9.

Risebrough, R.W., B.W. de Lappe, W. Walker II, B.R.T. Simoneit, J. Grimalt, J. Albaiges, J.A. Garcia Regueiro, A. Ballester i Nollaand M. Marino Fernandez. 1983. Application of the Mussel Watch Concept in studies of the distribution of hydrocarbons in the coastal zone of the Ebro Delta. Mar. Poll. Bull. 14:181-187.

Risebrough, R.W., W.M. Jarman, A.M. Springer, W. Walker II, and W.G. Hunt. 1986. A metabolic derivation of DDE from Kelthane. Environ. Toxicol. Chem. 5:13-19.

Risebrough, R.W. 1986. Pesticides and bird populations. Current Ornithology 3:397-427. Plenum Publishing Corporation. New York.

Risebrough, R.W. 1987. Distribution of organic contaminants in coastal areas of Los Angeles and the Southern California Bight. Report to the Los Angeles Regional Water Quality Control Board, Los Angeles. 114 pp.

Schaule, B.K., and C.C. Patterson. 1981. Lead concentrations in the Northeast Pacific: evidence for global anthropogenic perturbations. Earth Planet. Sci. Lett. 54:97-116.

Schaule, B.K., and C.C. Patterson. 1983. Perturbations of the natural lead profile in the Sargasso Sea by industrial lead. Pages 487-503 in S.C. Wong, E. Boyle, K.W. Bruland, J.D. Burton, and E.D. Goldberg, eds, Trace Metals in Sea Water. Plenum Press, New York.

Seifert, W.K., and J.M. Moldowan. 1978. Applications of the steranes, terpanes and monoaromatics to the maturation, migration and source of crude oils. Geochim. Cosmochim. Acta 42:77-95.

Settle, D.M., and C.C. Patterson. 1982. Magnitudes and sources of precipitation and dry deposition fluxes of industrial and natural leads to the North Pacific at Eniwetok. J. Geophys. Res. 87:8857-8869.

Sherwood, M. 1982. Fin erosion, liver condition, and trace contaminant exposure in fishes from three coastal regions. Pages 359-377 in Ecological Stress and the New York Bight: Science and Management. G.F. Mayer, ed. Estuarine Research Foundation, Columbia, South Carolina. 715 pp.

Simoneit, B.R.T., and I.R. Kaplan. 1980. Triterpenoids as molecular indicators of peleoseepage in recent sediments of the Southern California Bight. Mar. Environ. Res. 3:113-128.

Sladen, W.L.J., C.M Menzie, and W.l. Reichel. 1966. DDT residues in Adelie Penguins and a Crabeater Seal from Antarctica. Nature 210:670-673.

Steeman Nielsen, E. 1953. On organic production in the oceans. J. Cons. Perm. Int. Explor. Mer. 19:309-328.

Steeman Nielsen, E.S. 1960. Productivity of the oceans. Ann. Rev. Plant Physiol. 11:341-362.

Suszkowski, D.J., and E.D. Santoro. 1986. Marine monitoring in the New York Bight. Pages 754-759 in Oceans '86 Conference Record. IEEE Service Center, Piscataway, New Jersey, and Marine Technology Society, Washington, D.C.

Svensson, B.H., and R. Soderlund, eds. 1976. Nitrogen, Phosphorus and Sulphur - Global Cycles. SCOPE Report 7. Ecol. Bull. 22 (Stockholm). 192 pp.

Swanson, R.L., and C.J. Sindermann. 1979. Oxygen depletion and associated benthic mortalities in New York Bight. 1976. US Dept. of Commerce, NOAA Professional Paper No. 11. 345 pp.

Tatsumoto, M., and C.C. Patterson. 1963. The concentration of common lead in sea water. pp. 74-89 in Earth Science and Meteoritics North-Holland Publishing Co.

Tull, C.E., P. Germain, and A.W. May. 1972. Mortality of thick-billed murres in the West Greenland salmon fishery. Nature 237:42-44.

Wade, T.L., G.F. Oertel, and R.C. Brown. 1983. Particulate hydrocarbon and coprostanol concentrations in shelf waters adjacent to Chesapeake Bay. Can. J. Fish. Aquat. Sci. 40(Suppl.2):34-40.

Weaver, G. 1984. PCB contamination in and around New Bedford, Mass. Environ. Sci. Technol. 18:22A-27A.

Wiemeyer, S.N., and R.D. Porter. 1970. DDE thins eggshells of captive American Kestrels. Nature 227:737-738.

Yeats. P.A., and J.M. Bewers. 1983. Potential anthropogenic influences on trace metal distributions in the North Atlantic. Can. J. Fish. Aquat. Sci. 40:(Suppl. 2):124-131.

Transport of Marine Pollutants

EMİN ÖZSOY and ÜMİT ÜNLÜATA
Institute of Marine Sciences
Middle East Technical University
P.O.B. 28, Erdemli, İçel, Turkey

1. INTRODUCTION

A review of marine pollutant transport processes is given, with the objective of introducing the basic concepts. The subject has great extent and detail, and is a continuously developing area of research that is motivated by many practical interests. Marine pollutant transport processes are of great importance in many aspects of marine science, including engineering services related to marine waste disposal, the assessment of adverse effects of shipping and industrial sites, and the conservation of water quality in coastal and inland seas.

The basic concepts required in studying the diffusive/ dispersive transport of pollutants will be described. One is often forced, however, to use more complicated theory and/or numerical models to assess these effects in the presence of more complex geometry, current systems or in deep basins. There is also an often empirical element in the theory due to the need for realistic determinations of diffusion coefficients via experiments. This presentation precludes such variations to the theme, providing information on basic physics and the tools that one often needs. In addition to pollutants, the transport of heat, salt and other ecological quantities (e.g. plankton, detritus, nutrients, oxygen etc.) in the sea are also governed by similar laws. A number of basic books on fluid dynamics (Batchelor, 1967), and transport processes (Csanady, 1973; Fischer *et al.*, 1979; Kullenberg, 1982) are advisable to complement these notes.

The basics of hydromechanic theory is briefly summarized in Section 1. Simple solutions to the turbulent transport equations are reviewed in Section 2. The basics of shear flow dispersion are then provided in Section 3. Applications to transport of suspended sediments are considered in Section 4. Estuarine transport processes are reviewed in Section 5.

1.1 FLUID MOTION

The equations governing fluid motion are briefly reviewed. The *continuity equation*

35

$$\frac{D\rho}{Dt} + \rho \, \nabla \cdot \vec{u} = 0 \tag{1.1}$$

states the conservation of mass, where the *material derivative* is defined as

$$\frac{DX}{Dt} = \frac{\partial X}{\partial t} + \vec{u} \cdot \nabla X \; . \tag{1.2}$$

for any variable X. An *incompressible fluid* is defined as one in which the density of material elements of the fluid does not change, i.e. equation (1.1) simplifies to

$$\nabla \cdot \vec{u} = 0 \; . \tag{1.3}$$

The momentum equation

$$\frac{D\vec{u}}{Dt} + 2\Omega \times \vec{u} = \vec{g} - \frac{1}{\rho} \nabla p + \frac{1}{\rho} \nabla \cdot \underline{d} \; . \tag{1.4}$$

expresses *Newton's Second Law of Motion* in an inertial (rotating) coordinate system on the earth, where Ω is the earth's angular velocity, p is the pressure, $\vec{g}$ is the gravitational acceleration and $\underline{d}$ is the deviatoric stress tensor (Batchelor, 1967).

The *thermodynamic equation* is derived from the *First and Second Laws of Thermodynamics*. For seawater, which is assumed to be incompressible, the thermodynamic equation is expressed as

$$\frac{DT}{Dt} = \nabla \cdot K \nabla T + Q/c_p \tag{1.5}$$

where $K = \kappa/(\rho c_p)$ is the *thermal diffusivity coefficient*, with κ being the thermal conductivity and c_p the *specific heat at constant pressure* for the fluid, and Q stands for heat sources (e.g. internal heating due to solar radiation or frictional dissipation). Neglecting the latter term yields

$$\frac{\partial T}{\partial t} + \vec{u} \cdot \nabla T = \nabla \cdot K \nabla T \tag{1.6}$$

1.2 DIFFUSION IN A FLUID

In this Section we will derive the conservation equations governing the diffusion of dissolved substances that may be present in a fluid. Consider a *binary system* consisting of a mixture of two different fluids. The densities ρ_A and ρ_B represent the masses per unit volume of the mixture. The *concentration* of each constituent is defined as the mass of each constituent per unit mass of the mixture, $c_A = \rho_A/\rho$, and $c_B = \rho_B/\rho$.
Since the fluid density is $\rho = \rho_A + \rho_B$, we must have $c_A + c_B = 1$.

36

In a moving fluid mixture, we define the fluxes (passing through a fixed surface) of each constituent as

$$\vec{N}_A = \rho_A \vec{u}_A, \quad N_B = \rho_B \vec{u}_B \tag{1.7.a,b}$$

where $\vec{u}_A$ and $\vec{u}_B$, are hypothetical velocities that an infinitesimal group (or cloud) of particles would have on the average, representing the momenta of each constituent. The total momentum (or total flux) of the mixture is

$$\rho \vec{u} = \rho_A \vec{u}_A + \rho_B \vec{u}_B$$
$$= c_A \vec{u}_A + c_B \vec{u}_B, \tag{1.8}$$

where the hydrodynamic velocity is defined as **u**.

The fluxes defined in (1.7.a,b) are with respect to a fixed observer and involve both diffusion and bodily transport (convection) with the fluid velocity. For example, we can write (1.7.a) as

$$\vec{N}_A = \rho_A \vec{u}_A$$
$$= \rho_A (\vec{u}_A - \vec{u}) + \rho_A \vec{u} \tag{1.9}$$

so that the first term represents the transport relative to an observer moving with the fluid (i.e. diffusive transport) and the second term represents the convective (advective) transport.

The diffusive flux is commonly modelled by *Fick's Law* (an analogue of *Fourier's Law* in heat conduction) which relates this flux to the local gradients of the density of each constituent. In the general non-isotropic case, the flux vector can be expressed as the product of a tensor coefficient with the gradient vector. If the medium is assumed to be *isotropic*, the constant of proportionality is a scalar and we can write

$$\rho_A (\vec{u}_A - \vec{u}) = - D_{AB} \nabla \rho_A \tag{1.10}$$

for constituent A, where the diffusion coefficient D_{AB} characterizes the diffusivity of constituent A in medium.

We consider a *fixed* control volume **V** enclosed by a surface S with outward normal $\hat{n}$, and write a statement of the conservation of mass for each constituent:

$$\frac{\partial}{\partial t} \int_V \rho_A dV = - \int_S \vec{N}_A \cdot \hat{n} dS + \int_V r_A dV, \tag{1.11}$$

where r_A is the rate of production of constituent A due

37

to possible chemical interactions. Since the total mass of the mixture should be conserved,

$$r_A + r_B = 0,$$ (1.12)

i.e. the rate of production of either constituent must be at the expense of the destruction of the other.

Using (1.9), (1.10), (1.1), the divergence theorem, and assuming that $\rho \approx$ constant yields

and

$$\frac{Dc_A}{Dt} = \nabla \cdot D_{AB} \nabla c_A + \frac{r_A}{\rho}$$

$$\frac{Dc_B}{Dt} = \nabla \cdot D_{BA} \nabla c_B + \frac{r_B}{\rho} .$$ (1.13.a,b)

If the presence of each constituent influences the density, so that ρ is not constant, it can be verified that the equations are coupled through density, which then means that the corresponding diffusion equations must be solved together with the continuity (1.1), the momentum (1.4) equations, and an equation of state incorporating the effects of the two constituents and temperature, salinity of sea-water on density:

$$\rho = \rho(p, T, S, c_A, c_B)$$ (1.14)

Furthermore, the energy equation (1.5), and a diffusion equation (similar to (1.13.a)) for salinity (a third constituent) must also complement the above equations in order to be able to solve the system.

It is however, quite common that the mixture of interest is a weak (dilute) solution of one of the constituents (say $c_A \ll c_B$). Then, we can assume $c = c_A \ll 1$ (yielding $c_B \approx 1$, $\rho \approx$ constant), so that the second equation (1.13.b) becomes irrelevant and the conservation of mass for the dilute solution is expressed by the single equation

$$\frac{\partial c}{\partial t} + \bar{u} \cdot \nabla c = D\nabla^2 c + R$$ (1.56)

where R and D have replaced r_A/R and D_{AB} respectively. In this case, the convective diffusion equation (1.15) is decoupled from the remaining equations. Likewise, the influence of temperature and salinity on the density of seawater can often be neglected when the gradients of both properties are sufficiently small. To a good degree of approximation, the ocean can be assumed incompressible and homogeneous, as a result of which the energy and salt diffusion equations are decoupled from the continuity and momentum equations. Therefore, in principle, we first solve the hydrodynamics from (1.1) and (1.14) to determine the velocity field $\bar{u}(x,t)$. Consequently, for given velocity field, we seek solutions to equation (1.15).

1.3 TURBULENT MOTIONS

The equations derived in the preceeding sections are generally
for laminar (orderly) flows of fluids. Most fluids become
turbulent (by the generation of chaotic motions) due to various
reasons. Turbulence in fluids may be generated as a result of
instabilities with respect to fluctuations deriving their
energy from the mean motion, wind stirring or mechanical
stirring at the boundaries, etc. The result is the random
motion of fluid "particles" consisting of lumps (eddies) of
various sizes, superposed on the mean motion. Since the
motions are random and chaotic, a full description of turbulent
flows is in the realm of statistics, which on the other hand is
strongly dependent on the structure and generating mechanisms
of the turbulence activity.

In order to derive the turbulence equations, we proceed by
separating the flow variables into slowly varying and
fluctuating parts with respect to a time scale T, which is
assumed to be the upper limit of the turbulence time scales.
For the variables u and c in equation (1.15) we can
write

and

$$\vec{u} = \vec{\bar{u}}(x, \ T^{-1}t) + \vec{u}'(x, \ t)$$

$$c = \bar{c}(x, \ T^{-1}t) + c'(x, \ t)$$

(1.16.a,b)

where the quantities with overbars denote the long term (with
respect to T) averages, for example

$$\bar{c} = \frac{1}{T} \int_0^T c \ dt$$

(1.17)

and the primed quantities are the components with fluctuations
that are typically more rapid than the turbulence time scale T.
By definition, $\overline{c'} = \overline{c - \bar{c}} = 0$. The turbulence equations are
then obtained by averaging the respective equations over a time
period T. Since the conservation equations given in the
earlier sections, namely the continuity (1.1), momentum (1.4),
energy (1.5) and diffusion (1.15) equations are basically the
same types (i.e. have the similar time derivative, convective,
and diffusive terms), the averaging procedure results in
similar terms. It will therefore be illustrative to average
only one of these, that case being the diffusion equation.

After averaging, the linear terms in the equations will
preserve their form in that they will be the same differential
terms operating on the mean quantities (since the averages of
the fluctuating parts vanish). On the other hand, the
nonlinear terms give rise to additional terms arising due to
the averaging of the products of fluctuating variables, which
in general do not vanish since the individual fluctuations of
the variables can be correlated (arising due to the common
cause of turbulence). These turbulent products mainly
originate from the nonlinear convective fluxes and represent
the turbulent fluxes, of buoyancy in the case of the continuity
equation (1.1) [which vanishes for homogeneous, incompressible
fluids], of momentum (Reynolds' stresses) in the case of the

39

momentum equation (1.4), and of heat or concentration in the cases of the convective-diffusion equation (1.6 and 1.15).

We first put (1.15) into the flux form (making use of the continuity equation), substitute (1.16.a,b), and take averages. By making use of (1.17), we immediately obtain

$$\frac{\partial \bar{c}}{\partial t} + \nabla \cdot \bar{c}\bar{u} + \nabla \cdot \overline{c'\vec{u}'} = D\nabla^2 \bar{c} + R$$

(1.18)

i.e. the same as equation (1.15) with the exception of the term $\nabla \cdot \overline{c'\vec{u}'}$ arising due to the averaging of the nonlinear convective terms in the preceeding equation.

As we have noted above, the product $\overline{c'\vec{u}'}$ describes the statistical correlation of the fluctuating components of concentration and velocity, which are expected to be strongly correlated in a turbulent field. Because of the practical problems discussed above, these terms are often parameterized, using empirical formulations. The form of this term in (1.18) actually suggests that it may represent the divergence of a flux in much the same way that the molecular flux divergence appears in (1.13). We can therefore define

$$\vec{N}_T = \rho \overline{c'\vec{u}'}$$

(1.19)

as the *turbulent flux* of the matter represented by the concentration $\bar{c}$.

One way to parameterize this flux is to make an analogy to Fick's Law, and adopt it for turbulent flows (Further discussion of the *mixing length theory* on which the present approximation is based can be found in Schlichting (1968), and Tennekes and Lumley (1972)). With this analogy, we relate the turbulent fluxes to the local gradients of the mean concentration. For a turbulent fluid, the statistical properties of which we can not fully prescribe, it is imperative that we use a anisotropic version of the analogy,

$$\overline{c'u'}_i = E_{ij} \frac{\partial \bar{c}}{\partial x_j}$$

(1.20)

where i=1,2,3 are indices denoting the directions in three dimensional coordinates x_i, and E_{ij} the *turbulent diffusivity tensor*.

In the sea, the smallness of the vertical motions as compared to the horizontal motions (i.e. the shallow water approximation) results in the common situation that the vertical stratification is far greater than in the horizontal. Therefore, it is reasonable to expect that the vertical coordinate coincides with one of the *principal axes*, and the horizontal axes can (by choice) be aligned with the remaining principal coordinates of the diffusivity tensor E_{ij}, which then reduces (1.20) to the special form

$$N_i = \rho \overline{c'u'}_i = -\rho E_i \frac{\partial \bar{c}}{\partial x_i}$$

(1.21.a)

where $E_i=E_{ii}$ are the principal components of the diffusivity tensor. Writing in vector form this becomes

$$\vec{N}_T = \overline{\rho c'\vec{u}'} = -\rho(E\bullet\nabla)\bar{c}$$

(1.21.b)

where $E = (E_x, E_y, E_z)$ are the turbulent diffusivities in the principal coordinates (x,y,z).

The turbulent mechanism of mixing (turbulent diffusion) is in fact much more effective than the molecular diffusion, so that typically $E_x, E_y, E_z \gg D$, as a result of which the molecular diffusion term in (1.18) can be neglected. Substituting (1.21.b) in (1.18) and dropping the overbar notation, we obtain the turbulent diffusion equation

$$\frac{\partial c}{\partial t} + \vec{u}\bullet\nabla c = \frac{\partial}{\partial x}(E_x\frac{\partial c}{\partial x}) + \frac{\partial}{\partial y}(E_y\frac{\partial c}{\partial y}) + \frac{\partial}{\partial z}(E_z\frac{\partial c}{\partial z}) + R .$$

(1.22)

The sources or sinks of concentration are represented by R. In some cases, this term stands for the decay or decomposition of a non-conservative pollutant due to extraneous influences. A linear form representing first order decay processes is

$$R = -Kc .$$

(1.23)

In the case of homogeneous turbulence, the diffusivities are constants in space and time, and we have a simpler version of the turbulent diffusion equation:

$$\frac{\partial c}{\partial t} + \vec{u}\bullet\nabla c = E_x\frac{\partial^2 c}{\partial x^2} + E_y\frac{\partial^2 c}{\partial y^2} + E_z\frac{\partial^2 c}{\partial z^2} - Kc .$$

(1.24)

1.4 RELATIVE AND APPARENT DIFFUSION

In reality, turbulence constitutes a random field of motion, and therefore the *ensemble averaged* equations would be appropriate. We have so far circumvented this difficulty by empirically representing these effects within the bulk turbulent diffusivity coefficients in time averaged equations.

Even in the case of homogeneous, stationary turbulence, an initial release of concentration will diffuse at a rate which depends on the size of the patch, so that the constant coefficients in (1.24) are not appropriate. This is a consequence of the probabilty distribution of different length and time scales embodied in the turbulence field. At initial stages of the spreading, the material will be redistributed mainly by small scale eddies. As the patch size grows, larger eddies will begin to influence it, and distribute the material in a more efficient way.

In addition, each random realization of the ensemble will appear in an irregular form and will be different from other possible realizations as shown in Fig.s 1.1.a,b. The irregularities are only smoothed out if we take the ensemble

average of the process with respect to fixed coordinates and at
the same time intervals after the release; then the constant
concentration surfaces will be circular as shown in Fig. 1.1.c
(i.e. if the medium is isotropic). On the other hand, the
center of mass of the diffusing cloud of each realization may
be shifted randomly with respect to the ensemble mean center of
mass, due to the influence of eddies that are larger than the
cloud size (*meandering*) as shown in the first two figures.
If the ensemble averages were to be taken by shifting the
origin to the instantaneous center of mass of each realization,
then the average cloud would look smaller than that in fixed
coordinates as shown in Fig.1.1.d, for then we extract the
influence of meandering (Fischer *et al*., 1979). Note that
in fact, during the initial stages of development, diffusion by
small scale eddies, and advection by large scale eddies are
inseparable, making the definition of turbulent diffusion
somewhat arbitrary.

The *apparent diffusion* is that corresponding to Fig. 1.1.c,
and we should in principle use the apparent turbulent
diffusivity in the turbulent diffusion equation. On the other
hand, it is more convenient to obtain the *relative
diffusivity* through experiments, i.e. that corresponding to
Fig. 1.1.d, since individual clouds can be averaged
irrespective of their relative positions.

A good measure of the spreading of a cloud can be obtained by
calculating its variance (i.e. the normalized second moment of
the concentration distribution), defined as

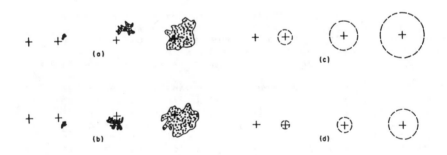

FIGURE 1.1 Turbulent diffusion from a small source. (a,b)
Random spread of two identical clouds, (c) Spread of the
ensemble mean, (d) Spred of the ensemble mean obtained by
shifting the origin to the centeroid of each realization.
(After Fischer *et al*. 1979)

42

$$\sigma_s^2 = \frac{\displaystyle\int_{-\infty}^{+\infty} cs^2 \, ds}{\displaystyle\int_{-\infty}^{+\infty} c \, ds} \qquad (1.25)$$

where c is the ensemble mean concentration and s stands for any of the coordinates (x,y,z) measured from the centroid of the cloud, and therefore σ_x, σ_y, σ_z are in essence the length scales (standard deviations) of the diffusing cloud. It can be verified, through multiplication of equation (1.22) with x^2, y^2, z^2 respectively and through integration by parts that

$$E_s = \frac{1}{2} \frac{d\sigma_s^2}{dt} \qquad (1.26)$$

for each of the coordinates $s = x,y,z$, i.e. the turbulent diffusivities are proportional to the rate of spreading.

The transformation between the ensemble mean values in fixed coordinates and those obtained by coinciding the centroids of different realizations is then obtained from (Csanady, 1973)

$$\sigma_s^2 = \sigma_{\hat{s}}^2 + m_s^2 \qquad (1.27)$$

where $\hat{s}$ refers to the coordinates with respect to the centroid in each relization and m_s^2 represents the variance due to the meandering.

For sufficiently long time after the injection (i.e. after the scale of diffusion becomes larger than the largest eddy sizes) m_s^2 becomes constant, so that it does not contribute to the diffusivities in (1.26). The time required for this to happen is typically the *Lagrangian time scale* (Fischer *et al.*, 1979).

For time larger than the Lagrangian time scale, the diffusivities are constant (i.e. the variance increases linearly with time in 1.26), and the solution of (1.24) with constant coefficients is appropriate in this case. On the other hand, for initial time after the release of a small patch, this approach is not valid, for then the diffusion is proportional to eddy sizes. Fischer *et al.* (1979) show that in this case the Fickian diffusion equation (1.23) is valid with respect to relative coordinates shifted to the instantaneous center of mass of the cloud, provided that the diffusivities are prescribed as

$$E = \frac{1}{2} \frac{d\sigma^2}{dt} = \alpha \, \sigma^{4/3} \,. \qquad (1.29)$$

This *"4/3 Law"* has experimentally been shown (Okubo, 1974) to apply to a wide range of diffusion problems (see Fig. 1.2); it simply states that the diffusivity increases as a power of the cloud size.

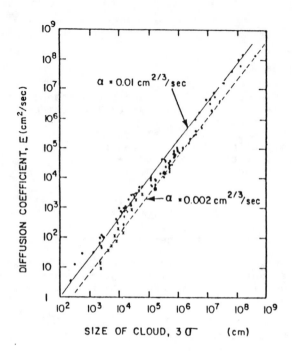

FIGURE 1.? The turbulent diffusion coefficient as a function of patch size (After Okubo,1974).

Variations to the thoery arising due to the consideration of eddy sizes, their stochastic bases and relation to statistical theory are discussed and interpreted with considerable latitude in Csanady (1973) and Fischer *et al.* (1979). In the following sections, we will mainly consider the cases in which the diffusivities are assumed to be constant.

2. SIMPLE MODELS OF TURBULENT DIFFUSION AND TRANSPORT

In this Section, solutions to the convective diffusion equation will be obtained under different initial and boundary conditions corresponding to typical simple situations that may be encountered in the environment. Although the real processes in the ocean can be more complex mainly due to the physical prescription of the yet undetermined turbulent diffusivity coefficients, these simple solutions will serve to illustrate the basic mechanics of diffusion. These classical solutions can be found in a number of basic references, including Csanady (1973) and Harleman (1970).

Equation (1.25) is in general not very easy to solve under general flow situations, often due to the variable coefficients introduced by the velocity field $u(x,t)$ determined by the equations of motion. An alternative approach is to lump the variability of the velocity field into the turbulent diffusivity coefficients, yielding *dispersion* equations to be demonstrated later in Section 3.

We consider a simple class of problems in which the velocity field is constant with speed U arbitrarily aligned with the x-axis. This case is analogous to the diffusion in a solid, when the equations are transformed to the coordinates fixed with respect to the uniform bodily motion of the fluid. We will assume homogeneous, non-isotropic turbulence and a non-conservative constituent which decays linearly. Namely, we consider the equation

$$\frac{\partial c}{\partial t} + U \frac{\partial c}{\partial x} = E_x \frac{\partial^2 c}{\partial x^2} + E_y \frac{\partial^2 c}{\partial y^2} + E_z \frac{\partial^2 c}{\partial z^2} - kc .$$

$$(2.1)$$

We define the coordinate transformations

$$X = \left(\frac{E}{E_x}\right)^{1/2}(x-Ut), \quad Y = \left(\frac{E}{E_y}\right)^{1/2}y, \quad Z = \left(\frac{E}{E_z}\right)^{1/2}z,$$

$$T = t,$$

$$(2.2.a\text{-}d)$$

so that the total advective rate of change (left hand side of 2.1) simplifies to

$$\frac{\partial c}{\partial t} + U \frac{\partial c}{\partial x} = \left(\frac{\partial X}{\partial t} \frac{\partial c}{\partial X} + \frac{\partial T}{\partial t} \frac{\partial c}{\partial T}\right) + U \left(\frac{\partial X}{\partial t} \frac{\partial c}{\partial X} + \frac{\partial T}{\partial t} \frac{\partial c}{\partial T}\right)$$

$$= -\left(\frac{E}{E_x}\right)^{1/2}U\frac{\partial c}{\partial X}+\frac{\partial c}{\partial T} + U\left(\frac{E}{E_x}\right)^{1/2} \frac{\partial c}{\partial X} = \frac{\partial c}{\partial T},$$

$$(2.3)$$

and the terms on the right hand side are transformed as

$$E_x \frac{\partial^2 c}{\partial x^2} = E_x \frac{\partial X}{\partial x} \frac{\partial}{\partial X} \left(\frac{\partial X}{\partial x} \frac{\partial c}{\partial X}\right) = E_x \frac{\partial^2 c}{\partial X^2}$$

$$(2.4)$$

We also set the yet undetermined constant E equal to

$$E^3 = E_x E_y E_z$$

$$(2.5)$$

i.e. one of the *invariants* of the anisotropic diffusivity tensor. With these transformations (2.1) becomes

$$\frac{\partial c}{\partial t} = E \left(\frac{\partial^2 c}{\partial X^2} + \frac{\partial^2 c}{\partial Y^2} + \frac{\partial^2 c}{\partial Z^2}\right) - kc .$$

$$(2.6)$$

We further make the transformation

$$c = \phi \, e^{-KT} \qquad\qquad (2.7)$$

upon which (2.6) is replaced by

$$\frac{\partial \phi}{\partial T} = E \left(\frac{\partial^2 \phi}{\partial X^2} + \frac{\partial^2 \phi}{\partial Y^2} + \frac{\partial^2 \phi}{\partial Z^2}\right) . \qquad\qquad (2.8)$$

This form of the equation in transformed variables is equivalent to the diffusion equation for a conservative substance in an isotropic field at rest. On the other hand, this equation is the familiar heat equation equivalently modelling heat conduction in an isotropic solid, for which the classical theory provides various solutions (e.g. Carslaw and Jaeger, 1959). We can in principle develop these classical solutions for (2.8), and back transform them by substituting (2.2.a-d), (2.5) and (2.7) to obtain solutions for (2.1).

We next consider the basic solution to the diffusion equation for a point source initial condition, and show how other solutions are developed from this basic solution.

2.1 THE BASIC SOLUTION TO DIFFUSION EQUATION (INSTANTANEOUS POINT SOURCE)

Consider the simple diffusion equation (isotropic conservative diffusion, stationary fluid)

$$\frac{\partial c}{\partial t} = E \, \nabla^2 c . \qquad\qquad (2.9)$$

We want to investigate the simple symmetric diffusion pattern in the case of an instantaneous point source, i.e. of some material injected at the point $x = x' = (x',y',z')$ released at some initial instant $t = 0$. We seek the solution to (2.9) with the initial condition

$$c(x,0) = \frac{M}{\rho} \, \delta(x-x') \, \delta(y-y') \, \delta(z-z')$$

$$= \frac{M}{\rho} \, \delta(\vec{x}-\vec{x}') \qquad\qquad (2.10)$$

where M is the total mass of the substance introduced and ρ is the density of the receiving fluid. Here $\delta(x-x')$ is the *Dirac delta "function"* with the important properties of

$$\int_{-\infty}^{+\infty} \delta(x) \, dx = 1$$

$$\int_{-\infty}^{+\infty} F(x')\delta(x-x')dx = F(x) \qquad\qquad (2.11.a-c)$$

$$\int_{-\infty}^{+\infty} e^{ikx} \, dk = 2\pi \, \delta(x) \ .$$

The delta function was introduced by the well known physicist Dirac in 1926, but it was later shown by Schwartz in 1950 not to be a "function", but rather a *generalized function* or *functional*, i.e. a set of functions which in some limiting case approach zero everywhere except at the relative origin $x=x'$, where its value becomes infinite. We can visualize a set of functions which monotonously decay away from the relative origin, such as in the case of the set constructed from $f_\alpha(x) = (\alpha\pi)^{-1/2} \exp[-(x-x')^2/\alpha]$ with varying values of α, as shown in Fig. 2.1. As $\alpha \to 0$, the peak at $x=x'$ becomes narrower and increases in height, approaching $\delta(x-x')$ in the limit.

The first property (2.11.a) of the delta function requires that the area under its curve be unity, the second (2.11.b) requires that its integral product with another function evaluates to the value of that function at the relative origin. The third property (2.11.c) states that it is the Fourier transform of unity.

If we integrate (2.10) in an infinite volume V_∞ enclosing the instantaneous source, we obtain from (2.11.a)

$$\int_{V_\infty} \rho c(x, 0) \, dV = M \iiint_{-\infty}^{+\infty} \delta(x-x') \delta(y-y') \delta(z-z') \, dxdydz = M$$
$$(2.12)$$

yielding the mass of the substance introduced, in agreement with the definition of concentration.

Since the source is located at an infinitely small point, the solution to (2.9) is expected to be radially symmetric. Writing (2.9) in the spherical coordinates (r, θ, ϕ) centered at the relative origin of the source and dropping the

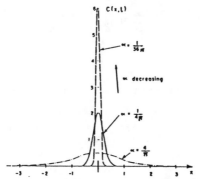

FIGURE 2.1 $f_\alpha(x) = (\alpha\pi)^{-1/2} \exp[-(x)^2/\alpha]$, the set of functions which reduce to the delta function as $\alpha \to 0$.

azimuthal and zonal terms (due to radial symmetry), the equation is

$$\frac{\partial c}{\partial t} = E \frac{1}{r^2} \frac{\partial}{\partial r} r^2 \frac{\partial c}{\partial r} \ . \tag{2.13}$$

where $r^2 = (x-x')^2 + (y-y')^2 + (z-z')^2$. The solution to (2.13) with the initial condition (2.10) can be obtained through various techniques, including Laplace transforms (Carslaw and Jaeger, 1959), or similarity transforms as will be presented here. We assume the solution is self-similar with the form

$$c = \zeta^{-m} f(\eta) \tag{2.14}$$

in the transformed coordinates

$$\eta = \frac{r^2}{4Et} \ , \qquad \zeta = 4Et \ . \tag{2.15}$$

The transformation to the original variables are

$$r = (\eta\zeta)^{1/2} \ , \quad t = \zeta/(4E) \ , \tag{2.16.a,b}$$

and the corresponding cross-derivatives are

$$\frac{\partial \zeta}{\partial t} = 4E \ , \qquad \frac{\partial \eta}{\partial t} = - \frac{r^2}{4Et^2} = - 4E \frac{\eta}{\zeta} \ ,$$

$$\frac{\partial \zeta}{\partial r} = 0 \ , \qquad \frac{\partial \eta}{\partial r} = \frac{2r}{4Et} = 2 \frac{\eta}{r} = 2(\frac{\eta}{\zeta})^{1/2}, \ . \tag{2.17.a-d}$$

The individual terms in (2.13) are then calculated as

$$\frac{\partial c}{\partial t} = \frac{\partial c}{\partial \zeta} \frac{\partial \zeta}{\partial t} + \frac{\partial c}{\partial \eta} \frac{\partial \eta}{\partial t}$$

$$= - 4E\zeta^{-m-1} (\eta f_\eta + mf) \tag{2.18}$$

and

$$\frac{1}{r^2} \frac{\partial c}{\partial r} (r^2 \frac{\partial c}{\partial r}) = \eta^{-1}\zeta^{-1} \frac{\partial \eta}{\partial r} \frac{\partial}{\partial \eta} (\eta\zeta \frac{\partial \eta}{\partial r} \frac{\partial c}{\partial \eta})$$

$$= \zeta^{-m-1} (\eta f_{\eta\eta} + \frac{3}{2} f_\eta) \tag{2.19}$$

where subscripts denote differentiation with respect to the transformed variable η.

With these substitutions we transform the partial differential equation (2.13) into the ordinary differential equation

$$\eta \, (f_{\eta\eta} + f_{\eta}) + (\tfrac{3}{2} \, f_{\eta} + mf) = 0 \qquad (2.20)$$

This equation is of second order, and general solutions can be obtained in series expansions. However, it is obvious that we can only obtain a similarity solution if we select the yet undetermined exponent in (2.14) as $m=3/2$. Then, by letting

$$g = f_{\eta} + f \qquad (2.21)$$

(2.20) directly integrates to

$$g = f_H + f = A \, \eta^{-3/2} \, . \qquad (2.22)$$

Integrating once more with use of integration factors yields

$$f = B \, e^{-\eta} + A \, e^{-\eta} \int \eta^{-3/2} \, e^{\eta} \, d\eta, \qquad (2.23)$$

where the first term represents the homogeneous solution and the second term is the particular solution.

For small values of $\eta = r^2/(4Et)$ (i.e. as $t \rightarrow \infty$), the second term grows with a trend of $t^{3/2}$. Since the initial concentration should decay with time through diffusion, this solution can not be accepted and therefore we set $A = 0$.

The constant B can be evaluated from the initial condition (2.10). In spherical coordinates we have

$$
\begin{aligned}
\frac{M}{\rho} &= \int_0^{2\pi} \int_0^{\pi} \int_0^{\infty} c(r) \, dr \, r \, d\phi \, r\sin\phi \, d\theta \\
&= 4\pi \int_0^{\infty} r^2 \, c(r) \, dr \\
&= 4\pi \int_0^{\infty} \eta \zeta \, \frac{f(\eta)}{\zeta^{3/2}} \, \frac{1}{2} \, (\frac{\zeta}{\eta})^{1/2} \, d\eta \\
&= 2\pi B \int_0^{\infty} \eta^{1/2} \, e^{-\eta} \, d\eta = \pi^{3/2} \, B \, .
\end{aligned}
\qquad (2.24)
$$

Therefore the solution of the problem becomes

$$c(r, t) = \frac{(M/\rho)}{(4\pi Et)^{3/2}} \, \exp(- \frac{r^2}{4Et}) \qquad (2.25)$$

or writing in Cartesian coordinates

$$c(x, y, z, t) = \frac{(M/\rho)}{(4\pi Et)^{3/2}} \exp- \, (\frac{(x-x')^2 + (y-y')^2 + (z-z')^2}{4Et}) \, . \qquad (2.26)$$

The instantaneous point source solution (2.25) decays as $t^{-3/2}$ and goes to zero everywhere as $t \to \infty$. At any fixed time the solution decays away from the relative origin as $\exp(-ar^2)$ (i.e. the spatial distribution is Gaussian). In fact, the behaviour of the solution can be visualized with the help of Fig. 2.1, replacing $\alpha = 4Et$. The shape of the function with respect to the radial distance measured from the source (instead of x in Fig. 2.1) is the same for any given time, although the time rate of decrease faster (as $t^{-3/2}$) than that would correspond to Fig. 2.1 as a result of the three-dimensionality of the problem. At any time, the constant concentration surfaces (of say $c=c_0$) are spheres.

The solution in the non-isotropic non-conservative, uniform flow case can be obtained by simply making use of (2.2.a-d), (2.5) and (2.7) as

$$c = \frac{M/\rho}{(4\pi Et)^{3/2}} \exp\left[-\left\{\frac{(x-x'-Ut)^2}{4E_x t} + \frac{(y-y')^2}{4E_y t} + \frac{(z-z')^2}{4E_z t} + kt\right\}\right]$$

(2.27)

where $E^3 = E_x E_y E_z$. In this case, the constant concentration surfaces are ellipsoids.

The variance (cf. equation 1.26) defined with respect to the center of the patch is

$$\sigma_s^2 = \frac{\int_{-\infty}^{+\infty} c\, \tilde{s}^2\, d\tilde{s}}{\int_{-\infty}^{+\infty} c\, d\tilde{s}}$$

(2.28)

where $\tilde{s}$ stands for any of the shifted coordinates $\tilde{x}=x-x'-Ut$ or $\tilde{y}=y-y'$ or $\tilde{z}=z-z'$ in equation (2.27). Noting that

$$\frac{\int_{-\infty}^{+\infty} r^2 \exp(-r^2/a)\, dr}{\int_{-\infty}^{+\infty} \exp(-r^2/a)\, dr} = \frac{(\pi a)^{1/2}\, a/2}{(\pi a)^{1/2}} = a/2$$

(2.29)

(2.28) evaluates to (for each axis)

$$\sigma_x = (2E_x t)^{1/2},$$
$$\sigma_y = (2E_y t)^{1/2},$$
$$\sigma_z = (2E_z t)^{1/2}$$

(2.30.a-c)

in the case of the instantaneous point source solution (2.31). If the medium is isotropic, the spread is obviously symmetric in all directions with $\sigma = (2Et)^{1/2}$. Note that these results are in agreement with equation (1.27).

2.2 CONSTRUCTION OF ELEMENTARY SOLUTIONS FROM THE BASIC SOLUTION

We can construct other elementary solutions from the basic solution obtained above, through *convolution* operations. For example consider the case of an initial concentration distribution C(x) at t=0

$$c(x,0) = C(x) = C(x,y,z) \quad . \tag{2.31}$$

Consider the simple diffusion equation (2.9) written as

$$L(c) = 0; \quad L = \frac{\partial}{\partial t} - E\nabla^2 \quad . \tag{2.32}$$

Let $\tilde{c}(x',t)$ be the basic solution of (2.32) for an instantaneous point source $\tilde{c}(x,0)=\delta(x-x')\delta(y-y')\delta(z-z')$ disregarding the dimensional coefficient M/ρ in (2.10). The solution to (2.32) with the initial condition (2.31) is constructed as

$$c(x,t) = \iiint_{-\infty}^{+\infty} C(x') \; \tilde{c}(x-x',t) \; dx'dy'dz' \tag{2.33}$$

where $\tilde{c}$ is (2.26) normalised with M/ρ. The proof is given as follows. Operating on (2.31) yields

$$L(c) = \iiint_{-\infty}^{+\infty} C(x') \; L(\tilde{c}) \; dx' \; dy' \; dz' = 0 \tag{2.34}$$

since L operates on x and t only and $\tilde{c}$ satisfies (2.32). Evaluating (2.33) at t=0, we obtain from (2.10)

$$c(x,0) = \iiint_{-\infty}^{+\infty} C(x') \; \delta(x-x') \; dx' \; dy' \; dz' = C(x) \tag{2.35}$$

by virtue of (2.11.b). The expression in (2.34) is therefore proved to be the solution. The method can be applied to arbitrary initial conditions, examples of which are given below.

2.3 INSTANTANEOUS LINE SOURCE

The concept of using the basic solution to construct other solutions is applied to the diffusion from an *instantaneous line source* with the initial condition

$$c(x,0) = (m'/\rho) \; \delta(x-x'') \; \delta(y-y'') \tag{2.36}$$

i.e. a point source in two-dimensions, located at $x=(x'',y'')$

with mass **m'**. Comparing with (2.10), the source strength **m'** is defined according to

$$m' = M \, \delta(z-z'),$$

$$\int_{-\infty}^{+\infty} m' \, dz = M,$$

(2.37.a-c)

$$m' \, dz = dM,$$

such that **m'** is distributed along the z-axis, with its total influence conceptually equalling **M**. Applying (2.38) we obtain

$$c(x,t) = \iiint_{-\infty}^{+\infty} \frac{m'}{\rho(4\pi Et)^{3/2}} \exp\left(-\frac{\bar{r}^2}{4Et}\right) \delta(x'-x'')\delta(y'-y'')\,dx'\,dy'\,dz'$$

(2.38)

where $\bar{r}^2 = (x-x')^2+(y-y')^2+(z-z')^2$.
Using (2.11.b) and integrating further yields the two-dimensional solution

$$c(x,t) = \frac{m'}{\rho 4\pi Et} \exp\left\{-\frac{(x-x'')^2+(y-y'')^2}{4Et}\right\}$$

(2.39)

The solution in the more general case (anisotropic, linear decay and uniform current) is easily obtained through the substitutions (2.2.a-d), (2.5) and (2.7).

An alternative interpretation of the convolution method rests in equations (2.38) and (2.37). We are equivalently summing up the influences of a sequence of point sources with strengths **dM** along the z-axis, to obtain the line souce solution.

Note that the instantaneous line source solution decays as t^{-1} as $t \to \infty$, at a slower rate compared to the point source solution. At any fixed time the spatial decay of the solution is again Gaussian.

A better measure of the spread in each direction is obtained from (2.32) yielding

$$\sigma_x = \sigma_y = (2Et)^{1/2}$$

(2.40)

i.e. the same as in the case of the point source solution.

2.4 INSTANTANEOUS PLANE SOURCE

We next consider an *instantaneous plane source*

$$c(x,0) = (m'''/\rho) \, \delta(x-x''')$$

(2.41)

where **m'''** represents the source strength

52

$$m'' = m' \ \delta(y-y''),$$
(2.42)

in comparison to (2.36). The solution can again be obtained by the same technique outlined in Section 2.2 as

$$c(x,t) = \iiint\limits_{-\infty}^{+\infty} \frac{m''}{\rho(4\pi Et)^{3/2}} \exp\left(-\frac{\tilde{r}^2}{4Et}\right) \ \delta(x'-x'') \ dx' \ dy' \ dz'$$
(2.43)

where $\tilde{r}^2 = (x-x')^2 + (y-y')^2 + (z-z')^2$.
Integrating further yields the solution

$$c(x,t) = \frac{m''}{\rho(4\pi Et)^{1/2}} \exp - \frac{(x-x'')^2}{4Et}$$
(2.44)

The one dimensional solution can also be obtained from a summation of the two dimensional solutions. The solution decays as $t^{-1/2}$ at large times, i.e. slower than the two and three dimensional instantaneous sources. The variance for the solution is again $\sigma_x = (2Et)^{1/2}$.

2.5 CONTINUOUS POINT SOURCE

We can use the above methods to construct solutions for *continuous sources* i.e. sources from which a substance is injected continuously, either for a certain period or for an infinite time. The construction technique of Section 2.2 is applicable in these cases, with summation of delta function inputs with respect to time. However, due to the complex time dependence of the basic solutions, we must apply the summation directly on the general case (2.27).

Consider the *continuous point source*, starting from an initial time $t=t_0$ and continuing up to time $t=t_1$, (Fig. 2.2) with a rate of mass injection $Q = dM/dt$. We idealize the situation as a summation of an infinite number of point sources progressing in time, each with a mass injection of dM per unit time increment $d\tau$.

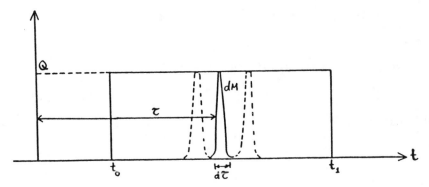

FIGURE 2.2 Idealization of the continuous point source as a series of instantaneous point sources.

53

We assume that the source is located at the origin $x=y=z=0$.
The solution can be constructed as

$$c = \int_{to}^{t\bullet} c_1(x, t-\tau) \, dM = \int_{to}^{t\bullet} q \, c_1(x, t-\tau) \, d\tau$$

(2.45)

where c_1 denotes the solution for the instateneous
solution (i.e. equation 2.27) divided by M, and t_x
the upper limit of integration in time. If we are interested
in the solution for the time interval $t_o < t < t_1$
(*during* continuous injection), we must integrate (2.45) up
to $t_x = t$. On the other hand, if we are interested in
$t > t_1$ (i.e. *after* the continuous source is stopped)
equation (2.45) must be integrated up to $t_x = t_1$.
Defining

$$\lambda^2 = \frac{E_x}{E_y} y^2, \quad \mu^2 = \frac{E_x}{E_z} z^2, \quad E^3 = E_x E_y E_z$$

(2.46)

and substituting (2.27), the solution (2.45) is written as

$$c = \int_{to}^{t\bullet} \frac{Q}{\rho [4\pi E(t-\tau)]^{3/2}} \exp{-\frac{[x-U(t-\tau)]^2 + \lambda^2 + \mu^2}{4E_x(t-\tau)}} d\tau$$

(2.47)

Note that a continuous point source with varying injection rate
$q(t)$ can also be taken into account by taking $q=q(t-\tau)$
in this equation. With the following definitions

$$r^2 = x^2 + \lambda^2 + \mu^2, \qquad \zeta = \frac{r}{2[E_x(t-\tau)]^{3/2}}$$

$$\nu = xU/(2E_x), \qquad \Omega = (U^2 + 4kE_x)^{1/2},$$

$$\beta = r\Omega/(4E_x)$$

(2.48.a-e)

equation (2.47) can alternatively be written as

$$c = \frac{Q \, e^{\nu}}{2\rho\pi^{3/2}(E_y E_z)^{3/2} r} \int_{\zeta_0}^{\zeta\bullet} \exp{-[\zeta^2 + \frac{\beta^2}{\zeta^2}]} \, d\zeta$$

(2.49)

where $\zeta_0 = \zeta(\tau=t_0)$ and $\zeta_\bullet = \zeta(\tau=t_\bullet)$.
Note that in the case $t_x = t$ (i.e. $t < t_1$),
$\zeta_\bullet = \infty$.

In order to integrate (2.49) we define

$$p = \zeta + \frac{\beta}{\zeta}, \qquad q = \zeta - \frac{\beta}{\zeta}$$

(2.50.a,b)

and consequently

$$p^2 - q^2 = 4\beta \ , \quad \frac{\beta}{\zeta^2} = \frac{p - q}{p + q}$$

(2.50.c,d)

Differentiating (2.50.a,b) gives

$$\frac{dp}{d\zeta} = 2 - \frac{p}{\zeta} = 1 - \frac{\beta}{\zeta^2} = \frac{2q}{p+q},$$

$$\frac{dq}{d\zeta} = 2 - \frac{q}{\zeta} = 1 - \frac{\beta}{\zeta^2} = \frac{2p}{p+q},$$

(2.51.a,b)

and therefore

$$p \, dp = q \, dq \ .$$

(2.52)

The integral in (2.56) can be formed into

$$\int \exp -(\zeta^2 + \frac{\beta^2}{\zeta^2}) \ d\zeta$$

$$= \frac{1}{2} \int [\exp(-p^2+2\beta)+\exp(-q^2-2\beta)] \ d\zeta$$

(2.53)

or, using (2.51.a,b)

$$\frac{1}{4} \int (\frac{p}{q}+1)\exp(-p^2+2\beta) \ dp + \frac{1}{4} \int (\frac{q}{p}+1)\exp(-q^2+2\beta) \ dq$$

(2.54)

which, with the help of (2.59) and (2.57.c) becomes

$$\frac{1}{2} \int \exp(-p^2+2\beta) \ dp + \frac{1}{2} \int \exp(-q^2-2\beta) \ dq$$

(2.55)

As a result, the solution (2.49) becomes

$$c = \frac{Qe^\nu}{8\pi\rho(EyEx)^{1/2}r}[e^{+2\beta}(erfc \ p)\Big|_{P_*}^{P_0}+e^{-2\beta}(erfc \ q)\Big|_{q_*}^{q_0}]$$

(2.56)

where evaluations are between the limits $p_0=p(\zeta_0)$, $q_0=q(\zeta_0)$, $p_*=p(\zeta_*)$, $q_*=q(\zeta_*)$.

Note that the transient solution (2.56) in the special case $U=K=0$ ($\beta=0$) becomes

$$c = \frac{Q}{4\pi\rho(EyEz)^{1/2}r}\{erfc\frac{r}{2[Ex(t-to)]}+erfc\frac{r}{2[Ex(t-t*)]}\}$$

(2.57)

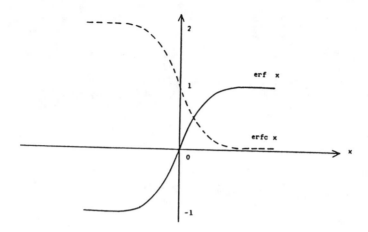

FIGURE 2.3 Error function and complementary error function.

In the above, the *error function* is defined as

$$erf\ u = \frac{2}{\pi^{1/2}} \int_0^u exp\ -s^2\ ds$$

(2.58.a)

and the *complementary error function* as

$$erfc\ u = 1 - erf\ u \qquad (2.58.b)$$

which are sketched in Fig. 2.3.

In the case of continuous injection, it is possible to have a steady solutions, which is obtained by letting $t_x=t$, $t_0 \to -\infty$ in (2.56). The lower limit of the integral (corresponding to $\tau=t_0$) becomes $\zeta_0=0$ and the upper limit $\zeta_\bullet \to -\infty$ in this case, and a definite integral results, yielding the steady state solution

$$c = \frac{Q}{\rho 4\pi(E_y E_z)^{1/2} r} exp\ -[2\beta-\nu] \qquad (2.59.a)$$

Consider the case $k=0$, then (2.59.a) becomes

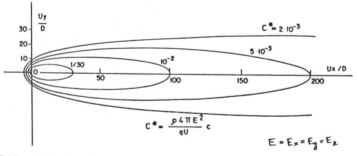

FIGURE 2.4 Solution for a continuous point source

$$c = \frac{Q}{\rho 4\pi (E_y E_z)^{1/2} r} \exp -[\frac{U(r-x)}{2E_x}] \qquad (2.59.b)$$

These solutions are shown in Fig 2.4. At long distances away from the origin along the x-axis, i.e. for $x \gg \lambda$ and $x \gg \mu$, we can simplify and (2.48.a) to

$$r = (x^2 + \lambda^2 + \mu^2)^{1/2} = x (1 + \frac{\lambda^2 + \mu^2}{x^2})^{1/2} \approx x + \frac{1}{2} \frac{\lambda^2 + \mu^2}{x} \qquad (2.60)$$

so that (2.59.b) can be approximated as

$$c = \frac{Q}{\rho 4\pi (E_y E_z)^{1/2} x} \exp - \{\frac{U}{4x}(\frac{y^2}{E_y} + \frac{z^2}{E_z})\} \qquad (2.61)$$

i.e. the distribution becomes two dimensional at large x.

At large distances along the x-axis the solution has *boundary layer* structure, with diffusion occuring transverse to the flow, and negligible diffusion along x. A particle released at the origin is swept to any point x in a time duration of $\bar{t} = x/U$; which upon substitution into (2.68) yields the familiar (two dimensional) instantaneous line source solution (2.44) with $m'U = Q$, $\bar{t} = t$, in this case the line source being oriented along x. It can be verified that this is equivalent to the solution of the system

$$U \frac{\partial c}{\partial x} = E_y \frac{\partial^2 c}{\partial y^2} + E_z \frac{\partial^2 c}{\partial z^2} \qquad (2.62)$$

2.6 CONTINUOUS LINE SOURCE

Solutions for the case of a *continuous line source* aligned with the z-axis (at x=0, y=0) with an injection rate of $q' = dm'/dt$ can be obtained either by summation of continuous point source solutions in space or the summation of instantaneous line source solutions in time. With the definitions

$$r^2 = x^2 + \lambda^2, \qquad \lambda^2 = \frac{E_x}{E_y} y^2, \qquad \varepsilon = \Omega^2 (t - \tau) \qquad (2.63.a-c)$$

and ν, Ω, β as defined in (2.48.c-e), the summation of instantaneous line sources yields (Harleman,1970)

$$c = \frac{q' e^{\nu}}{4\pi \rho (E_x E_y)^{1/2}} \int_{\varepsilon_\bullet}^{\varepsilon_0} \frac{1}{\varepsilon} \exp -[\varepsilon + \frac{\beta^2}{\varepsilon}] d\varepsilon \qquad (2.64)$$

where $\varepsilon_0 = \varepsilon(\tau = t_0)$ and $\varepsilon_\bullet = \varepsilon(\tau = t_\bullet)$.

This general form can be integrated numerically. Note that in the case $t_x = t$, $\varepsilon_* = 0$.

The steady state solution is obtained by letting $t_o \to \infty$. In the case $t_x = t$, this is

$$c = \frac{q'e^{\nu}}{2\pi\rho(E_x E_y)^{1/2}} K_0(2\beta) \qquad (2.65)$$

where K_0 is the modified Bessel function of the second kind and order zero. For large values of r (i.e. $\beta >> 1/2$) we can approximate (2.65) as

$$c = \frac{q'}{4\rho(\pi\beta E_x E_y)^{1/2}} \exp -[2\beta - \nu] \qquad (2.66)$$

2.7 CONTINUOUS PLANE SOURCE

Consider a continuous plane source (an assemblage of continuous line sources or instantaneous plane sources) in the y-z plane positioned at the origin x=0. Let the strength of the source be $q'' = dm''/dt$. Defining

$$r = x, \qquad y = \Omega \left(\frac{t-\tau}{4E_x}\right)^{1/2} \qquad (2.67.a,b)$$

and y, Ω, β as in (2.48.c-e) the solution can be obtained by the methods outlined above, resulting in

$$c = \frac{2q''e^{\nu}}{\rho\pi^{1/2}\Omega} \int_{y_*}^{y_0} \exp -[y^2 + \frac{\beta^2}{y^2}] \, dy \ . \qquad (2.68)$$

Comparing with (2.49) and (2.56) the solution is evaluated as

$$c = \frac{q''e^{\nu}}{2\rho\Omega}[\exp(+2\beta)(\mathrm{erfc}\ p)\Big|_{p_*}^{p_0} + \exp(-2\beta)(\mathrm{erfc}\ q)\Big|_{q_*}^{q_0}] \qquad (2.69)$$

where p, q, p_0, q_0, p_x, q_x are defined the same way as in (2.57.a,b) replacing ζ by y.

2.8 INFLUENCE OF FINITE SOURCE DIMENSIONS

The influence of finite source dimensions can in principle be accounted for through the superposition techniques outlined above. As an example, we consider the case of an instantaneous line source confined in $-h < z < h$ at $x'=0$, $y'=0$. For brevity, we consider the isotropic, conservative non-convective case. In this case, the solution (2.38) is modified accordingly, to

$$c = \int_{-h}^{+h} \frac{m'}{\rho(4\pi Et)^{3/2}} \exp - \frac{\hat{r}^2}{4Et} \, dz' \qquad (2.70)$$

where $\hat{r}^2 = x^2 + y^2 + (z-z')^2$. Making the substitutions

$$\mu(z') = \frac{(z-z')}{(4Et)^{1/2}}, \qquad dz' = -(4Et)^{1/2} \, d\mu$$

$$\qquad\qquad (2.71.a,b)$$

yields the solution

$$c = \frac{m'}{\rho\pi^{3/2} 4Et} \exp - \frac{x^2+y^2}{4Et} \int_{\mu(+h)}^{\mu(-h)} \exp - \mu^2 \, d\mu$$

$$\qquad\qquad (2.71.c)$$

$$= \frac{m'}{8\rho\pi Et} \exp - \frac{x^2+y^2}{4Et} [\mathrm{erf} \frac{z+h}{(4Et)^{1/2}} - \mathrm{erf} \frac{z-h}{(4Et)^{1/2}}] .$$

Note that the solution is symmetric about z=0, since
$\mathrm{erf}(-r) = \mathrm{erf}(+r)$.

2.9 INFLUENCE OF BOUNDARIES

On impervious boundaries the diffusive flux must vanish normal to the boundary surface,

$$\hat{n} \cdot \nabla c = 0 \qquad\qquad (2.72)$$

where $\hat{n}$ in the unit normal to such a surface.

In some simple cases, the solution with this boundary condition is equivalent to superposition of mirror images with respect to the boundary of the unbounded solutions. To illustrate this method consider an instantaneous point source positioned at $y'=z'=0$ and $x=L$ where x is measured perpendicular to a boundary at x=0 and in the y,z plane. The solution is

$$c = \frac{M}{\rho(4\pi Et)^{3/2}} \exp - \frac{y^2+z^2}{4Et} [\exp - \frac{(x-L)^2}{4Et} + \exp - \frac{(x+L)^2}{4Et}]$$

$$\qquad\qquad (2.73)$$

where the second term is due to an image at x=-L. Note that the solution is symmetric with respect to x=0 and if the source is located next to the boundary, the concentration is increased to twice the value of the unbounded solution. In more complicated cases, the solution is obtained through superposition or other mathematical techniques. Consider an initial vertical distribution $c_0(z)$ in a two dimensional uniform flow with finite depth as shown in Fig 2.5. The initial condition is

$$c(x,z,0) = c_0(z) \, \delta(x) \qquad\qquad (2.74)$$

The diffusion pattern is governed by a convective diffusion equation, which, upon transforming the x coordinate by

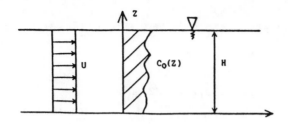

FIGURE 2.5 Diffusion in confined flow

$$X = x - Ut \qquad (2.75)$$

can be written as

$$\frac{\partial c}{\partial t} = E_x \frac{\partial^2 c}{\partial X^2} + E_y \frac{\partial^2 c}{\partial z^2} . \qquad (2.76)$$

The boundary conditions are

$$\frac{\partial c}{\partial z} = 0, \qquad \text{on} \quad z = 0, H, \qquad (2.77)$$

by virtue of (2.72). A Fourier cosine series solution

$$c = \sum_{n=0}^{\infty} \phi_n(X, t) \cos \frac{n\pi z}{H} \qquad (2.78)$$

is assumed, satisfying the boundary conditions (2.77). The solution satisfying (2.76) is then obtained as follows

$$c = \frac{1}{(4\pi E_x t)^{1/2}} \exp{-\frac{(x-Ut)^2}{4E_x t}} \sum_{n=0}^{\infty} a_n \exp{-(\lambda_n^2 t)}\cos\frac{n\pi z}{H} \qquad (2.79.a)$$

where

$$\lambda_n^2 = (n\pi/H)^2 E_z \qquad (2.79.b)$$

and

$$a_0 = \frac{1}{H} \int_0^H c_0(z) \, dz, \quad n=0 \qquad (2.79.c, \partial)$$

$$a_n = \frac{2}{H} \int_0^H c_0(z) \cos \frac{n\pi z}{H} \, dz, \quad n=1,2,\ldots.$$

Note that each term in the series solution (2.79.a) decays

in a time of $T_n = 1/\lambda_n^2 = (H/n\pi)^2/E_z$
approximately, except the term $n=0$, which survives as a
dominant term. Therefore, at large times
$(t >> H^2/(\pi^2 E_z))$, the solution is

$$c = \frac{a_Q}{(4\pi E_x t)^{1/2}} \exp - \frac{(x-Ut)^2}{4E_x t} \qquad (2.80)$$

i.e. uniformly distributed with depth and equivalent to that of
an instantaneous plane source with strength $a_0 = m''/\rho$
(cf. equation 2.44).

2.10 INFLUENCE OF VARIABLE DIFFUSION COEFFICENTS

As noted in Section 1.4, the assumption of constant diffusivity
is actually inappropriate shortly after the release of a small
source. Another reason for variable diffusivities can be the
presence of a solid boundary, because near the boundaries the
structure of the turbulence is modified and eddies decrease in
size. Furthermore, the texture of the boundaries can also be
important, since the turbulence field near a flat surface will
differ from that near a rough surface. Modifications of the
solutions which describe the initial growth stages and
diffusion near boundaries have been obtained by various
investigators, for example by Joseph and Sendner (1958), and
Okubo(1962) in the case of instantaneous line sources,
Walters(1962), Sutton(1953) and Smith (1957) in the case of
continuous point souces, and Pasquill (1962) in the case of
continuous line sources located on solid boundaries.

For these more advanced diffusion theories, the reader can
consult Slade(1968), Frenkiel and Munn(1974), Csanady(1973) and
Fischer *et al.* (1979).

3. SHEAR FLOW DISPERSION

3.1 INFLUENCE OF VELOCITY SHEAR

The dramatic effects of velocity shear will first be
demonstrated by a simple solution due to Okubo and Karweit
(1969) who considered the linear velocity profile

$$u = U + \alpha y + \beta z \qquad (3.1)$$

for the x-component of velocity and obtained a solution to
equation (1.25) for an instantaneous point source at $x=0$, $y=0$,
$z=0$ and time $t=0$. The solution was obtained as

$$c = \frac{M}{\rho(4\pi E t)^{3/2}} \exp -(\frac{\varepsilon^2}{4E_\varepsilon t} + \frac{y^2}{4E_y t} + \frac{z^2}{4E_z t} + kt) \qquad (3.2)$$

where

$$\varepsilon = x - Ut - \frac{1}{2}(\alpha y + \beta z)t, \qquad (3.3.a)$$

$$E_\xi = E_x (1 + \Phi^2 t^2)$$

(3.3.b)

$$E^3 = E_\xi E_y E_z$$

(3.3.c)

and

$$\Phi^2 = (\alpha^2 E_y + \beta^2 E_z)/(12 E_x)$$

(3.3.d)

It can be seen from the above that shortly after the release $(\Phi t)^2 \ll 1$ the influence of shear is unimportant and the solution is very similar to the case without shear in (2.27). On the other hand, for large time $(\Phi t)^2 \gg 1$, the peak concentration decays as $t^{-5/2}$, much faster than the uniform flow solution with $t^{-3/2}$ decay. In effect and by virtue of (3.3.b) the effective diffusivity is considerably increased for large time, due to the elongating influence of shear. Also note that for large time we can approximate (3.3.b) as

$$E_\xi \approx E_x (\Phi t)^2 = \frac{1}{2} \frac{\partial \sigma_\xi^2}{\partial t}$$

(3.4)

which yields

$$\sigma_\xi^2 = \frac{2}{3} E_x \Phi^2 t^3$$

(3.5)

and (3.4) can alternatively be expressed as

$$E_\xi \approx (\frac{3\Phi}{2E_x})^{2/3} \sigma_\xi^{4/3}$$

(3.6)

which is analogous to the "4/3 Law" (cf. 1.29) for relative diffusion but arises in the much different context of shear flow.

The above example displays the important and convenient result that the effect of shear can be incorporated into *"dispersion coefficients"* in anology to the diffusivities.

The shear flow dispersion behaviour differs considerably when the flow is confined between boundaries, for example in a shallow sea, where the no flux condition (2.72) applies at the bottom and the free surface. The following method is based on the early analyses of Taylor (1953, 1954), Aris (1956), Elder (1959), Bowden(1965) and Fischer(1967). The formulation of the shallow water equations will closely follow that of Nihoul and Adam(1974). We start with equation (1.25), or alternatively,

$$\frac{\partial c}{\partial t} + \nabla \cdot \vec{u} c + \frac{\partial}{\partial z} wc = \frac{\partial}{\partial z} E_z \frac{\partial c}{\partial z} + T - kc$$

(3.7)

where use has been made of the continuity equation (1.3) and $\nabla = (\partial/\partial x, \partial/\partial y)$ and $\vec{u} = (u, v)$ stand for the horizontal components of the gradient and the velocity vectors, w is the vertical velocity and T represents the

horizontal diffusion terms $T=\nabla \cdot E_H \cdot \nabla c$. We
assume that the horizontal velocity and concentration can be
separated into vertically averaged and fluctuating (deviation
from the vertical average) components

$$c = \bar{c} + c"$$
$$u = \vec{\bar{u}} + u" \qquad\qquad (3.8.a,b)$$

where for instance

$$\bar{c} = \frac{1}{H} \int_{-h}^{\eta} c \, dz . \qquad\qquad (3.9)$$

The free surface and the bottom are respectively defined at
$z=\eta$ and $z=-h$, bounding a total depth $H=\eta+h$.
Integrating the continuity equation (1.3) in the vertical
yields

$$\frac{\partial H}{\partial t} + \nabla \cdot H\vec{\bar{u}} = 0. \qquad\qquad (3.10)$$

Then, integrating (3.7) in the vertical and making use of
(3.10), we have

$$\frac{\partial \bar{c}}{\partial t} + \vec{\bar{u}} \cdot \nabla \bar{c} = - \frac{1}{H}\nabla \cdot H\overline{\vec{u}"c"} + \bar{T} - k\bar{c} \qquad\qquad (3.11)$$

Subtracting (3.11) from (3.7) gives

$$\frac{\partial c"}{\partial t} + \vec{\bar{u}} \cdot \nabla c" + \vec{u}" \cdot \nabla c" + w\frac{\partial c"}{\partial z} + \vec{u}" \cdot \nabla \bar{c}$$

$$= \frac{\partial}{\partial z}E_z\frac{\partial c"}{\partial z} + T" + \frac{1}{H}\nabla \cdot H\overline{\vec{u}"c"} - kc" . \qquad (3.12)$$

We assume $\vec{u}$ (and therefore $\vec{\bar{u}}$ and $\vec{u}"$) are known;
equation (3.12) can then be solved for $c"$. With the
correlation $\overline{\vec{u}"c"}$ determined, the dispersion equation (3.11)
can then be solved for $\bar{c}$, using some simplifying
assumptions first introduced by Taylor (1953). Through an
order of magnitude analysis it can be shown that the basic
balance in (3.12) is

$$\vec{u}" \cdot \nabla \bar{c} = \frac{\partial}{\partial z}E_z\frac{\partial c"}{\partial z}, \qquad\qquad (3.13)$$

which is much easier to integrate.

The basic assumptions are that $c"<<c$, $|\vec{u}"|=O(|\vec{\bar{u}}|)$,
$w=O(|\vec{u}"|H/L)$, where H is the depth and L is the
horizontal scale. Defining

$$c"/\bar{c} = O(\epsilon), \qquad\qquad |\vec{u}"/\vec{\bar{u}}| = O(1)$$
$$w/|\vec{u}"| = O(\mu), \qquad\qquad H/L = O(\delta) \qquad (3.14.a-d)$$

where ϵ, μ, δ are small numbers $<< O(1)$. The orders of each term (written in the same sequence as 3.12) are

$$
O(1) + O(1) + O(1) + O(\frac{\mu}{\delta}) + O(\frac{1}{\epsilon})
$$

$$
= O(\frac{E_z}{HU} \frac{1}{\delta}) + O(\frac{E_H}{LU}) + O(1) + O(\frac{kL}{U}),
$$

(3.15)

where U is a velocity scale and E_H stands for the horizontal diffusivites E_x, E_y. Shear dispersion effects obviously become important only when diffusion time scales are comparable with convection time scales i.e. E_z and $E_H = O(HU)$. The decay term $O(kL/U)$ is often small or at most $O(1)$. Therefore the last term on the left hand side and the first term in the right hand side (the terms of $O(\epsilon^{-1})$ and $O(\delta^{-1})$) dominate, yielding (3.13).

Equation (3.13) can be integrated twice, resulting in

$$
c''(z) - c''(0) = (\int_{-h}^{z} \frac{1}{E_z} \int_{-h}^{z} \vec{u}'' \, dz' \, dz'') \cdot \nabla \bar{c} .
$$

(3.16)

Then the dispersive flux term on the right hand side of (3.11) becomes

$$
S = - \frac{1}{H} \nabla \cdot \overline{H \vec{u}'' c''}
$$

(3.17)

$$
= - \frac{1}{H} \nabla \cdot \int_{-h}^{\eta} \vec{u}'' \int_{-h}^{z} \frac{1}{E_z} \int_{-h}^{z''} \vec{u}'' \, dz' \, dz'' \cdot \nabla \bar{c} \, dz
$$

$$
= - \frac{1}{H} \frac{\partial}{\partial x_i} \int_{-h}^{\eta} u_i'' \int_{-h}^{z} \frac{1}{E_z} \int_{-h}^{z''} u_j'' \, dz' dz'' dz \frac{\partial \bar{c}}{\partial x_j},
$$

the last expression being written in indical notation, with x_i for $i=1,2$ representing x and y directions respectively. Note that the integration of the second term on the left hand side of (3.16) does not contribute to (3.17) since $c''(0)$ is constant.

The form of (3.17) suggests that we can express this term in linear proportion to the local gradients of mean concentration in analogy to the Fickian expression (1.21), i.e.

$$
S = \frac{1}{H} \frac{\partial}{\partial x_i} H K_{ij} \frac{\partial \bar{c}}{\partial x_j}
$$

(3.18)

so that we can write

$$K_{ij} = -\frac{1}{H} \int_{-h}^{\eta} u_i'' \int_{-h}^{z} \frac{1}{E_z} \int_{-h}^{z''} u_j'' \, dz' \, dz'' \, dz$$

(3.19)

where K_{ij} is the horizontal *dispersion tensor*.

Note that the dispersion terms in (3.18) are generally anistropic, depending on the three dimensional structure of the integrated velocity components. Making the substitution (3.17) and dropping overbars, equation (3.11) describing the dispersion of vertically averaged concentration c becomes

$$\frac{\partial c}{\partial t} + \vec{u} \cdot \nabla c = \frac{1}{H} \nabla \cdot H \underline{\hat{E}} \cdot \nabla c - kc$$

(3.20)

where

$$\underline{\hat{E}} = \hat{E}_{ij} = K_{ij} + E_i \, \delta_{ij}$$

(3.21)

for i=1,2 where K_{ij} is the dispersion tensor, E_i stand for the vertical averages of the horizontal turbulent diffusivities Ex and Ey (in principal coordinates), and $\hat{E}_{ij}$ is the total dispersion tensor. The dispersion coefficients K_{ij} depend on the variable current distributions, and therefore it is often not possible to write either K_{ij} or $\hat{E}_{ij}$ in principle coordinates, since the orientation of these coordinates are subject to change with horizontal position. We must therefore use the full anisotropic form of (3.20). Since the vertical assymmetries of the current and concentration profiles contributing to (3.19) are predominant, it is natural to expect that $\hat{E}_{ij} \approx K_{ij}$ and therefore to neglect the turbulent diffusivity. Note that if the depth variations are small, we can write (3.20) as

$$\frac{\partial c}{\partial t} + \vec{u} \cdot \nabla c = \nabla \cdot \underline{\hat{E}} \cdot \nabla c - kc$$

(3.22)

The reader is referred to Fischer *et al.* (1979) and Nihoul and Adam (1974) for examples of two-dimensional dispersion.

3.2 LONGITUDINAL DISPERSION

If we consider unidirectional steady flows in the x-direction that are bounded in a cross sectional area of A, and neglect the decay term, it can be verified that the corresponding equation (3.20) becomes

$$\frac{\partial \bar{c}}{\partial t} + \bar{u} \frac{\partial \bar{c}}{\partial x} = \frac{1}{A} \frac{\partial}{\partial x} A(K_x + \bar{E}_x) \frac{\partial \bar{c}}{\partial x}$$

(3.23)

where u and c are respectively the sectionally averaged velocity and concentration, $\bar{E}_x$ the sectional averaged longitudinal turbulent diffusivity (in the flow direction).

$$K_x = - \frac{1}{A\frac{\partial \bar{c}}{\partial \varepsilon}} \int_A \overline{u''c''} \, dA$$

(3.24)

is defined as the *longitudinal dispersion coefficient* where

$$\varepsilon = x - \bar{u} \, t$$

(3.25)

and c" is solved from

$$u'' \frac{\partial \bar{c}}{\partial x} = \frac{\partial}{\partial y} E_y \frac{\partial c''}{\partial y} + \frac{\partial}{\partial z} E_z \frac{\partial c''}{\partial z} \, ,$$

(3.26)

in analogy to (3.13).

With this approach, Taylor (1953) solved the dispersion problem for shear flow in a pipe, taking the laminar equivalent of (3.26) written in cylindrical coordinates, from which c" is calculated for laminar velocity profiles. Then he determined the longitudinal dispersion coefficient from (3.24), yielding

$$K_x = \frac{R^2 u_m^2}{192 D_r}$$

(3.27)

where R is the radius of the tube, u_m the centerline velocity and D_r the radial molecular diffusivity. In his later work, Taylor (1954) extended his analysis to the turbulent shear flow in a pipe. Using the *Reynolds analogy*, which states that the transfer of mass, heat, momentum and turbulence are exactly analogous, he was able to relate the concentration and velocity profiles to the turbulent diffusivity, and obtained

$$E_L = K_x + E_x = 10.1 \, R \, u_*$$

(3.28)

where u_* is the *shear velocity* (defined as $u_* = (\tau/\rho)^{1/2}$, τ being the shear stress).

Note that (3.27) and (3.28) constitute two different ways of writing the dispersion coefficient. The difference implies (considering the turbulent equivalent of (3.27)) that the cross-stream turbulent diffusivity can be expressed as $E_r = \alpha R u_*$, where α is constant. In fact Taylor (1954) obtained (3.28) as $E_L = (10.04 + 0.06) h u_*$ ($\alpha = 0.06$), showing the negligible contribution of the turbulent diffusivity.

Elder (1959) applied the same technique to two dimensional unidirectional flow with infinitely wide horizontal extent and constant depth h. In this case (3.24) becomes

$$\frac{\partial c}{\partial t} + u \frac{\partial c}{\partial x} = E_L \frac{\partial^2 c}{\partial x^2} \, ,$$

(3.29)

where $E_L = K_x + E_x$ and

$$K_x = -\frac{1}{h} \int_0^h u'' \int_0^z \frac{1}{E_z} \int_0^{z''} u'' \, dz' \, dz'' \, dz$$

(3.30)

and for a logarithmic velocity profile

$$E_L = 5.93 \, hu_*\, .$$

(3.31)

is obtained. Note that if we use the non-dimensional variables

$$\eta = z/h, \qquad \phi = u''/\sigma_u, \qquad e = E_z/\bar{E}_z,$$

(3.32.a-c)

where

$$\sigma_u^2 = \overline{u''^2} = \frac{1}{h} \int_0^h u''^2 \, dz,$$

$$\bar{E}_z = \frac{1}{h} \int_0^h E_z \, dz \, ,$$

(3.32.d,e)

we can also express (3.30) as

$$K_x = \frac{h^2 \sigma_u^2}{\bar{E}_z} I$$

(3.33)

where

$$I = -\int_0^1 \phi \int_0^\eta \frac{1}{e} \int_0^{\eta''} \phi \, d\eta' \, d\eta'' \, d\eta$$

(3.34)

Fischer *at al.* (1979) note that the dimensionless integral
I has values of 0.05-0.1 for most practical flows,
so that it may suffice to take I=0.1. The rms
amplitude of the velocity deviation from the mean is lumped
into the parameter σ_u, and E_z is the mean
vertical diffusivity in (3.33).
Bowden (1965) considered various velocity and vertical eddy
diffusivity distributions and showed that for those cases
considered, values of E_L/hu_* (cf. equation 3.29)
ranged between 5.9-25. On the other hand, observations in
natural water courses of limited width indicated considerably
higher values: For example Fischer (1967) reported values of
E_L/hu_* in the range 50-700, and to account for
the large difference with theory, it was proposed that lateral
(transverse) shear effects not considered so far could have
caused the disperancy. Fischer(1967) has in fact argued that,
since the transverse mixing time scale in wide channels should
be larger than the vertical mixing time scales, transverse
shear could have a more predominant effect on longitudinal
dispersion. This is an apparent paradox, since with increasing
widths we do not apparently obtain the case of the infinitely
wide channel (the two dimensional case), but has since been

resolved through both theory and experiments.

With this motive, Fischer (1967) considered the lateral shear acting on the vertically averaged velocities and first, averaged the velocity distribution vertically as

$$\tilde{u}^{"}(y) = \frac{1}{h(y)} \int_{-h(y)}^{0} u^{"}(y,z) \, dz = \frac{q^{"}(y)}{h(y)} \qquad (3.35)$$

where $u^{"}$ is the deviation of the velocity field from the cross-sectional average u, and $h(y)$ is the lateral depth variations. By requiring that the $T^{"}$ terms in equation (3.12) balance the left hand side of (3.13) instead of the vertical diffusion terms, the longitudinal dispersion coefficient is then calculated as

$$E_L = -\frac{1}{A} \int_0^W q^{"}(y) \int_0^y \frac{1}{E_y h(y^{"})} \int_0^{y^{"}} q^{"}(y') \, dy' \, dy^{"} \, dy \qquad (3.36)$$

where E_y is the transverse eddy diffusivity, w is the width and A the cross-sectional area of the channel.

In actual water courses, a number of other effects modify the dispersion, such as the actual three-dimensional channel cross-sections, secondary flows, channel irregularities etc., so that some empirical judgement enters the dispersion formulations. Considering these effects, Fischer (1975) gives an estimate of the longitudinal dispersion coefficient in analogy to (3.31) as

$$E_L = 0.0011 \frac{u^2 w^2}{h u_x} \qquad (3.37)$$

where u, w, h are the mean velocity, width and depth of the channel and u_x the friction velocity. Reasonable agreement with observations is reported (Fischer et al., 1979). In laterally confined flows such as in rivers, estuaries and continental shelves, the transverse mixing effects are important and should be taken into consideration.

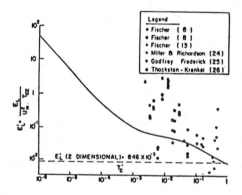

FIGURE 3.1 Longitudinal Dispersion coefficient in a rectangular channel (After Taylor, 1974).

While Fischer's analysis accounting for transverse mixing indicates that these effects can increase the dispersion, it provides little insight into the problem of longitudinal dispersion due to three-dimensional velocity and concentration distributions. Taylor (1974) has considered the turbulent flow constrained by both horizontal and vertical boundaries of rectangular cross-section, and has obtained exact solutions for velocity and concentration distributions, from which the longitudinal dispersion is calculated (from 3.25). Although an oversimplifying assumption of constant turbulent diffusivities has been used, Taylor's(1974) results show increasing dispersion effects for increasing aspect ratios as displayed in Fig. 3.1. Here, the non-dimensional variables are defined as

$$E_L' = \frac{E_L}{U_m^2 T_{cz}} = \frac{E_L}{U_m^2 h^2/E_z}$$

$$T_c' = \frac{T_{cz}}{T_{cy}} = \frac{h^2/E_z}{w^2/E_y} = \frac{h^2 E_y}{w^2 E_z}$$

(3.38.a,b)

where U_m is the maximum (centerline) velocity, h is the height and w is the width of the channel, and T_{cz} and T_{cy} are respectively the vertical and lateral mixing time scales.

The infinitely wide channel case (two-dimensional flow) is shown by the dotted lines Fig. 3.2 and corresponds to $E_L = 8.46\times10^{-3} U_m^2 h^2/E_z$ (which is in analogy to (3.31), but has a different form due to the assumption of constant E_z). Note that the dispersion coefficient for the rectangular section (solid line) does not approach the two-dimensional solution as $w\to\infty$ ($T_c'\to0$) and in fact differs by a large factor from this case. The three-dimensional problem includes the lateral shear effect, which is present no matter how wide the channel, while the two-dimensional problem has no such effect by definition. Comparison with various field and laboratory data indicates the increasing trend with increasing T_c' values, in spite of the different flow geometries and subjective evaluations of the diffusion coefficients.

3.3 DISPERSION IN OSCILLATORY SHEAR FLOW

The analysis of dispersion in oscillatory shear flow is more complex than the steady unidirectional flows considered above, mainly due to two reasons. Firstly, the unsteady and convective terms in (3.12) must be kept in addition to those already appearing in (3.13). Secondly, oscillatory motions create phase lags between concentration and velocity distributions both in space and time. However, assuming a single frequency of oscillation and by averaging the equations both vertically and in time, a mean dispersion coefficient in analogy to Taylor's(1954) hypothesis for unidirectional flows can be defined as

$$\langle E_L \rangle = \frac{1}{\partial c/\partial \varepsilon} \langle \overline{u^* c^*} \rangle \qquad (3.39.a)$$

where the velocity and concentration are decomposed into vertically averaged and deviational components

$$c = \bar{c}(t) + c^*(z, t)$$
$$u = \bar{u}(t) + u^*(z, t) \qquad (3.39.b,c)$$

following the notation of the earlier section, and where

$$\varepsilon = x - \int_0^t \bar{u}(t') \, dt' \qquad (3.39.d)$$

represents a coordinate transformation to the time and space averaged center of the patch. The angled brackets imply time averaging

$$\langle X \rangle = \frac{1}{T} \int_0^T X \, dt \qquad (3.39.e)$$

where T is the period of the oscillation.

Bowden (1965) investigated the dispersion coefficient in two dimensional oscillatory flow, but because he used equation (3.13) without due concern for the unsteady terms (for the limit $T \to \infty$), he obtained a longitudinal dispersion coefficient that is one half the value for steady unidirectional flow. The ratio of 1/2 arises because of the phase shift in time between velocity and concentration.

Okubo (1967) investigated the same problem, specifying a shear profile with velocity increasing linearly in the vertical and which has both fluctuating and steady components

$$u = \left(\frac{z}{h}\right) \left(U_0 + U \sin \frac{2\pi}{T} t\right) \qquad (3.40)$$

where U_0 and U are surface velocity amplitudes and T the period of the oscillation. Okubo obtained solutions through Aris' (1956) method of moments, and for this case he expressed the dispersion as a functional representation of

$$\langle E_L \rangle = f(U_0, U, h, T, T_c) \qquad (3.41.a)$$

where

$$T_c = h^2/E_z \qquad (3.41.b)$$

is the time scale of vertical mixing. He showed that the effects of steady and oscillatory parts of the motion on the longitudinal dispersion are additive (superposed) such that

$$\langle E_L \rangle_t = (E_L)_s + \langle E_L \rangle_0 \qquad (3.42)$$

70

where the subscripts t, s and o denote the total, steady and oscillatory contributions respectively. Okubo (1967) obtained two limits for his solution:

$$\langle E_L \rangle_t = \frac{U_0^2 T_c}{120} [1 + \frac{120}{4\pi^2} (\frac{U T}{U_0 T_c})^2], \text{ for } T \ll T_c$$

and

$$\langle E_L \rangle_t = \frac{U_0^2 T_c}{120} [1 + \frac{120}{236} (\frac{U}{U_0})^2], \text{ for } T \gg T_c \qquad (3.43.a,b)$$

When the flow is steady (U=0) the equivalent value is

$$(E_L)_s = \frac{U_0^2 T_c}{120} \qquad (3.44)$$

(the solution for this case of steady flow with linear profile can also be found in Fischer *et al.*, 1979, p.85). On the other hand, the oscillatory flow dispersion coefficient in the case of equal amplitudes with the steady case (U=U$_0$) as related to (3.44) are

$$\frac{\langle E_L \rangle_o}{(E_L)_s} = 3.04 (\frac{T}{T_c})^2, \text{ for } T \ll T_c,$$
$$\qquad (3.45.a)$$

and

$$\frac{\langle E_L \rangle_o}{(E_L)_s} = 0.51, \qquad \text{for } T \gg T_c \qquad (3.45.b)$$

This result indicates that for $T \ll T_c$, the dispersion coefficient is proportional to T^2, whereas for $T \gg T_c$ it is a constant about one half the value of the steady case.

This behaviour is expected, since for long periods of oscillation, the diffusion process is similar to that in steady flow, where an initial patch has sufficient time to diffuse before the flow reverses. On the other hand, in the limit T→0 (rapid oscillations), the diffusing patch returns to its original position rapidly before any diffusion can take place, and therefore cannot respond to the shear in the velocity profile, making the oscillatory dispersion coefficient vanish in this limit. (A discussion of these limits is given in Fischer *et al.*, 1979, p.95).

Later, Holley *et al.* (1970) considered the same problem with the linear velocity profile (3.40) (without the steady component, U_0=0) and obtained an analytical solution from which an expression for $\langle E_L \rangle$ is obtained:

$$\langle E_L \rangle = E_0 f(T') \qquad (3.46)$$

where

$$T' = T/T_c = T \; E_z/h^2$$

and   (3.47)

$$E_o = \frac{U^2 T_c}{240}$$ (3.48)

which is the constant value of $\langle E_L \rangle$ for the limit
$T \gg T_c$ (i.e. one half of 3.44). The function (3.46) is
shown in Fig.3.2.a, where the ratio $\langle E_L \rangle/E_o$ is
plotted against T'. In applying the results to estuaries,
Fischer *et al.* (1979, p.235), make an analogy to (3.31),
and use some empirical judgement to replace E_o on the
left hand side of (3.46) by

$$E_o = \alpha \; I\sigma_u^2 h^2/E_z$$
$$= \alpha \; I\sigma_u^2 T$$ (3.49)

assuming E_o, the limit of (3.46) for $T \gg T_c$, to
be proportional to the steady dispersion coefficient. In fact
for a linear velocity profile (3.40) it is found that (Fischer,
et al., 1979, p.93, Table 4.1)

$$\sigma_u^2 = U^2/24 \text{ and } I = 1/10$$

(3.50)

so that (3.44) results in the case of steady flow. Comparing
with (3.48) the proportionality constant is found as $\alpha = 1/2$,
and therefore (3.49) reduces to (3.48). It should be noted
that in their analogy, Fischer *et al.* (1979) have
erroneously omitted this proportionality constant, which should
be included. Nevertheless, by combining (3.46), (3.49) and
(3.50) we obtain

$$\langle E_L \rangle = \alpha I\sigma_u^2 T \; \frac{f(T')}{T'} = \frac{U^2 T}{240} \frac{f(T')}{T'} \; ,$$

(3.51)

which is in fact the same as (3.46), written differently.
Fischer *et al.* (1979) suggest using the first equality in
(3.51) empirically for velocity distributions other than the
case considered above. For example in a wide and shallow

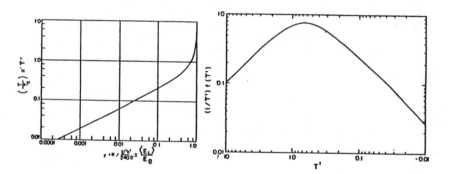

FIGURE 3.2 (a) Oscillatory flow dispersion coefficient,
function $f(T')$ (Holley *et al.*, 1970), and (b) the
normalized function $g(T')$ due to Fischer *et al.* (1979).

72

estuary, they suggest using the time scale $T_c = w^2/E_H$ corresponding to lateral mixing, rather than that for vertical mixing, since then the lateral shear is expected to dominate the dispersion. The function $g(T') = (T')^{-1} f(T')$ is shown in Fig. 3.2.b.

In all of the analyses discussed above, it has commonly been assumed that the velocity profile at each instant is the same as an equivalent steady flow. In reality, the velocity distribution is also subject to a convective-diffusion equation of its own, where the turbulent diffusion of momentum in the direction transverse to the flow must be taken into account. In fact, the diffusion of the momentum in oscillatory flow gives rise to *shear waves* in the fluid, just as concentration waves in the case of diffusion equation which propagate in the transverse direction causing phase shifts which depend on position in the fluid, and which are important in the correlations of u'' and c'' in equation (3.39.a).

These influences of simultaneous diffusion of momentum and concentration were accounted for the first time by Taylor (1974), who solved both equations and rigorously constructed the oscillatory longitudinal dispersion coefficient from (3.39.a). In his analyses, Taylor used constant turbulent diffusivity coefficients for both momentum and concentration.

Taylor's (1974) oscillatory flow dispersion coefficient is analogous to (3.51), although the dependence on T' is modified compared to the Holley *et al.* (1970) solution. The results are plotted in Fig.3.3 where it is shown that a maximum value of $\langle E_L \rangle$ is obtained at certain values of T_c depending on the period T.

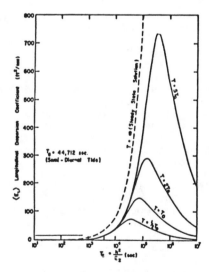

FIGURE 3.3 Oscillatory flow dispersion coefficient for different values of the period T (Taylor, 1974).

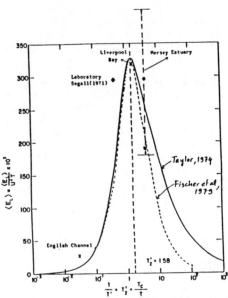

FIGURE 3.3 Normalized oscillatory flow dispersion coefficient (After Taylor, 1974 and Fischer *et al*., 1979).

Taylor (1974) also showed that it is not appropriate to normalise the oscilatory flow dispersion coefficient with respect to the steady flow dispersion, for they are two different processes. When the variables are normalized as

$$\langle E_L' \rangle = \frac{\langle E_L \rangle}{U^2 T} \quad \text{and} \quad T_z' = (T')^{-1} = \frac{T_c}{T} \tag{3.52}$$

a single curve results as shown in Fig.3.5. The maximum dispersion occurs for $T=T_c/1.58=0.63T_c$, i.e. when the oscillation period is of the same order as the transverse mixing time $(T'=0.63)$. Experimental verification of the results as obtained by Taylor (1974) are also superposed.

It is quite interesting to plot the empirical formulation of Fischer *et al*. (1979) based on the solution of Holley *et al*. (1970) in comparison to Taylor's (1974) results, as shown in Fig. 3.4. It is noted that the agreement of the two versions of the oscillatory dispersion coefficient is quite good for $T>T_c$ $(T_z'<1)$. On the other hand, the two solutions differ considerably for the range $T<T_c$ $(T_z'>1)$, for the phase distribution of concentration and velocity profiles begin to play important roles, which are not accounted for in the former solution.

Taylor (1974) has further considered the oscillating flow in a channel of rectangular cross-section. The main results are the shifting of the period of oscillation for which maximum dispersion occurs and rather small modifications in the functional form displayed in Fig. 3.4 for different values of the ratio T_{cz}/T_{cy}. Taylor has found that by varying $T_c'=T_{cz}/T_{cy}$ (cf. equation 3.38) in the

74

range 10^{-5} to 1, the peak value of the dispersion cofficient corresponding to Fig. 3.4 changes by about 15 % and the value of $T_z'=1/T'$ at which the peak occurs varies between 1.58 - 4.5.

Oscillatory shear flow dispersion with applications on horizontal mixing in the ocean have been investigated by Young *et al.* (1982). Considering a periodic shear flow velocity

$$u = u_0 \sin(mz) \cos(\omega t) \qquad (3.53)$$

in an infinite domain, they solve for the concentration distribution and obtain a dispersion coefficient

$$\langle E_L \rangle = \frac{1}{4} \left(\frac{u_0^2}{\omega}\right) \left(\frac{\kappa_\bullet}{1+\kappa_\bullet^2}\right) \qquad (3.54.a)$$

where

$$\kappa_\bullet = E_v m^2/\omega \; . \qquad (3.54.b)$$

Since the velocity field is periodic in z, the results can also be interpreted for an equivalent flow between horizontal boundaries placed at $z=0$ and $z=\pi/m$, where the velocities vanish, i.e. a flow with a vertical extent and oscillation period of

$$h = \pi/m \quad \text{and} \quad T = 2\pi/\omega \qquad (3.55)$$

respectively. The dispersion coefficient in (3.54.a) can then be put into the form

$$\frac{\langle E_L \rangle}{u_0^2 T} = \frac{1}{16} \left(\frac{T'}{1+(\pi T'/2)^2}\right) \qquad (3.56)$$

where $T'=E_v T/h^2=T/T_c$ as defined in (3.52.b). This solution also gives a maximum value at $T'=2/\pi=0.64$ where the value of the function is $\langle E_L \rangle/u_0^2 T$ =0.04. It can be observed that the form of the solution is similar to those presented earlier in Fig.3.4, coinciding better with the functional form of Holley *et al.* (1970), but the magnitude of the calculated values are about one order larger than that plotted in the same figure. This is because the original solution is obtained for unconfined flow, and the characteristic velocity u_0 largely differs from that defined earlier for confined flows, the similarity only being established through heuristic arguments.

The important result in the case of infinite domain, just like in the confined flow case, is that two different limits are obtained for oscillatory shear flow dispersion, i.e.

$$\langle E_L \rangle \approx \left(\frac{mu_0}{2\omega}\right)^2 E_z, \qquad \text{for } \kappa_\bullet \ll 1 \qquad (3.57.a)$$

and

$$\langle E_L \rangle \approx (\frac{u_0}{2m})^2 \, E_z^{-1}, \quad \text{for } \kappa_* \gg 1.$$

(3.57.b)

The first case corresponds to rapid oscillations with high
vertical wavenumber and vanishes in the limit $\kappa_* \to 0$.
Young and Rhines (1982) note the similarity of this case to the
"Okubo (1967) mechanism". In this limit, the dispersion is
directly proportional to E_z. The second case
corresponds to long period oscillations and is analogous to
Taylor's (1953) initial theory of steady flow dispersion, where
the dispersion effect is inversely proportional to E_z.

Young et al. (1982) also construct dispersion coefficients
for a random velocity field, from observed and empirical models
of the shear spectrum in the ocean. They conclude that shear
dispersion by an internal-wave field is dominated by the Okubo
(1967) mechanism, rather than the Taylor(1953) mechanism, since
they show a dependence on E_z. The transition from the
internal-wave shear dispersion regime to the meso-scale
stirring regime caused by eddying motions in the ocean is also
discussed by Young et al. They find the important result
that meso-scale stirring begins influencing the dispersion at
horizontal scales as small as 100m.

4. SUSPENDED SEDIMENTS

4.1 TURBULENT DIFFUSION OF SUSPENDED MATTER

In natural water bodies, such as estuaries, rivers, lakes and
the ocean, suspended matter is quite common. The terms
suspended matter, suspended solids, suspended sediments,
gelbstoff or seston are widely applied to refer to these
concentrations of solids. Since the concentrations are often
smaller than that of the main constituent of water, suspended
sediment often does not influence the density of the mixture
so that, we can use the previous approximations of Section 1.2
in its definition. However, the distinguishing property of
suspended matter is that individual particles are often heavier
(denser) than water. As a result, they sink in the vertical,
characterized by the settling velocity $\mathbf{w}_s$, which
differs from the vertical velocity $\mathbf{w}$ of the fluid
particles. We modify (1.7.a), (1.9) and (1.10) to write

$$\vec{N}_A = \rho_A (\vec{u}_A - w_s \hat{k})$$
$$= \rho_A (\vec{u}_A - \vec{u}) + \rho_A (\vec{u} - w_s \hat{k})$$
$$= -D_{AB} \, \nabla \rho_A + \rho_A (\vec{u} - w_s \hat{k}),$$

(4.1)

where the first term describes the diffusive flux and the
second term the convective flux of suspended matter. Following
the earlier developments of Section 1.2, we derive the
turbulent diffusion equation

$$\frac{\partial c}{\partial t} + \vec{u} \cdot \nabla c - \frac{\partial w_s c}{\partial z} = \nabla \cdot \ddot{E} \cdot \nabla c - kc$$

(4.2)

in analogy with (1.25) or (3.22). The boundary conditions
at a solid boundary are also modified as compared to (2.72).
Since the velocity and the flux of material normal to the
surface must vanish ($u=0$ and $\mathbf{N}\cdot\hat{n}=0$) in (4.1), we
have

$$(\mathbf{E}\cdot\nabla c)\cdot\hat{n} + c\ (\mathbf{w}_s\hat{k})\cdot\hat{n} = 0$$

(4.3.a)

which, for a horizontal surface ($\hat{n}=\hat{k}$) becomes

$$E_z \frac{\partial c}{\partial z} + w_s c = 0\ .$$

(4.3.b)

Note that in the above, we have assumed that no sediment can
pass across a solid boundary. In free surface flows, (4.3.b)
is valid at the surface, if no sediments are input from the
atmosphere. In applying (4.3.b) to the bottom boundary, we must
account for the bottom deposition loss of sediments. While the
flow in the interior is often turbulent, there exists a viscous
sub-layer near the boundary. If the size of the settling
particles is larger than the thickness of this layer they are
reflected from the bottom. On the other hand, particles
smaller than the viscous layer thickness tend to stay near the
bottom to form a layer of *fluid mud* and are eventually
deposited on the bottom. The following bottom boundary
condition has therefore been suggested by Sayre (1969) and
Jobson and Sayre (1970):

$$E_z \frac{\partial c}{\partial z} + (1-\alpha)w_s c + yq = 0$$

(4.3.c)

where α represents the probability that a particle
settling to the bottom is deposited there, and yq is
the average rate of entrainment into the flow, q being the
storage at the bed. Sayre (1969) and Jobson and Sayre (1970)
have obtained analytical and numerical solutions to the two-
dimensional version of equation (4.2) with the surface and
bottom boundary conditions (4.3.b,c) respectively, and an
initial condition of a vertical line source. The solutions are
functions of $\eta=z/h$, $\tau=tE_z/h^2$, α and a
settling velocity parameter $\beta=w_s/\kappa u_*$ where

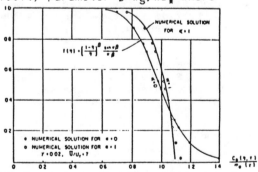

FIGURE 4.1 Vertical distribution of suspended matter for
$\beta=0.1$ and at $\tau=0.5$ (After Sayre, 1970).

77

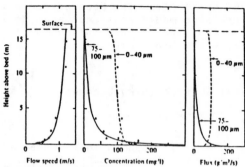

FIGURE 4.2 Profiles of velocity, concentration and flux of
suspended sediments in the Thames estuary (After McCave, 1979).

κ=0.41 is Von Karman's constant.   The concentration
profile becomes time-independent (in the convected coordinates)
after an initial time of τ>0.5, and is self similar.
Sayre's solutions for τ>0.5, β=1, α=0 and α=1
are shown in Fig. 4.1.   Jobson and Sayre (1970) verified their
solutions with experimental data.   Later, Sumer

(1974) has obtained analytical solutions for the various ranges
of parameters, and has shown that some of the special cases
reduce to Sayre's solutions.

In the above descriptions the settling velocity w_s is a
function of sediment density, size and eddy viscosity,
empirical values of which can be found in the literature.   In
situ values can be obtained through methods outlined in McCave
(1979).   Note, however, that w_s is different for each
type of sediment (fine, coarse sand, silt, detritus, organic
debris etc.) and separate equations with appropriate values of
w_s are required to describe the diffusion of each size
fraction.   In reality, the settling of sediments in sea water
is often influenced by flocculation (combining of small
particles into larger aggregates through electrodynamic
attraction).   The probability of flocculation is a function of
particle type, electrolytic strength (i.e. salinity), and
velocity shear (Dyer, 1979).

Example measurements of suspended sediment profiles in an
estuary are shown in Fig. 4.2.   Coarser sediments are usually
concentrated near the bottom, whereas fine sediments in
suspension are more uniformly distributed in the vertical.

4.2 SHEAR FLOW DISPERSION

Considering the diffusion equation (4.2) in the presence of
sediments, and vertical averaging following section 3.1 yields
(Nihoul and Adam, 1974)

$$\frac{\partial c}{\partial t} + \vec{u} \cdot \nabla c = \frac{1}{H} \nabla \cdot H\hat{\underline{E}} \cdot \nabla c + Q - kc \qquad (4.4)$$

where ∇ represents the two dimensional gradient operator as in section 3.1, and

$$Q = \frac{1}{H} (E_z \frac{\partial c}{\partial z} + w_s c) \Big|_{z=-h}^{z=\eta} \qquad (4.5)$$

represents the total flux through the surface and the bottom. Subtracting the averaged equation (4.4) from (4.2) yields the same equation as (3.12) with the addition of the following terms on the right hand side

$$lhs(3.12) = rhs(3.12) + w_s \frac{\partial c''}{\partial z} - Q \qquad (4.6)$$

In estimating the dispersion tensor $\underline{E}$, Nihoul and Adam (1974) assume low concentrations of fine sediment and therefore neglect the influence of these terms in (4.6) and use the basic balance in (3.13) to derive the expression for K_{ij} in (3.19). Therefore the dispersion is assumed to be the same as that for neutrally buoyant concentrations.

On the other hand, the settling of suspended matter influences the concentration profiles as shown earlier and produce nonuniform distributions. It should therefore be expected that, in general the dispersion coefficient should be a function of settling velocity. Sayre(1969) and Sumer (1974) have taken the settling terms into account and have calculated the dispersion coefficient (normalized with respect to the neutrally buoyant case) as a function of the settling velocity as shown in Fig.4.3. With increasing settling velocity (e.g. sediment size), the dispersion is increased with respect to the neutrally buoyant case.

For completing the description of horizontal dispersion in (4.4), the flux term Q must be specified. This is often done

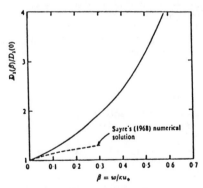

FIGURE 4.3 Longitudinal dispersion coefficient for suspended matter (normalized with respect to neutrally buoyant substances) as a function of β (After Sumer, 1979).

79

empirically and, neglecting the surface fluxes, Q represents
the deposition losses to the bottom or reentrainment from the
bottom into the flow. A model of practical importance
(Sayre, 1969) is given by (4.3.c) and (4.5).

The flux terms should be different for the different processes
of deposition and resuspension. For the deposition of cohesive
sediments Krone (1962,1976) suggests (after McCave, 1979)

$$Q = - w_s c \ r \ (\frac{1-\tau/\tau_c}{1+t/t_c})$$ (4.7)

for $\tau<\tau_c$, where τ is the bed shear stress,
τ_c the critical shear stress below which deposition
occurs, t the time, t_c the "coagulation time" (the
mean time between collisions of particles), and r the mean
number of particles in a floc. This formula is actually of
little practical use since the time t from the beginning of
flocculation cannot be easily determined in nature. A further
complication arises because the settling velocity is a function
of concentration; when the sediment concentration is
sufficiently high, $w_s=Kc^n$, where K and n
are coefficients depending on sediment type and occasion, with
n=4/3 suggested by Krone(1962) and n=1 or 2
suggested by Owen(1971). For low concentrations of sediments,
w_s can often be taken as constant (n=0).

When both the rate of flocculation and the sediment
concentration are low, the approximations r=1 and
$t<<t_c$ can be made, upon which (4.7) reduces to

$$Q = - w_s c \ (1 - \tau/\tau_c)$$
   (4.8)

In many estuaries, a turbidity maximum and a corresponding
region of high deposition is found in the mid-reaches of the
estuary, where the bed shear stress decreases due to opposing
effects of river and open sea waters (e.g. near the tip of a
salt wedge). In this bottom convergence region, sediment
concentration increases and bottom shear vanishes, yielding
high deposition rates according to (4.8).

In the case of resuspension of sediments from the bottom
(erosion), a different formula applies according to
Partheniades (1968)

$$Q = M \ (\tau/\tau_e - 1)$$
   (4.9)

where M is an erosion rate constant and τ_e the
minimum required bed shear stress for erosion to take place.

The above relations are often difficult to use in modelling
practice, mainly because they require the switching on and off
of the deposition and erosion processes according to situation.
Nihoul and Adam (1974) have adopted (4.8) for general modelling
application, assuming that it applies for both deposition
($\tau<\tau_c$) and erosion($\tau>\tau_c$). Replacing
$\tau=(\rho f/8)u^2$, where f is the Darcy-Weisbach

80

bottom friction coefficient, they write

$$Q = w_s c \left(1 = \frac{u^2}{u_c^2}\right)$$ (4.10)

Note that this source/sink function formulates the deposition/erosion as a completely reversible process.

Nihoul and Adam (1974) have used (4.10) in (4.2) to model dispersion and settling of sediments near a dump site in a shallow sea with tidal flows. The mass m of sediments deposited on the bottom are calculated from

$$\frac{\partial m}{\partial t} = Q .$$ (4.11)

The source/sink function (4.10) has been utilized by Ozsoy (1977, 1986) to model suspended sediment transport and deposition on the seaward side of a tidal inlet. During the ebb tide, the flow is in the form of a quasi-steady jet, with pronounced lateral diffusion. Then, using the jet velocity distribution obtained by Ozsoy and Unluata(1982), equation (4.4) with (4.10) is solved for the horizontal diffusion within the jet (Fig.4.4.a). Ambient concentrations, lateral entrainment into the jet, depth variations, bottom friction and settling velocity are taken into account. The contours of bottom deposition rate seaward of the inlet are computed via (4.11). In Fig.4.4.b, the inlet velocity is critical

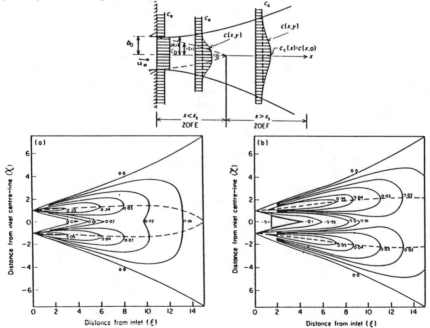

FIGURE 4.4 (a) Jet diffusion, and (b,c) bottom deposition of sediments near a tidal inlet (After Ozsoy, 1986).

$u_o=u_c$, so that no deposition takes place at the jet
core, where no diffusion or settling occurs. In the diffusion
regions of the jet, diffusion and settling processes compete
and yield maximum deposition rates at lateral lobes. In the
second case (Fig. 4.4.c), the inlet velocity exceeds the
critical velocity ($u_o=1.4u_c$), and the material
eroded at the jet core is deposited in the bar system
encircling the mouth. The implications on tidal inlet
morphology and river deltas are discussed in Ozsoy (1986).
Wang (1984) has applied Ozsoy's model (although without due
reference to the original solution provided by Ozsoy, 1977) to
the dynamics and growth of a river delta.

Sediment diffusion and dispersion is an emerging field of study
deserving much attention. It must however be stressed that the
subject is a complicated one requiring considerable empirical
guidance. Many aspects of sediment transport, such as the bed-
load mode of transport have not been described within the
limited scope of this course. An expedient summary of marine
sediment transport, its relationships with shelf circulation
and implications on morphology can be found in Stanley and
Swift (1976). The modelling of sediment transport on the
continental shelf requires special attention, an introduction
to which can be found in Smith (1977).

5. ESTUARINE TRANSPORT

5.1 INTRODUCTION

An estuary is a semi-enclosed coastal water body communicating
with the sea through a mouth or entrance region and which is
diluted considerably by the influence or river runoff in the
interior region. Although this definition is quite general, it
does not sufficiently describe an estuary, since the physical
nature of each estuary differs considerably from another with
respect to the varying influences of geometrical shape (depth
and area distribution, sand bars, islands, channels, ruggedness
of coasts etc.), amount of freshwater inflow, the nature of the
restricted exchange at its connection with the sea, the degree
of tidal influence, the weather conditions etc. As a result of
these varying influences, each estuary has a different
personality and the stratification and circulation in one
estuary may differ greatly from another. There has been
various attempts to classify estuaries, for example by
Pritchard (1967) and Hansen and Rattray (1966), basically
utilizing the salinity and the velocity ratios of surface
values to mean cross sectional values. Generally, an estuary
can be of *salt wedge type*, where fresh water on the surface
and sea water at the bottom are sharply separated by a wedge,
or *partially mixed type*, where vertical stratification is
strong but an interface is not formed, or *well mixed type*,
where vertical stratification is small.

Since estuarine processes are quite complicated, they are the
subjects of detailed theory in their own right. However, the

general hydrodynamic, thermodynamic and mass conservation
laws of Section 1 can, in principle, be applied to estuaries,
with further specific assumptions and reductions required. The
various aspects of estuarine processes can be found in a number
of specialized books such as Ippen (1966), Dyer (1973), Officer
(1976), Kjerfve (1978), McDowell and O'Connor (1977).

Our purpose here is not to describe in detail the hydrodynamic
and mixing characteristics in estuaries, but rather how these
characteristics influence the transport and dispersion of a
substance in solution, e.g. a pollutant. On the other hand,
transport processes in an estuary are highly dependent on the
hydrodynamic and mixing characteristics, and therefore we
venture for a brief review of the influencing factors.

5.2 ESTUARINE MIXING

The processes of estuarine mixing will be briefly summarized,
following Fischer *et al.* (1979), but keeping the scope much
more concise within the present context. Various mechanisms
are considered, which are often superposed in real estuaries.

Wind mixing:
Wind drift and mixing is often important in shallow and wide
estuaries. The surface stress exerted by the wind constitutes
a force at the surface, which is redistributed over the water
column through the vertical diffusion of momentum. In salt-
wedge type estuaries (two layer stratification), the wind
induced driving force influences only the upper layer, and
causes entrainment processes at the interface. On the other
hand, in well-mixed estuaries, the wind force is distributed
over the whole depth, so that it influences shallow regions
more than deep regions. A residual wind-induced circulation
can therefore be driven in estuaries with large depth
variations, which can influence the dispersion patterns
(Fischer *et al*, 1979).

Influence of stratification on mixing:
One of the most important factors to be considered in estuaries
is the inhibiting influence of stratification on turbulence,
and hence on vertical mixing. As compared to the homogeneous
cases considered earlier, an extra amount of energy is required
for vertical mixing in order to overcome the potential energy
of stratification. In estuaries, this energy is derived from
boundary and internal shear. According to a well known
formula due to Munk and Anderson (1948), the vertical
diffusivity decreases with increasing stratification and
increases with increasing vertical shear. Fischer *et al.*
(1979), however, caution for indiscriminate use of such
formulas, since many other processes that need empirical
definition can influence the diffusion processes.

Longitudinal Dispersion:
Assuming an estuary with longitudinal variations of cross
sectional area $A(x)$, and a flow induced by fresh water
inflow $u=Q_f/A$, where Q_f is the river discharge,

equation (3.23) should in principle be applicable to describe
the longitudinal dispersion in an estuary, i.e. considering
steady flows due to river discharge alone, we have

$$A(x)\frac{\partial c}{\partial t} + Q_f\frac{\partial c}{\partial x} = \frac{\partial}{\partial x} A(x)(K_x+E_x)\frac{\partial c}{\partial x} \ . \tag{5.1}$$

Here, K_x is the longitudinal dispersion coefficient
which must now be evaluated from (3.24) based on the different
conditions of stratification, velocity distribution, transverse
mixing etc., as summarized above.

In principle, the use of equation (5.1) with appropriate values
of the longitudinal dispersion coefficient, should describe the
dispersion processes in an estuary. However, as noted above,
K_x is modified due to a number of influences. A
method often suggested was to obtain K_x from the
observed longitudinal salinity distributions, since the cross
sectionally averaged salinities also obey (5.1) with the
unsteady term omitted for equilibrium conditions, and hence
could be used as a tracer. On the other hand, Fischer *et
al* (1979) admit that, in spite of the considerable
developments in the last 25-30 years since these ideas were
suggested, there is still no general predictive method to
obtain the dispersion coefficient in estuaries. Nevertheless,
equation (5.1) has often been used in estuaries with
experimentally determined values of the dispersion
coefficients, with examples provided by Officer (1976) and
Fischer *et al* (1979).

Tidal dispersion:
In the above sub-sections, the influence of oscillatory shear
flows, such as that occurs due to tidal propogation in
estuaries have not been considered. In the presence of
stratification and residual circulations, such analyses are
tedious and produce little of practical use, although an
understanding of various contributions can be reached (cf. Dyer
(1973) and Fischer *et al* (1979)).

On the other hand, the longitudinal dispersion in well mixed
estuaries due to tidal oscillations alone can be estimated
through the methods outlined for oscillatory shear flows in
section 3. Fischer *et al* (1979) have taken this route, but
considering the dominant influence of transverse mixing have
formulated (3.51) such that the transverse mixing time have
been used instead of the vertical mixing time. We have already
discussed these aspects of the applications in section 3.

Tidal pumping:
The tidal oscillatory flow in estuaries often gives rise to a
net steady circulation, which only becomes apparent after
averaging the currents over the period of oscillation. These
residual circulations arise mainly due to the nonlinear terms
in the equations of motion which yield mean currents when
averaged: convection and turbulent bottom friction and their
interactions with bottom topography. Examples of residual

circulations in estuaries are given by Stommel and Farmer
(1952), Bowden and Gilligan (1971), Van de Kreeke (1975,1978)
etc. These residual circulations contribute effectively to the
longitudinal dispersion and exchange processes.

Stommel and Farmer (1952) have considered the residual
circulations near the mouth of an estuary. As shown in Fig.
5.1.a, the ebb flow in the estuary is in the form of a sink
flow converging towards the mouth, and the volume of water
ejected out of the estuary is in the form of a semi-circle.
During flood flow, the water entering from the sea is idealized
as a rectangular plug intruding the estuary. Over one tidal
cycle, only a proportion of material introduced on the ocean
side during flood will return to the ocean during the
subsequent ebb-flow, leading to trapping within the estuary.
The residual circulation represented for the estuary side in
Fig. 5.1.a is in fact also valid for the ocean side of the
estuary mouth, reversing the roles of flood and ebb (mirror
image of Fig.5.1.a with respect to the mouth region).

The flood flow represented in Fig.5.1.a (or alternatively the
ebb-flow on the ocean side) is actually in the form of a

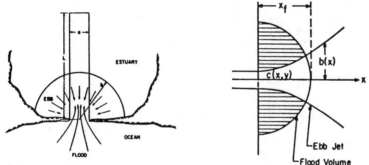

FIGURE 5.1 Idealizations of tidal residual flow near
entrances (a) Stommel and Farmer (1952), (b) Ozsoy (1977)

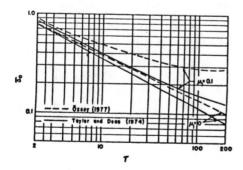

FIGURE 5.2 Ocean mixing coefficient as a function of bottom
friction and tidal excursion (After Ozsoy, 1977).

turbulent jet as shown in Fig.5.1.b, rather than the idealized form of a slug. The hydrodynamic and mass transport characteristics of such jets have been investigated by Ozsoy and Unluata (1982) and Ozsoy (1986), allowing the calculation of exchange. An ocean mixing coefficient y_o, defined as the ratio of the average concentrations passing through the mouth during the respective flood and ebb phases can therefore be defined and calculated as a function of bottom friction, mouth geometry and the ratio τ $=Tu_o/2b_o$ of the tidal excursion length u_oT to the inlet width $2b_o$ (u_o is the mouth flow velocity, T the period of the tide). The ocean mixing coefficient thus calculated by Ozsoy (1977) and Mehta and Ozsoy (1978) for the case of constant depth is shown in Fig.5.2 as a function of a bottom friction parameter $\mu=fb_o/8h_o$ (f is the Darcy-Weisbach bottom friction coefficient and h_o is the depth) and the excursion length ratio τ. Taylor and Dean (1974) have considered the same problem earlier, but have found a different expression since they neglect lateral entrainment in the jet.

These concepts of tidal exchange at an entrance region has been applied by Ozsoy (1977) to the exchange of a pollutant between a bay and the ocean. The tidal flow is idealized as a series of quasi-steady flows (with inlet velocity u_o during ebb and $-u_o$ during flood). The average concentrations at the inlet (entrance) during the ebb and flood flows are related as

$$c_{if}{}^n = y_o \, c_{ie}{}^{n-1}$$

(5.2)

where the subscripts denote i=inlet, f=flood, e=ebb and the superscript n represents the n th tidal cycle starting with flood. The mixing on the bay side is assumed to be more complex due to its confined nature, where it is assumed that

$$c_{ie}{}^n = y_b c_{if}{}^n + (1-y_b) c_{be}{}^{n-1}$$

(5.3)

where b=bay, c_{be} the volume averaged bay concentration during ebb, and y_b a coefficient describing the bay mixing, and varying in the range (0,1), so that the concentration of the ebb flow at the inlet is always between the values c_{if} and c_{be}, representing the inlet (flood) and bay (previous ebb) concentrations. Considering further the mass balance of the bay during the flood and ebb phases, Ozsoy (1977) obtained the recursion formula

$$c_{be}{}^n = A_1 c_{be}{}^{n-1} + A_2 c_{be}{}^{n-2}$$

(5.4.a)

where

$$A_1 = 1 + y_o y_b - \frac{(1-y_b)k}{(1-k/2)}$$

(5.4.b)

$$A_2 = \frac{(1-y_b)y_o(1+k/2)}{(1-k/2)} - y_o$$

(5.4.c)

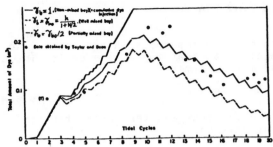

FIGURE 5.3 Total dye in Card Sound, experimental values (points) by Taylor and Dean (1974), and calculations by Ozsoy (1977)

and where

$$k = \Omega/V = 2a_0/h_b \qquad\qquad (5.4.d)$$

is the ratio of the tidal prism Ω to the mean bay volume V, a_0 being the tidal amplitude and h_b the mean depth of the bay.

Ozsoy (1977) applied this method to Card Sound in Florida, where a dye injection study had earlier been made by Taylor and Dean (1974). Using numerical values of the parameters and the recursion formula (5.4.a), reasonable estimates of the dye remaining in the bay were obtained, as shown in Fig. 5.3. In the case of no mixing in the bay, it is sufficient to take $y_b=1$; on the other hand, if the bay waters are completely mixed with the incoming tidal waters during flood, it is shown that y_b should have the value $y_b=k/(1+k/2)=y_{bo}$. These two limits bound the possible solutions that can be obtained for specific cases of bay mixing. An assumption of partial mixing in the bay with $y_b=y_{bo}/2$ have yielded reasonable agreement with observations of Taylor and Dean (1974), which were obtained by integrating the dye concentration over the bay volume.

Tidal Trapping:
In estuaries with storage basins, or relatively stagnant regions of branching waterways or embayments along the coasts, tidal currents can cause a subtle and additional dispersive effect called tidal trapping. A patch of pollutant released in such a system may get partially trapped at the surrounding embayments or shallow banks during a certain phase of the tide and gets released into the mainstream flow some time later. This influence results in increased dispersion, since the material in the mainstream flow and the fraction caught in the trap zones are seperated from each other. Shijf and Schonfeld (1953) and Okubo (1973) have studied tidal trapping, and have found that it may significantly contribute to dispersion as compared to shear effects alone. Fischer *et al.* (1979) estimate that the trapping mechanism may play a major role in many estuaries.

5.3 CHARACTERISTIC TIME SCALES

There are various time scales characterising the various mechanisms of exchange and transport in estuaries. We have already seen in section 4, that two of the basic time scales are the *transverse mixing times*

$$T_{cv} = h^2/E_V, \quad \text{and} \quad T_{ch} = w^2/E_H$$

(5.5.a,b)

the former being for vertical and the latter for transverse horizontal (lateral) mixing.

Fischer *et al.* (1979) suggest another time scale based on empirical judgement and in analogy to the above, namely the *replacement time*, representing the time required for a slug of material initially concentrated at one end of the basin to reach approximately uniform concentration throughout the basin, given as

$$T_r = 0.4 \ L^2/E_L$$

(5.6)

where L is the length of the basin and E_L the longitudinal dispersion coefficient.

An important concept is the *flushing time* which is the average time spent by a tracer particle in the estuary, defined as the ratio of the fresh water volume in the estuary to the fresh water flux (Officer, 1976; Fischer *et al.*, 1979):

$$T_f = V_f/Q_f$$

(5.7)

where Q_f is the fresh water volume flux and V_f is the total volume of fresh water in the estuary, calculated from

$$V_f = \int_V \frac{S_0-S}{S_0} \ dV = \int_V f dV = fV \ .$$

(5.8)

Here, S_0 is the ocean salinity, S the salinity in the basin and V the volume of the basin, and f is the freshness defined as the fraction of fresh water at any point, and $f=(S_0-S)/S_0$ is the mean freshness of the basin. Note that the above flushing time is defined for an estuary influenced by a fresh water inflow alone.

For a tidal estuary, the *tidal prism flushing time* (Officer, 1976) is obtained by letting V_P and V_R respectively represent the volumes of ocean and river water entering the estuary during a tidal cycle, and writing the salt balance at high tide

$$(V_P + V_R)S = V_P \ S_0$$

(5.9.a)

where S is the mean salinity in the estuary, and the mean freshness is therefore $f=V_R/P$ where $P=V_P+V_R$ is the tidal prism. Then the tidal prism flushing time is

$$T_t = \frac{fV}{Q_f} = \frac{fV}{V_R/T} = \frac{V}{P} T \qquad (5.9.b)$$

with T being the tidal period. Since neither the entire
estuary, nor the ebb-water on the ocean side are not usually
completely mixed during each tidal cycle, T_t is
generally smaller than T_f.

If we perform a dye experiment in an estuary we need another
measure of *pollutant flushing time*. Considering a
continuous release of rate q and steady-state conditions to
prevail, this is given as (Officer,1976)

$$T_p = \rho cV/q \qquad (5.10)$$

where ρ and c are the mean estuarine density and
concentration.

Instead of the flushing time the term *residence time* is
also often employed. However there seems to be a confusion
with respect to the terminology applied to the various time
scales of exchange.

Realizing the often confused and misleading terminology, Bolin
and Rodhe (1973) have reviewed these concepts, and have derived
the basic time scales. Basing their analyses on rigorous
foundations, they have defined the time scales based on the
age τ of any fluid element in the reservoir (i.e. the
time elapsed since the entry of that element in the reservoir).
The total mass of the basin is $M_o = \rho V$. A
cumulative age distribution function $M(\tau)$ gives the
mass that has spent a time less or equal to τ in the
reservoir. All material elements spend an infinite time or
less in the basin, so that

$$\lim_{\tau \to \infty} M(\tau) = M_o \qquad (5.11)$$

An age frequency distribution functon $\Psi(\tau)$ can then be
defined and normalized such that

$$\int_o^\infty \Psi(\tau) \, d\tau = 1 \qquad (5.12)$$

which is related to the cumulative function through

$$\Psi(\tau) = \frac{1}{M_o} \frac{dM(\tau)}{d\tau} . \qquad (5.13)$$

Secondly, consider a steady state volume flux F_o of
material entering the basin or equivalently leaving the basin.
A cumulative transit time function $F(\tau)$ is defined,
giving the mass leaving the basin per unit time of those fluid
elements which has spent a time of τ or less in the
basin. Obviously,

$$\lim_{\tau \to \infty} F(\tau) = F_o \qquad (5.14)$$

and again we define a frequency distribution of transit time $\Phi(\tau)$ such that

$$\int_0^\infty \Phi(\tau) \ d\tau = 1 \ . \tag{5.15}$$

This frequency function is then

$$\Phi(\tau) = \frac{1}{F_0} \ \frac{dF(\tau)}{d\tau} \ . \tag{5.16}$$

In the case of a steady-state balance, the two sets of functions are related through

$$F_0 - F(\tau) = M_0 \ \Psi(\tau) = \frac{dM(\tau)}{d\tau} \ , \tag{5.17}$$

or with the aid of (5.14)

$$\Phi(\tau) = - \ \frac{M_0}{F_0} \ \frac{d\Psi(\tau)}{d\tau} \ . \tag{5.18}$$

Since $F(0)=0$, it follows from (5.18) that

$$\Psi(0) = \frac{F_0}{M_0} \ . \tag{5.19}$$

Equipped with the above tools, Bolin and Rodhe (1973) defined the various time scales as follows:

The *turn-over time* is the ratio of the total mass of the reservoir to the total flux

$$\tau_0 = \frac{M_0}{F_0} \ . \tag{5.20}$$

The *average transit time* of particles leaving the basin (the expected life time of newly incorporated particles) is given by

$$\tau_t = \int_0^\infty \tau \Phi(\tau) \ d\tau = \frac{M_0}{F_0} = \tau_0 \tag{5.21}$$

which is integrated by making use of (5.18). Therefore, the average transit time and turn-over time equivalent. An alternative name for both time scales is *residence time* as suggested by Bolin and Rodhe (1973), who note that this last term has often been misused.

Another time scale that can be defined is the *average age* of particles in the reservoir at any time, given by

$$\tau_a = \int_0^\infty \tau \Psi(\tau) \, \partial\tau = \frac{1}{M_0} \int_0^\infty \tau \, dM(\tau) . \qquad (5.22)$$

Since it is shown that $\tau_t = \tau_0$, there are
basically two time scales τ_t and τ_a.
The relation between these two time scales is determined by the
form of the frequency functions $\Psi(\tau)$ and $\Phi(\tau)$.
Three cases can be distinguished according to the ranges of
these time scales:

$\tau_a < \tau_t$:
A reservoir with modest transport velocities and source and
sink regions placed far apart belongs to this case (for
example, a well-mixed, wide and elongated estuary).

$\tau_a = \tau_t$:
A well-mixed reservoir with isolated source/sink regions, such
that all elements in the reservoir have equal probability of
exiting at any time is characterised by this condition (for
example, a well-mixed estuary of very small volume). Bolin and
Rodhe (1973) note, however, that since any element in the basin
is comprised of particles of all ages, it is impossible in this
case to establish the frequency functions by direct
observation. In this case, the sufficient condition (from
5.19, 5.21 and 5.22) is

$$\Psi(\tau) = \Phi(\tau) = \frac{1}{\tau_a} \exp(-\tau/\tau_a) . \qquad (5.23)$$

$\tau_a > \tau_t$:
This case represents the situation in which most of the fluid
particles entering the reservoir exit in a short time and those
remaining particles stay in the reservoir for a much longer
time. Such a case is possible if the source and sink regions
are close to each other (short circuiting), so that any
particles diffusing in the relatively stagnant major part of
the basin are trapped in these regions (for example a salt-
wedge or partially mixed estuary with stagnant regions).

Takeoka (1984a) developed these concepts further and redefined
the residence time differently from Bolin and Rodhe
(1973), producing two different residence times, one for the
reservoir and one for the inlet. Takeoka's residence time is
not the same as the average transit time, since he defined it
as being the average time required for the particles to reach
the outlet, which becomes a complement of the average age.
These results were then applied to coastal seas (1984a,b).

While the earlier definitions of time scales in this section
apply to specific situations in estuaries, the latter more
rigorous definitions outlined above apply to more general
situations, involving larger basins with more structural
variations. On the other hand, they require the determination
of frequency functions through direct observations or various
models.

REFERENCES

Ariathurai, R. and R.B. Krone, 1976. Finite element model for cohesive sediment transport, Journal of the Hydraulics Division, American Society of Civil Engineers, 102 (HV3), 323-338.

Aris, R. 1956. On the Dispersion of a Solute in a Fluid Flowing Through a Tube, Proc. Roy. Soc., A, vol.235, pp.67-77.

Batchelor, G.K. 1967. *An Introduction to Fluid Dynamics*, Cambridge University Press.

Bolin, B. and H. Rodhe, 1973. A note on the concepts of age distribution and transmit time in natural reservoirs, Tellus, 25, 58-62.

Bowden, K.F., 1965. Horizontal Mixing in the Sea due to a Shearing Current, J. Fluid Mech., vol.21, pp.83-95.

Bowden, K.F., and R.M. Gilligan, 1971. Characteristic features of estuarine circulation as represented in the Mersey Estuary, Limnol. Oceanogr. 16, 490-502.

Carslaw, H.S. and J.C. Jaeger, 1959. *Conduction of Heat in Solids*. 2nd ed. Oxford Univ. Press (Clarendon).

Csanady, G.T. 1973. *Turbulent Diffusion in the Environment* D. Reidel Publishing Company.

Dyer, K.R. 1973. *Estuaries: A Physical Introduction*, Wiley.

Dyer, L.R. 1979. *Estuarine hydrography and Sedimentation*, Cambridge University Press.

Elder. J.W., 1959. The Dispersion of Marked Fluid in Turbulent Shear Flow, J. Fluid Mech., vol.5, pp.544-560.

Fischer, H.B., 1967. The Mechanics of Dispersion in Natural Streams, J. Hyd. Div., ASCE, vol.93, No.HY6, pp.187-216.

Fischer, H.B. 1975. Discussion of "Simple method for predicting dispersion in streams" by R.S. McQuivey and T.N. Keefer, J. Environ. Eng. Div. Proc. Am. Soc. Civ. Eng. 101, 453-455.

Fischer, H.B. 1978. On the tensor form of the bulk dispersion coefficient in a bounded skewed shear flow, J. Geophys. Res. 83, 2373-2375.

Fischer, H.B., List, E.J., Koh, R.C.Y., Imberger, J. and N.H. Brooks, 1979. *Mixing in Inland and Coastal Waters*, Academic Press.

Frenkiel, F.N. and R.E. Munn, ed.s, 1974. *Turbulent Diffusion in Environmental Pollution*, Adv. in Geophys., 18A, Academic Press.

Hansen, D.V., and M. Rattray, 1966. New dimensions in estuary classification. Limnol. Oceanogr. 11, 319-325.

Harleman, D.R.F. 1970. *Transport Processes in Water Quality Control*, Massachusetts Institute of Technology Department of Civil Engineering. Unpublished Lecture Notes.

Holley, E.R., Harleman, D.R.F. and H.B. Fischer, 1970. Dispersion in Homogeneous Estuary Flow, J. Hyd. Div., ASCE< No.HY8, pp.1691-1709.

Ippen, A.T., ed., 1966. *Estuary and Coastline Hydrodynamics*., McGraw Hill.

Jobson, H.E., and W.W. Sayre, 1970. Vertical transfer in open channel flow, J. Hydraul. Div. Proc. Am. Soc. Civ. Eng. 96, 703-724.

Joseph, J. and H. Sendner, 1958. Uber die Horizontale Diffusion im Meere, Deut. Hydrogr. Zeit., v. 11, No. 2, pp. 49-77.

Kjerfve, B. ed., 1978. *Estuarine Transport Processes*, University of South Carolina Press.

Krone, R.B., 1962. *Flume Studies of the Transport of Sediment in Estuarial Shoaling Processes*, report, Hydraul. Eng. Lab., Sanit. Eng. Res. Lab., Univ. Calif., Berkeley.

Krone, R.B. 1972. *A Field Study of Flocculation as a Factor in Estuarial Shoaling Processes*, U.S Army Corps of Engineers, Committee on Tidal Hydraulics Technical Bulletin, 19.

Krone, R.B. 1976. In: McCave, I.N., ed., *Engineering Interest in the Benthic Boundary Layer*, pp.143-56. Plenum Press.

Kullenberg, G. (ed.) 1982. **Pollutant Transfer and Transport in the Sea**, CRC Press, volumes 1 and 2.

McCave, I.N. 1979. Suspended Sediment, in Dyer, K.R. (ed.), *Estuarine Hydrography and Sedimentation*, Cambridge University Press.

McDowell, D.M. and O'Connor, B.A. 1977. *Hydraulic Behaviour of Estuaries*, Wiley.

Mehta, A.J. and E. Ozsoy, 1978. Inlet Hydraulics, Flow Dynamics and Nearshore Transport, In: Bruun, P., ed., *Stability of Tidal Inlets Theory and Engineering*, Elsevier Scientific Publishing Company.

Munk, W. and E.R. Anderson, 1948. Notes on a theory of the thermocline. J. Mar. Res. 7, 276-295.

Nihoul, J.C.J. and Y. Adam, 1975. Dispersion and Settling Around a Waste Disposal Point in a Shallow Sea. Journal of Hydraulic Research 13, No.2.

Odd, N.V.M. and M.W. Owen, 1972. A two-layer model of mud transport in the Thames Estuary. Proc. Inst. Civil Eng. Supplement, 9, 175-205.

Officer, C.B., 1976. *Physical Oceanography of Estuaries (and Associated Coastal Waters)*, Wiley.

Okubo, A., 1962. *A Review of Theoretical Models of Turbulent Diffusion in the Sea*, Chesapeake Bay Inst., Johns Hopkins Univ., Technical Report No. 30.

Okubo, A., 1967. The Effect of Shear in an Oscillatory Current on Horizontal Diffusion from an Instantaneous Source, Int. J. of Oceanology and Limnology, vol.1, No.3, pp.194-204.

Okubo, A. and M.J. Karweit, 1969. Diffusion from a Continuous Source in a Uniform Shear Flow, Limnology and Oceanography, vol.14, pp.514-520.

Okubo, A. 1973. Effect of shoreline irreularities on streamwise dispersion in estuaries and other embayments. Neth. J. Sea Res. 6, 213-224.

Okubo, A. 1974. Some speculations on oceanic diffusion diagrams. Rapp. P., V. Reun. Cons. Int. Explor. Mer. 167, 77-85.

Owen, M.W. 1971. The Effect of Turbulence on the Settling Velocities of Silt Flocs. Proc. 14th Congress int. Ass. Hydraulics Res., Paris, Paper D-4, 27-32.

Owen, M.W., 1977. Problems in the modelling of transport, erosion and deposition of cohesive sediments, In: Goldberg, E.D., McCave, I.N., O'Brien J.J. and J.H. Steele, *The Sea, v. VI, Marine Modelling*, pp.515-37. Wiley-Interscience.

Ozsoy, E., 1977. *Flow and Mass Transport in the vicinity of tidal inlets*. Coastal and Oceanographic Engineering Laboratory, University of Florida, Report No. TR-036, 196 pp.

Ozsoy, E. and U. Unluata, 1982. Ebb-tidal flow characteristics near inlets. Estuarine, Coastal and Shelf Science, 14, 251-263.

Ozsoy, E., 1986. Ebb-tidal Jets: A Model of Suspended Sediment and Mass Transport at Tidal Inlets, Estuarine, Coastal and Shelf Science 22, 45-62.

Partheniades, E., 1965. Erosion and Deposition of Cohesive Soils. Proc. Amer. Soc. Civil Eng., J. Hydraulics Division, 91, 105-39.

Pasquill, F., 1962. *Atmospheric Diffusion*, Van Nostrand.

Pritchard, D.W., 1967. Observations of circulation in coastal plain estuaries. In: G.H. Lauff, ed., *Estuaries* pp.37-44. AAAS Publ. No.83, Washington, D.C.

Sayre, W.W., 1969. Dispersion of silt particles in open channel flow, Journal of the Hydraulics Division, Proceedings of the American Society of Civil Engineers, vol.95, No.HY3.

Schijf, J.B. and J.C. Schonfeld, 1953. Theoretical considerations on the motion of salt and fresh water, Proc. Minnesota Int. Hydraul. Conf., Minneapolis, Minnesota pp.321-333.

Schlichting, H., 1968. *Boundary Layer Theory*, 6 th ed., Mc Graw Hill.

Slade, D.H., ed., 1968. *Meteorology and Atomic Energy*, US Atomic Energy Comission.

Smith, J.D., 1977. Modelling of Sediment Transport on Continental Shelves, in Goldberg, E.D., McCave I.N., O'Brien J.J. and J.H. Steele, ed., *The Sea*, volume 6, Wiley.

Stanley, D.J. and D.J.P. Swift, ed.s, 1976. *Marine Sediment Transport and Environmental Management*. Wiley.

Stommel, H. and H.G. Farmer, 1952. *On the nature of estuarine circulation*, Woods Hole Oceanographic Inst., References No. 52-51, 52-63, 52-88 (3 vols. containing chapters 1-4 and 7).

Sumer, B.M., 1974. Mean velocity and longitudinal dispersion of heavy particles in turbulent open-channel flow. J. Fluid Mech. vol.65, part 1, pp.11-28.

Sumer, S.M., and Fischer, H.B. 1977. Transverse mixing in partially stratified flow. J. Hydraul. Div. Proc. Am. Soc. Chem. Eng. 103, 587-600.

Sutton, O.G., 1953. *Micrometeorology*. McGraw-Hill, New York, 333 pp.

Takeoka, H., 1984a. Fundemental concepts of exchange and transport time scales in a coastal sea, Continental Shelf Research. vol.3, No.4, pp.311-326.

Takeoka, H., 1984b. Exchange and transport time scales in the Seto Inland Sea, Continental Shelf Research. vol.3, No.4, pp.327-341.

Taylor, G.I., 1953. Dispersion of a soluble matter in solvent flowing slowly through a tube. Proc. R. Soc. London Ser. A 219, 186-203.

Taylor, G.I., 1954. The dispersion of matter in turbulent flow through a pipe. Proc. R. Soc. London Ser. A 223, 446-468; (1960). Sci. Pap. 2, 466-488.

Taylor, R.B., 1974, *Dispersive Mass Transport in Oscillatory and Unidirectional Flow*, Technical Report No.24, Coastal and Oceanographic Engineering Laboratory University of Florida.

Taylor, R.B., and R.G. Dean, 1974. Exchange Characteristics of Tidal Inlets, Proc. 14th Coastal Engineering Conference, ASCE, pp.2268-2289.

Tennekes, H. and J.L. Lumley, 1972. *A First Course in Turbulence*, M.I.T. Press, Cambridge. Massachusetts.

Van de Kreeke, J. and R.G. Dean, 1975. Tide-induced mass transport in lagoons, Journal of the Waterways, Harbors and Coastal Engineering Division, ASCE 101, 393-402.

Van de Kreeke, J., 1978. Mass Transport in a Coastal Channel, Marco River, Florida, Estuarine and Coastal Marine Science 7, 203-214.

Walters, T.S., 1962. Diffusion into a Turbulent Atmosphere from a Continuous Point Source, and from an Infinite Across-Wind Line Source, at Ground Level, Int. J. Air and Water Poll., vol.6, 1962, pp.349-352.

Wang, F.C., 1984. The dynamics of a River-Bay-Delta System, Journal of Geophysical Research, vol.89, No.C5, pp. 8054-8060.

Young, W.R., Rhines, P.B. and C.J.R. Garrett, 1982. Shear-Flow Dispersion, Internal Waves and Horizontal Mixing in the Ocean, Journal of Physical Oceanography, vol.12, No.6.

■
Transformation of Pollutants in the Marine Environment

J. ALBAIGÉS
Department of Environmental Chemistry
CID (CSIC)
Jorge Girona Salgado, 18-26
08034–Barcelona, Spain

GENERAL CONTEXT

It has already been mentioned that the marine environment is not a unique and homogeneous system (Sect. 1). The oceans can be subdivided into regions both vertically (i.e., air-sea water interface, mixed layer, deep water, sediment-water interface and sediment) and horizontally (i.e., estuaries, coastal zone and open ocean), the boundaries being defined by gradients of some physical and/or chemical properties of the water masses. Despite of this, the composition of the sea water is under a considerable degree of control and defines certain conditions under which the reactions involving pollutants may occur.

These conditions consist, mainly, on the pH, the redox potential and the organic matter of the sea water. The pH varies normally from 7.8 to 8.2, the major regulator being the CO_2 system, described in Figure 1. Therefore, if a reaction does not happen at pH values close to 8, it does not happen in sea water. The main implications of this constraint are in the partitioning equilibria or speciation of many organic or inorganic species as well as in the occurrence of hydrolytic reactions.

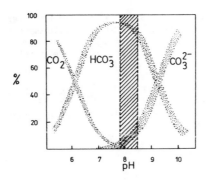

FIGURE 1.   Distribution of the three CO_2 forms in water as a function of pH.

The buffering capacity of the CO_2 system in the water column is relatively small; it has been calculated as approximately 0.25mM. However, if reactions involving complexation and the formation of ion pairs are considered, the calculated buffer capacity increases to 0.72mM. If, in addition, the sedimentary solid phases or the suspended particles (which are relevant in estuaries and shallow waters) are taken into account this capacity may increase by a factor of 400 (Wangersky, 1980).

The redox potential of sea water is fixed by the presence of dissolved oxygen Despite some exceptions, such as those of the anoxic deep waters (e.g. the Baltic Sea or the Cariaco trench) or the mid-depths in upwelling zones (e.g.the Walvis Bay or the Peruvian coast), the redox reactions in the sea are dominated by the presence of molecular oxygen. These reactions may be mediated by abiotic factors such as light (photochemical decompositions) or by the oxygenic biological activity (metabolic oxydation).

The third and not so obvious factor controlling marine reactions is set by the presence of organic matter in disolved and particulate forms. In fact, the major pool of organic carbon in our planet is that of dissolved organic matter in the sea (DOM) (Bolin et al. 1979). Although its precise structure is still unknown their main features indicate that it is strongly involved in the distribution of many trace metals in the water column through chelation and complexation (Mantoura, 1981) as well as in photochemical reactions because of the presence of chromophore moieties (Zafiriou et al. 1984).

In this context, the marine pollutants undergo their biogeochemical cycle, the nature and the rates of the reactions determining the ultimate fate of the components in the marine environment.

These reactions consist in chemical and biological processes, which can be classified, according to their relative time scales, in a group of relatively rapid processes that generally can be treated with a thermodynamic approach and a group of slower processes that require a kinetic treatment. This classification is shown in Table I. By extension we include in this list some physicochemical processes which are determinant for the availability of components to reaction.

These processes exhibit a great variety of understanding and their study can be attempted through three different strategies:

1) Laboratory scale experiments, such as the determination of physicochemical parameters (i.e. partition coefficients).

2) Mesocosms experiments, utilizing systems containing several levels of interaction between physical, chemical and biological components.

3) Field studies, which are essential for conclusive confirmation of the above tests or experiments.

The description of the processes that follows is based on results obtained from these approaches. The presentation is focused on the photochemical and microbial degradation processes, which are by far the most significant and the most efficient in transforming marine pollutants.

TABLE I

Description of processes governing the behavior of marine pollutants (adapted from Baughman and Burns, 1980).

Approach	Process	Status
Kinetic	Hydrolysis	Reliable studies in aqueous phase. Effects of sediment poorly understood. Recent interest on the effect of binding or interaction of humic substances with the reactants.
	Microbial transformation	Vast body of knowledge on bacterial metabolism. Few rate constants in literature
	Photolysis	Models for direct photolysis available. Indirect processes have been observed. Sensitized reactions poorly understood.
	Oxidation	Very few published data on free radical reactions on natural conditions except in the case of light-mediated processes. Influence of free radical scavengers and particulate matter poorly known.
Thermodynamic	Sorption and ligand exchange	Extensive data available on equilibrium constants (K_{oc}) mostly from lab tests. Influence on other processes (i.e. photochemistry) difficult to predict.
	Bioconcentration	Assessment from physical properties (i.e. K_{ow} in the case of hydrophobic organics). Recent application of the fugacity concept.

PHOTOCHEMICAL DEGRADATION OF POLLUTANTS

Background

The major portion of the solar energy which penetrates the atmosphere is absorbed in the oceans that cover 70% of the earth surface. In fact, energy from the sun is the primary driving force for the biochemical processes in the marine environment. Much of it is used in photosynthesis and heating of the sea surface. The remainder, which is absorbed, can initiate photochemical reactions which may have significant effects upon the dissolved organic and inorganic compounds, both natural and anthropogenic. The study of these reactions make the object of the modern marine photochemistry (Zika, 1981).

Figure 2 illustrates the fate of solar radiation on seawater. About 10-20% of the incident energy is lost by reflection or scattering and the remaining 80-90% is available for initiating a series of processes which are listed in Table II.

Approximately one half of the energy corresponds to the infrared region and is absorbed rather easily in the upper one meter and dissipated as thermal energy. The remainder of the radiation is composed of the visible (400-700 nm) and the near-ultraviolet (290-400 nm) which is able to penetrate the photic zone over to 100 m inducing photoprocesses. In a photophysical process such as fluorescence, the molecule goes from its ground state (So) through the excited singlet (S1;electron spins paired) and triplet (T1; electron spins unpaired) states back to ground state. These light-absorbing entities are called chromophores. In a photochemical process the same course is followed, except that during the lifetime of the excited species in either a singlet or triplet state, a chemical alteration occurs resulting in a formation of a new ground state product. In some circumstances the new product is reactive and undergoes either intra or intermolecularly further reactions in the system. Such reac-

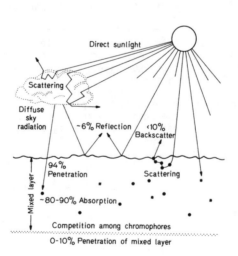

FIGURE 2.   The fate of solar radiation on sea water. (Roof, 1982; with perm.).

TABLE II

Process concerned with excitation and deactivation of a molecule (pollutant)
by solar radiation

Excitation	S_0	$\xrightarrow{h\nu}$	S_1
Internal conversion	S_1	$\dashrightarrow$	S_0 + heat
Intersystem crossing	S_1	$\dashrightarrow$	T_1 + heat
Fluorescence	S_1	$\dashrightarrow$	$S_0 + h\nu'$
Chemical reaction	S_1	$\dashrightarrow$	stable products, radicals (secondary reactions)
	T_1	$\longrightarrow$	id.
Phosphorescence	T_1	$\dashrightarrow$	$S_0 + h\nu'$
Energy transfer	T_1 + A	$\longrightarrow$	$S_0 + A_1^T$

The subscripts (i.e. 0, 1) indicate ground and first activation states.
S = singlet; T = triplet; A = acceptor molecule

tions are known as secondary reactions (see Table II). A compound does not
need to absorb radiation energy directly to undergo photochemical processes.
If an excited species does not react, bur proceeds to transfer the energy to
another molecule, then this becomes the reactive species. This process is
referred as photosensitization and the donor molecule is called photosensitizer.

A good sensitizer functions following the reaction sequence (see Table II):
excitation, intersystem crossing and energy transfer. Generally, the transfer
of triplet excitation energy is more efficient than that from singlet. If the
sensitizer is converted to a new compound having a low-lying triplet state,
then may act as a quencher and the reaction is stopped completely. It should
be emphasized, however, that sensitizers need not always to be absolutely
photostable because energy transfer will often reduce by itself photoreactivity.

Although photochemical reactions have energy thresholds below which the reaction is not energetically feasible, it appears that sunlight may promote several reactions at all wavelengths in the ultraviolet and visible region. The
sea surface solar spectrum and energy distribution is shown in Fig. 3. The
lowest wavelength recorded on the earth's surface is 286 nm. Therefore, compounds in the marine environment absorbing at longer wavelengths are candidates for primary photochemical processes.

Initially, a workable approach for the study of environmental photoreactions
should consist in the use of laboratory model systems. However, such studies
have often been criticized because do not reflect actual environmental conditions. Laboratory studies usually include a limited number of adjustable para-

meters (i.e. solvents, reactants, phases present, ...) and for convenience (to obtain appreciable photoconversion) numerous investigators have used radiation below 290 nm, making the results not easily extrapolable to natural conditions. Experimental approaches for conducting photoproduct studies relevant for the environment have been reviewed (Zepp, 1982). Despite their limitations, the results obtained with these systems provide useful information for understanding the processes, namely the degradation rates and the identity of the products (Sundstrom and Ruzo, 1978; Zepp and Baughman, 1978).

Sunlight can transform marine pollutants by two mechanisms:

1. direct absorption of light by the pollutant itself followed by reaction
2. adsorption of light by a photosensitizer that reacts with the pollutant or mediate a reaction pathway that includes the pollutant.

In Table III a diversity of natural seawater photo processes is summarized, taken from a report of the NATO-ARI "Photochemistry of Natural Waters" (Zafiriou et al, 1984) and other references from the literature. A more extensive list of reactions encountered in the aquatic environment has been reported by Roof (1982).

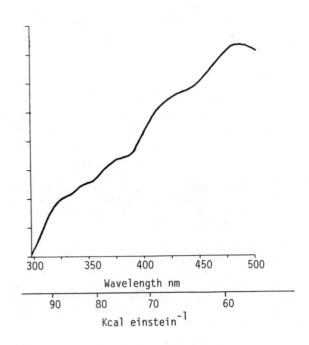

FIGURE 3.   The sea surface solar spectrum and energy distribution.

TABLE III

Natural photoprocesses observed in the marine environment

Substrate	Products	Probable mechanism and effects
NO_2^-	$.NO + .OH$	Direct photolysis. Induces NO air-sea flux.
CH_3I	$.CH_3+ .I$	Direct photolysis. Air-sea exchanges.
Natural organic chromophores	$.C + HO_2$ $C^+ + .O_2^-$ $.C + O_2 + H_2O_2$	Transfer (H, electron, energy) to O_2. Initiation of numerous reactions.
Petroleum hydrocarbons (RH, ArH, $ArBH_2R$, R_2S)	$ROO.$ $R=O$, RCO_2^-, $ArOH$ $ArCOR$, $ArCHO$, R_2SO	Oxygen addition. Oxidizes organic radicals. Direct or sensitized photolysis. Changes in oil properties and toxicity.
Pesticides (Disulfoton)	Disulfoton sulfoxide	Singlet oxygen. Changes in properties and toxicity.
Herbicides (2,4-D, 2,4,5-T)	Oxidation, reduction products	Direct photolysis. Complex.
Pentahclorophenol	Phenols, quinones, acids	Direct photolysis. Changes in properties and toxicity.
PCBs, DDT	Complex	Direct or sensitized photolysis Dehydrochlorination.
Cu(II)	$Cu(I)^+$	Alteration of cycling-toxicity.
Domestic wastes Fe(III)-NTA	Fe(II) + amine + + CO_2	Charge transfer to metal. NTA degradation.

Direct photolysis

Direct photolysis of organic pollutants has been the subject of intensive laboratory investigation.   Chromophores may undergo photodecomposition through the formation of radical species and further reactions involving isomerizations, rearrangements, additions or substitutions. Examples are given in Table III and Fig. 4.

FIGURE 4. Direct photolysis of:

A. PCBs (Cl atoms not depicted)
B. Pentachlorophenol
C. Metoxychlor
D. Triazine
E. Carbamate

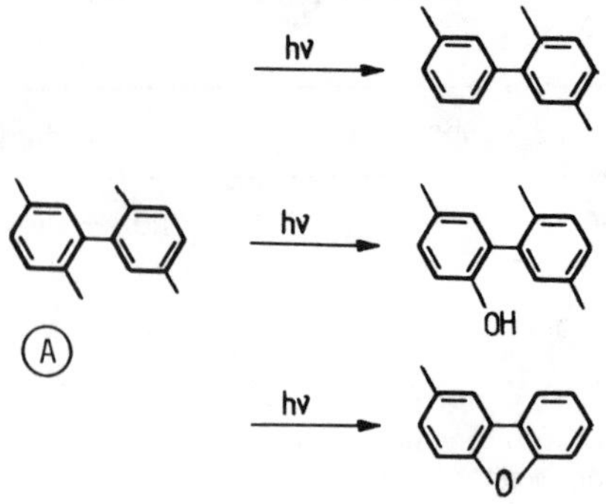

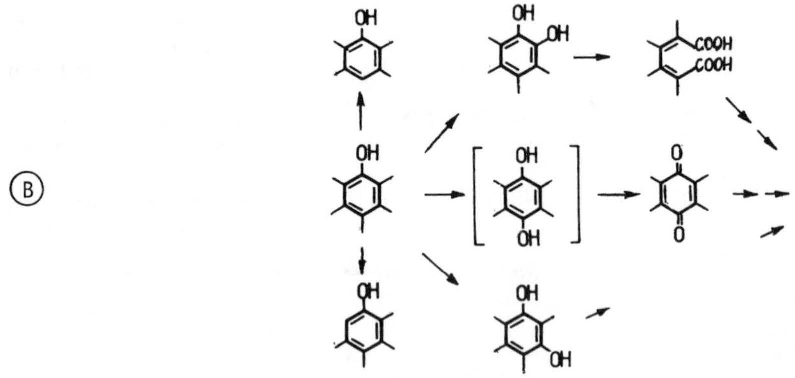

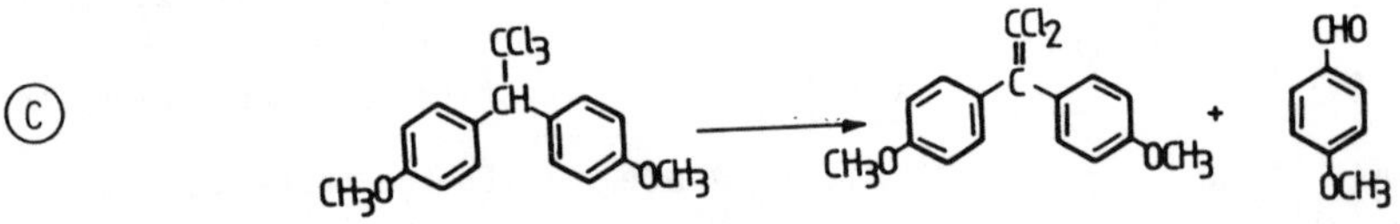

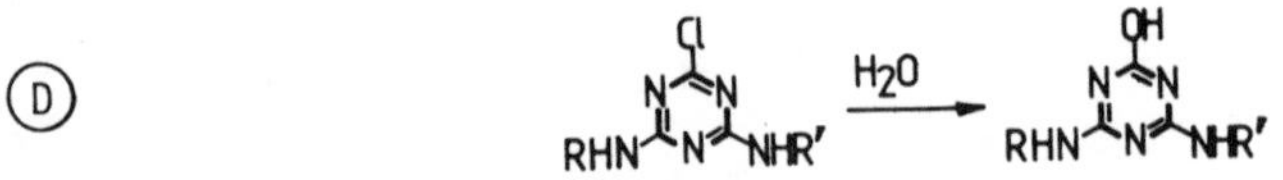

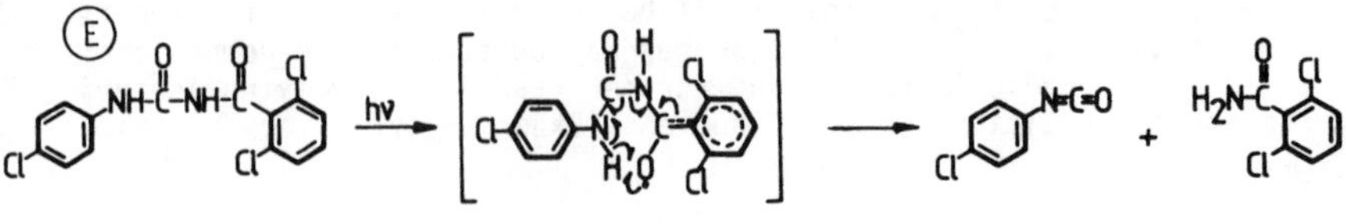

In an oxygenated solution such as seawater the most likely reactions of the formed radicals will be those associated with autoxidation. At this respect chlorinated biphenyls have been widely investigated. The photolytic C-Cl bond cleavage
in water at $\lambda > 310$ nm leads to three different reaction pathways namely dechlorination, nucleophilic substitution and dibenzofurane formation (Hutzinger and
Roof, 1980) (Fig. 4). Photochemical degradation of PCBs is influenced by the
degree of chlorination and the position of chlorine substitution in the ring.
After 24 hours of exposure to irradiation in hexane ($\lambda > 310$ nm) the tetrachloro-
biphenyls remaining were about 30%, whereas only 4% and 1% remained, respectively, of the hexa- and octachloro congeners.

The reaction was also decreasing from o- >m - > p-chlorine substituents, probably because of the release of steric hindrance in the dechlorination of the
orto position (Ruzo et al, 1974).

Direct photolysis of pentachlorophenol seems to follow through nucleophilic
substitution (Wong and Crosby, 1979). By dimerisation there is some evidence
of the formation of dioxin precursors, although TCDD has not been really detected, probably because it is rapidly decomposed.

The insecticides DDT and methoxychlor undergo dehydrochlorination, yielding DDE
and the analogue DMDE (Fig. 4), the latter much more rapidly. When the degradation was performed in natural waters instead of distilled water the half-life
was two orders of magnitude less pointing to a photosensitized reaction, as will
be discussed in the next section (Zepp et al, 1976). It is worth to mention
here that the DDT-DDE association in the environment is no more univocous.
Risebrough et al (1986) have demonstrated that other precursors (e.g. Kelthane) may also yield DDE.

Chlorophenoxyacetic acids (2,4-D and 2,4,5-T) photodecompose in water by cleavage of the ether linkage as well as by the substitution of chlorine atoms with
hydroxyl groups (Crosby and Wong, 1973).

Triazine herbicides photolyse in water to give mainly substitution products.
Carbonyl compounds (e.g. carbamates and ureas) react by the well known Norrish
type II pathway, which involves intramolecular hydrogen abstraction followed
by bond cleavage (Fig. 4) (Ruzo et al, 1974)

Given the absorption spectrum and quantum yield compound (the fraction of absorbed light that results in the excitation process, $\emptyset$), methods are available
for predicting the rate of reaction as a function of the environmental variables (Zepp and Baughman, 1978). Thus, the near-surface half-life, $t_{\frac{1}{2}}$, is defined as:

$$t_1 = \frac{1.81 \times 10^{20}}{\emptyset \, \Sigma \, \epsilon_\lambda \, Z_\lambda}$$

where, ϵ_λ is the molar absorptivity of the pollutant (in $mol^{-1} \, cm^{-1}$) and
Z_λ is the solar irradiance (in photons $cm^{-2} \, sec^{-1}$) both at the wavelength
centered at λ. Then, $\Sigma \, \epsilon_\lambda \, z_\lambda$ indicates summation over the wavelengths of sunlight absorbed by the pollutant. All these parameters can readily be obtained
from the absorption spectrum of the pollutant in water (or methanol when the
solubility of the compound is a problem), and from tabulated values (Zepp et
al, 1977 and 1978).

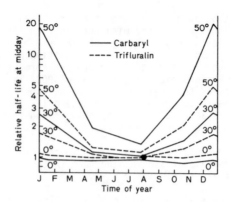

FIGURE 5. Midday half-lives for direct photolysis of carbaryl and trifluralin
for several northern latitudes relative to midsummer at 30ºN (•)
(Zepp et al., 1977; with perm.).

This calculation is exemplified in Figure 5 , where it is shown the dependence
of the photolysis half-lives of the insecticide Carbaryl and the herbicide
Trifluralin on the time of the year and for several latitudes. As a comparison,
predicted half-lives for direct photolysis of PAHs and pesticides (Mid-summer
at latitude 40° N) are 25 days (quinoline), 12 days (2,4-D), 10 days (parathion)
5.7 days (dibenzothiophene) and 0.1 days benzo (a) pyrene and benzo (a) anthracene (Zepp and Baughman, 1978). A general question arises from all these data.
How close are the assumed rates calculated from spectra in air-saturated pure
water from the actual photolysis rate in a water body?.

Of course, if the pollutant undergoes a reaction in natural water that changes its chemical nature or its physical disposition, then its photochemistry
will change. For this reason, laboratory experiments may give unrealistic
rates and product distributions ascribable to medium effects. Examples of
such reactions are the dissociation of phenol to phenoxide ions, forming
phenoxy radicals and hydrated electrons. Likewise, NTA and EDTA become photo-
activated and degraded when they form complexes with metal ions, such as Fe
(III) (Zika, 1981).

There is a variety of reactions involving photoderived unstable oxidation
states of metals (e.g. Cu^+ or Fe^{++}), when they are complexed with natural organic or inorganic ligands, which can alter their environmental cycling and toxicity. These reactions often follow charge transfer ligand to metal or vice-
versa or ligand exchanges (Zika, 1981):

$$\left[M^{x+} \ L^{y-} \right]^z \xrightarrow{h\nu} M^{(x-1)+} + L^{(y-1)-}$$
$$\longrightarrow M^{(x+1)+} + L^{(y+1)-}$$

Sorption of pollutants onto suspended sedimens should have also important effects upon photolysis rates. Photoreactions of adsorbed pollutants are environmentally important but difficult to study. Therefore, the information available at this respect is not conclusive.

Among the effects observed we may refer to those derived from the bathochromic spectral shifts occurred for sorbed molecules, the quenching of excited states of adsorbed pesticides, or the enhancement of electron transfers between charged species upon adsorption on silica colloids (Zafiriou et al, 1984).

Finally, direct photoreactions of dissolved inorganic chromophores may have important environmental implications, particularly concerning the air-sea exchanges and the formation of secondary products of environmental concern.

Nitrogen oxides and methane are the result of photodissociation of the seawater natural components nitrite and methyliodide, respectively. A net loss by this process of about 10% of the sea surface nitrate per day has been estimated (Zafiriou et al., 1984). The transfer of volatile materials to the atmosphere and particularly of photoreactive gases has received attention recently for their possible climatological implications. At the same time, the formation of hydroxyl radicals (see Table III) may accelerate the photolysis of organic pollutants even as refractory as the silicone oligomers (Anderson et al., 1987).

On the other hand, very recently, Suzuki et al. (1987a and b) have shown evidence of the formation of mutagenic nitroarenes in the photoreaction of aromatic hydrocarbons in waters containing nitrite or nitrate ions. Experiments have been conducted with pyrene and biphenyl and the reaction pathway for the latter is exemplified as follows:

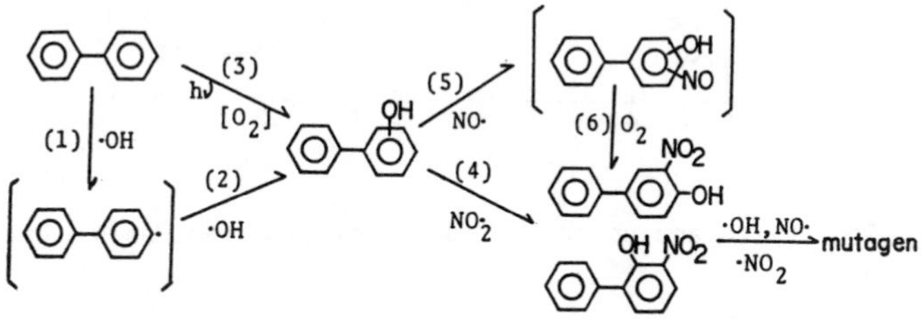

An important conclusion from these results is that the aquatic environment and particularly the marine one cannot be considered only as a sink for pollutants but also as a source.

Indirect photolysis

Indirect photolysis has not been studied so extensively, although it is common in seawater and is especially important because may alter molecules that resist direct photolysis.

Several pionering experiments demonstrated that photoreactions of pollutants not or heavily occurring in distilled water underwent in natural waters (Ross and Crosby, 1973 and 1975; Zepp et al., 1975 and 1976) (Fig. 6). The nature of these transformations has been attributed to the initial excitation of an "unknown seawater chromophore", followed either by energy transfer or by transfer of electrons or hydrogen atoms to other components of the system, including the pollutants (see Table II). Extensive studies have been carried out in order to identify the structural features of this "unknown chromophore", described in the literature with different terms such as marine humic and fulvic acids, Gelbstoffe (or yellow substance), etc., mainly because the different isolation techniques used (Ehrhardt, 1984).

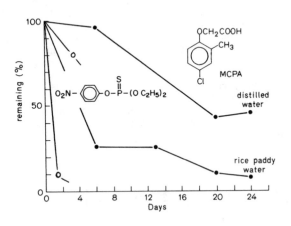

FIGURE 6. Sensitized and unsensitized photolysis of MCPA (●) and parathion (○)

In general, these substances, that account for 65-90% of the total organic carbon present in seawater, are complex mixtures of macromolecules with several functional groups, recognizing the major classes of plankton byproducts dissolved in seawater, namely carbohydrates, aminoacids and fatty acids. In coastal areas and particularly estuaries, this fraction may contain other constituents of terrigenous origin as well as a larger quantity of particulate and colloidal materials. Hypothetical structures, based on different analytical and synthetic approaches are shown in Fig. 7 (Stuermer and Harvey,1978; Harvey et al.,1983).

These materials absorb UV light and can transfer their excitational energy to chemicals present in the environment, thereby acting as photosensitizers. The wavelengths of sunlight absorbed by the sensitizers correspond to triplet energies ranging from 40-80 Kcal/Mole, in the same range as many chemicals.

These materials absorb UV light and can transfer their excitational energy to chemicals present in the environment, thereby acting as photosensitizers. The wavelengths of sunlight absorbed by the sensitizers correspond to triplet energies ranging from 40-80 Kcal/Mole, in the same range as many chemicals. These chemicals include most aromatic compounds with two or more aromatic rings, nitroaromatic compounds and chemicals with conjugated double bonds such as di-ketones and polyenes. Some triplet state energies are listed in Table IV. Thus, such pollutants should be able to participate in photoreactions, even those which do not absorb UV light of $\lambda > 300$ nm appreciably. At this respect, the degradation of p-cresol is increased by factors of 2 to 12 (Smith, 1978).

An important difference between direct and sensitized photolysis is that the latter proceeds most rapidly in the most opaque water bodies to sunlight, whereas the opposite is true for direct photolysis.

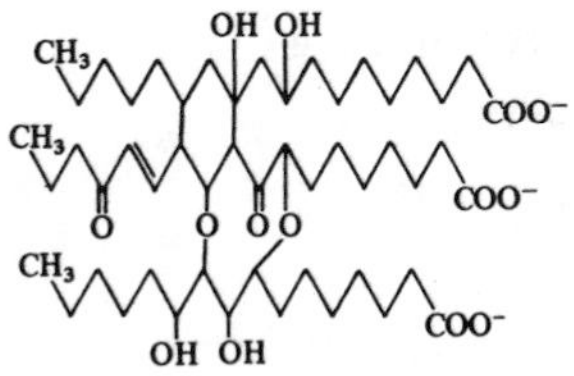

Marine Fulvic Acid 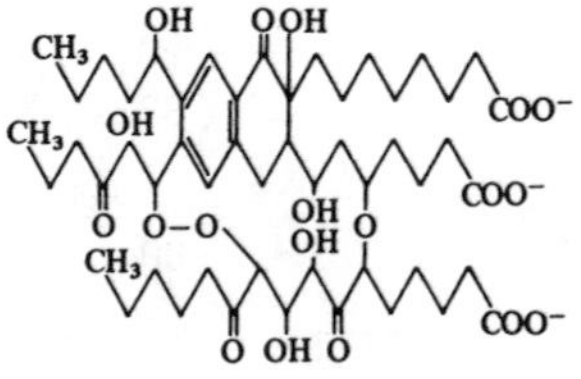   Marine Humic Acid

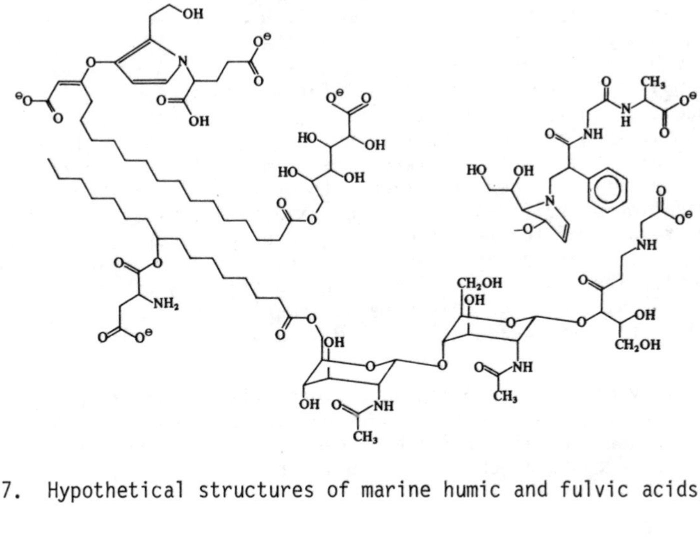

FIGURE 7. Hypothetical structures of marine humic and fulvic acids.

109

TABLE IV

Triplet state energies of selected synthetic chemicals
(from Zepp and Baughman, 1978)

Substrate	T. Energy kcal/mole	Substrate	T. Energy kcal/mole
DDT	79	Benzophenone	69
Methoxychlor	80	Biphenyl	65
DDE	54	Quinoline	62
Parathion	58	Naphthalene	61
Diazinon	76	Nitrobenzene	60
Carbaryl	60	Chrysene	57
2,4-D ester	77	Pyrene	49
p-Nitrophenol	58	Benz(a)anthracene	47
Hexachlorobenzene	70	Anthracene	42
Benzene	85	Benzo(a)pyrene	42
Phenol	82	Naphthacene	29
Aniline	77	Oxygen	23
Acetone	80		

Energy transfer sensitization is even more probable in the case of energy acceptors with lower triplet energies such as oxygen, the most important dissolved gas in seawater. The ground-state triplet of molecular oxygen is sufficiently low that triplet sensitizers may transfer energy to it at the maximum rate, resulting the singlet oxygen, a more powerful oxidant species than the oxygen itself. In this way, it has been estimated that about 1% of the solar energy absorbed by humic substances in surface waters (Fig. 7) produces "singlet oxygen" ($^1O_2$). Although it has only a lifetime of about 4 μs, its steady state concentration in sunlit waters can exceed a value of $10^{-14}M$ (Haag and Hoigné, 1986), enabling reaction with nucleophilic organic solutes, present as micropollutants in surface waters. The main reactions involved are the "ene reaction" with alkenes, reaction with cyclic 1,3-dienes to form endoperoxids, oxidation of sulfides to form sulfoxides, and electron-transfer reactions. The results of the measurements of rate constants for a series of model compounds revealed that these reactions are environmentally significant only for molecules undergoing 1,3-addition reactions with $^1O_2$ (e.g. furan type molecules). Also, phenoxide anions (e.g. trichlorophenols at natural pH) exhibit similarly high reactivities (Scully and Hoigné, 1987).

A great variety of oxygenated compounds identified in seawater, such as aldehydes, phenylalkanones and alkanols and aromatic ketones (Ehrhardt et al., 1982; Gschwend et al., 1982) have been proposed as sensitized photoreaction products of hydrocarbons dissolved in the sea.

The importance of polycyclic aromatic hydrocarbons in initiating photoreactions of marine pollutants and of alkyl aromatics in the radical propagation of oxidation chains has been recently shown by Thominette and Verdu (1984). The effects of photosensitizers in the fate of oil spills in the marine environment has been extensively studied. At this respect, the rate of photooxidation of petroleum films spread on seawater was greatly increased by the addition of naphthalene derivatives (Pilpel, 1975). These derivatives apparently acted as photosensitizers and caused an increase in solubilization of the films through a reaction mechanism involving the formation of peroxides and alcohols, these accelerating the formation of emulsions and their availability for degradation.

On the other hand, photooxidation of dimethylnaphthalenes accomodated in seawater, in the presence of oil and particularly of sulfur-containing compounds, has been shown to occur more efficiently in the aromatic ring due to the abundant formation of singlet oxygen (Sydnes et al., 1985). In the absence of oil the photooxidation was fairly inefficient and took place exclusively in benzylic positions.

The following scheme has been proposed for the oxidation of sulfur components of petroleum in the sea and their implication on further oxidation of hydrocarbon components (Burwood and Speers, 1974):

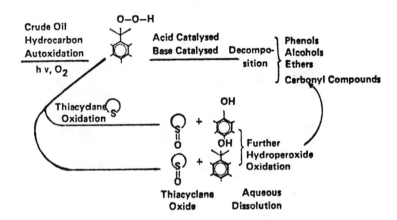

The partially oxidized hydrocarbons may also undergo polymerization, increasing the concentration of heavy components (asphalthenes) in oil spills. A detailed investigation of the structural changes occuring at this stage has been published (Albaigés and Cuberes, 1980; Tjessem and Aaberg, 1983).

111

Erhardt et al (1984) were able to simulate the reaction in the laboratory using phenylalkanes as substrates for photooxidation and anthraquinone as photosensitizer. Although the results of the experiment were insuficient for a rigurous justification of the reaction mechanisms it is suggested that a photooxidized polycyclic aromatic hydrocarbon (in this case anthracene) could undergo an intersystem crossing reaction to the relatively long-lived first triplet state (see Table I), which is able to abstract a benzylic hydrogen atom from an alkylbenzene. Momzikoff et al. (1983) pointed out that this direct action of an excited photosensitizer on a highly diluted substrate is unlikely in the presence of dissolved oxygen, an effective triplet quencher. However, it was proved that molecular singlet oxygen, generated by energy transfer from excited anthraquinone, does not carry sufficient energy to attack the benzylic positions of the substrates. As a result the mechanism shown in Fig. 8 was proposed.

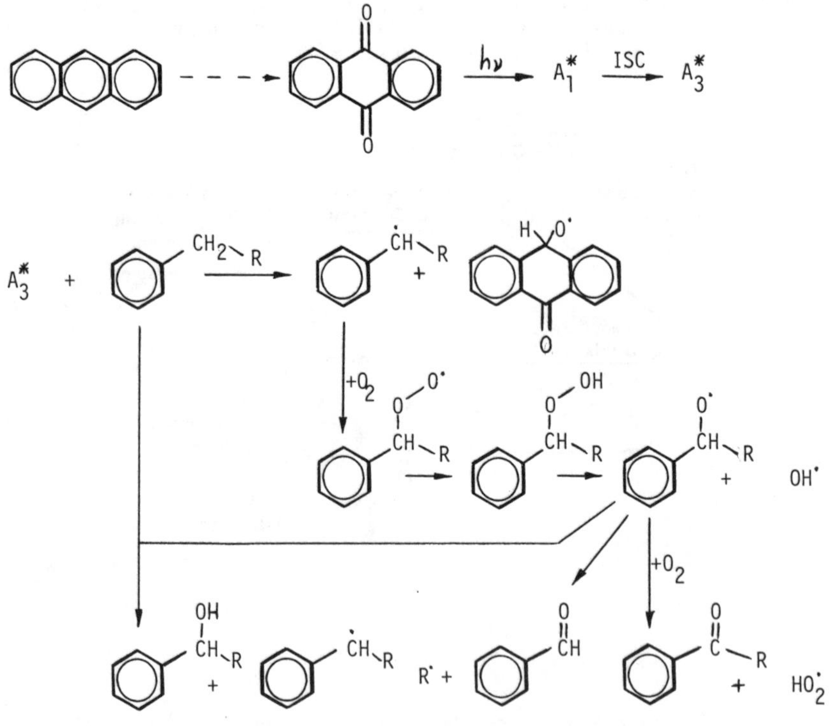

FIGURE 8. Photosensitized decomposition of alkylbenzenes in seawater.

112

The partially oxidized hydrocarbons may also undergo polymerization increasing the concentration of heavy components (asphalthenes) in oil spills.

A detailed investigation of the structural changes occurring at this stage has been published (Albaigés and Cuberes, 1980; Tjessem and Aaberg, 1983).

A generalization of the sequence of photosensitized reactions that a marine organic pollutant may undergo may be exemplified as follows:

$$X \xrightarrow{\ h\nu\ } X_1^* \xrightarrow{\ ISC\ } X_3^*$$

$$X^* + RH \longrightarrow XH\cdot + R\cdot \xrightarrow{+R\cdot} R\text{-}R$$

with branches:

$$XH\cdot \xrightarrow{O_2} X + HO_2^{\cdot}$$

$$R\cdot \xrightarrow{O_2} RO_2^{\cdot} \xrightarrow{\ RH\ or\ XH\ } RO_2H \longrightarrow RO\cdot + HO\cdot$$

$$RO_2H \xrightarrow{R\cdot} RO\cdot + ROH$$

$$RO\cdot + HO\cdot \xrightarrow{RH} ROH + R\cdot$$

where X is the photosensitizer
and RH is a hydrocarbon

The photochemical induction of metal-organic interactions in seawater which may contribute to their transport and mobilization has also been observed, however, the process has not yet been completely understood (Zika, 1981).

Finally, it should be considered that since humic substances contain aromatic polycyclic structures, they may act as quenchers by accepting the excitational energy of the environmental chemicals, if that is higher than that of the excited state of the humic material, thus retarding the degradation of the pollutants. An example has been described in which the half-life of aqueous atrazine was three times longer when irradiated in presence of fulvic acids (Khan and Schnitzer, 1978). A similar behaviour was observed in a series of PAHs, among them benzo(a)pyrene (Smith, 1978).

The environmental implications of this feature are evident, especially in estuarine areas, by the concurrence of land-based pollutant discharges and the formation of colloidal phases in the mixing waters zone.

BIOCHEMICAL TRANSFORMATIONS

Background

Microorganisms play the most important role in the biological transformation of marine pollutants, due to their widespread colonization of the marine environment and their enzymatic activity. Uptake and metabolism by higher organisms, although ecotoxicologically significant is irrelevant from the standpoint of the pollutants budget. Nevertheless, encapsulation in faecal material, particularly zooplanctonic, may contribute significantly to the transfer of conservative pollutants from the water column downward to the sediment.

The microorganisms distribution in the marine environment depends on many properties of the water column such as light penetration, temperature, oxygen, etc. In Fig. 9 it is shown the vertical distribution of habitats that determine specific populations of microorganisms. It is obvious that the dominant situation is the aerobic environment, the most reducing zones being restricted to the deep waters or highly producing areas in the ocean. Phototrophs dominate the illuminated euphotic zone, contributing substantially to the planktonic biomass. Chemoorganotrophic bacteria are confined to oxygenated waters and sediments. They include contributions from aerobes above the oxic/anoxic boundary and from NO_3^- and $SO_4^=$ - respiring anaerobes below it. Sedimentary chemoorganotrophs are ultimately dependent on the down-ward fluxes of organic matter (i.e., algal debris, faecal pellets, ...), being able to alter the chemical composition of the overlying waters to the extent of originating an anoxic layer.

On conventional media 90% of the bacterial isolates correspond to Achromobacter, Alcaligenes, Alteromonas, Flavobacterium, Pseudomonas and Vibrio genera (Grant and Long, 1985).

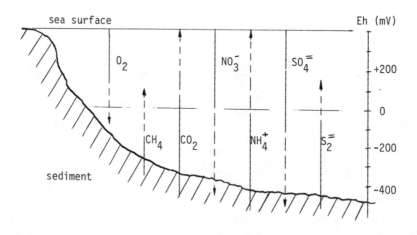

FIGURE 9. Physicochemical gradients in the water column (depths not on scale)

Much of the information presently available on the biological transformation of marine pollutants has been generated in laboratory studies with pure substrates and microbial cultures (mainly Pseudomonas and Alcaligenes sp.), making the extrapolation to environmental conditions sometimes delicate. See, for example, the reviews of Gibson (1978), Watkinson (1978) and Furukawa (1986). However, it should be difficult to envisage another approach for this study. Details on the mechanisms of the microbial transformations are unlikely to be obtained by direct experimentation, because of the diversity of interrelated variables upon which microbial activity depends. Bollag (1979) has discussed in detail the various types of biological transformations -mineralization, cometabolism, bioaccumulation and polymerization- that can occur. More recently, Richards and Shieh (1986) have reviewed the literature on the biological fate of organic priority pollutants in the aquatic environment, including biological waste water treatment systems. In the following section the main processes will be exemplified.

Microbial metabolism

Microbial metabolism in the seawater has been recognized under aerobic and anaerobic conditions, affecting differently the cycling of pollutants. Under aerobic conditions the degradative metabolism involve oxygen as a substrate for a class of enzymes known as oxygenases. These enzymes incorporate molecular oxygen into the organic molecules to form products that can follow more readily metabolic processes.

In anaerobic respiration, inorganic electron acceptors are utilized by the microbes to metabolize the organic pollutant. If no external electron acceptors are used but the organic molecules themselves, then the process is known as fermentation, which also occurs under anaerobic conditions. These processes are illustrated in Fig. 10.

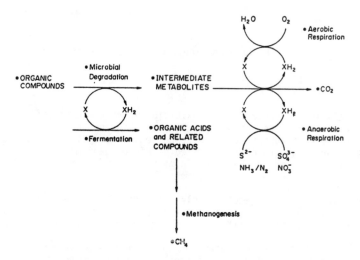

FIGURE 10. Microbial metabolic routes for organic pollutants.

If light is present, then photometabolism is also possible. An interesting example of biologically mediated photodecomposition has been reported by O'Kelly and Deason (1976), where the insecticide malathion was not degraded by the action of sunlight when dissolved in pure water, but was rapidly converted to the monoacid derivative when the aqueous solution contained algae (either Chlorella sp or Nitzschia sp.) (in Zepp and Baughman, 1978).

In the marine environment there are other beneficial associations of environmental factors, namely microorganisms, in the sense that often metabolites produced by one species becomes toxic and inhibit the further growth of the organism, but may be used by another as a growth substrate.

Examples of these comunities have been described in the literature both in aerobic and anaerobic environments. Nocardia, for example, can only grow with hydrocarbons (cyclohexane) as the sole source of carbon, in the presence of Pseudomona species (Watkinson, 1978). Another interesting example of a beneficial microbial association is that described by Slater (1978) consisting on the coexistance of Chlorobium and Desulfovibrio species in anaerobic environments accessible to light. The close relationship between both metabolic pathways is summarized in Fig. 11 .The hydrogen sulfide produced by the latter species serves as an electron donor for the photosynthetic activity of Chlorobium which lead to the regeneration of sulfate, thus closing the cycle.

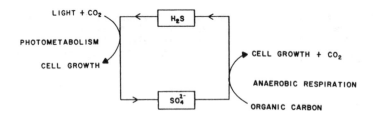

FIGURE 11. Photometabolic and anaerobic microbial association.

Parallel to this is the concept of co-oxidation where a microorganism is able to oxidize a substrate not used for growth when another cosubstrate is furnished for it. This technique has widened considerably the number of compounds that are known to be metabolized by microorganisms in the laboratory (Raymond et al., 1971) and probably explains the degradation as the ultimate fate of petroleum in the marine environment.

In fact, petroleum is an extremely complex mixture of hydrocarbons, both in number and types, thus exposing the bacterial populations to a variety of structural components and facilitating the occurrence of such processes.

The ubiquitous occurrence of hydrocarbons in the marine environment has prompted an extensive investigation of their metabolic processes (Watkinson, 1978; Atlas, 1984).

More than 25 genera of hydrocarbonoclastic bacteria have been identified (i.e. Pseudomonas, Vibrio, Acinetobacter, Alcaligenes, etc...) and used in laboratory studies. Particular attention has deserved the understanding of the fate and weathering of petroleum spills in the sea, because of their large impact on the biological resources and human activities. A number of general reviews have been published on this subject (e.g., Malins, 1977; Albaigés, 1980; Jordan and Payne, 1980) as well as reports on the fate of oil in different spillages (e.g. Torrey Canyon, Amoco Cadiz, Ixtoc) (Patton et al., 1981; Gundlach et al, 1983) and the reader is referred to them. The present section will be focused on the microbial metabolism of individual components or families and the mechanisms of the reactions involved.

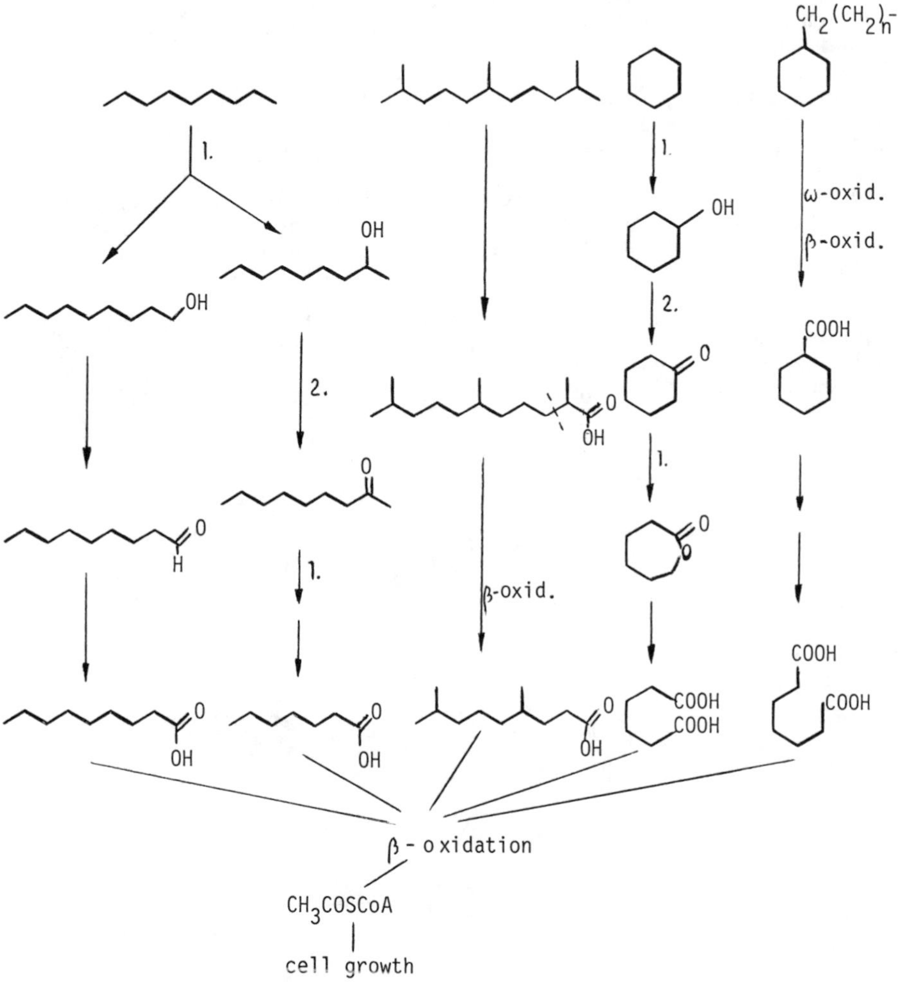

FIGURE 12.   Microbial degradation of aliphatic hydrocarbons (1. monoxygenase; 2. dehydrogenase).

The major pathways used by aerobic microorganisms for the metabolism of alkanes implies α-, β- or ω-oxidation (Fig. 12), usually by the cytochrome P-450 oxygenase. These reactions are directed towards the production of acetyl-coenzyme A which can be readily assimilated by the organisms for energy production and cell growth. Anaerobic degradation may also occur, but in this case the first step is a dehydrogenation, the alkene formed being hydroxylated afterwards (Watkinson, 1978).

Alicyclic hydrocarbons are more resistant to degradation, particularly those containing fused-rings, such as the known molecular markers, steranes and hopanes. This is why these components, that can be recognized even in heavily degraded oils, have been proposed as indicators of pollutant sources (Albaigés and Albrecht, 1979). The degradation pathway is shown in Fig. 12 and offers the particularity of an intermediate biological Baeyer-Villiger reaction, reaction which is frequently encountered in the microbial degradation of ketone substrates.

Alkyl-substituted cycloalkanes (Fig. 12) follows a pathway more like that of n-alkanes. The anaerobic degradation goes through the ring aromatization in the step of the cyclohexylcarboxylic acid (benzoic acid formation).

Interest in the degradation of aromatic compounds stems from the known toxicity (and carcinogenicity in some cases) of some of them. Many insecticides, herbicides, solvents, detergents, etc. which are considered as environmental pollutants contain aromatic rings in their molecular structures.

The degradation of the aromatic ring follows a different pathway (Fig. 13) involving the introduction of hydroxy groups, usually two in ortho position, by a dioxygenase enzyme. Gibson has shown that this did exhibit a cis configuration, differently to higher organisms which generate the trans isomer.

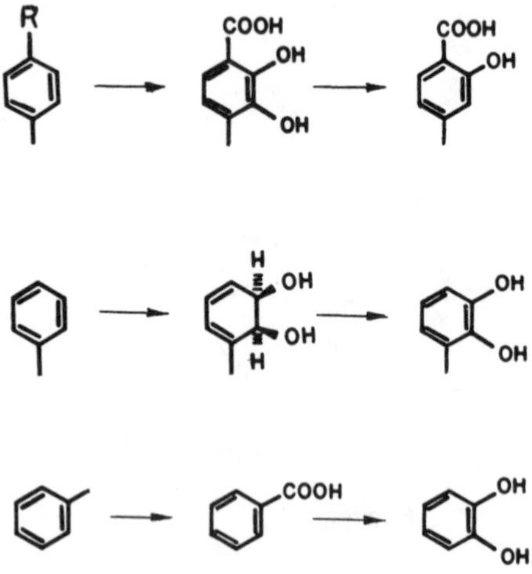

FIGURE 13. Initial reactions in the aerobic degradation of alkylaromatic hydrocarbons.

Further oxidation leads to the formation of catechols which are the substrates for ring fission. Therefore, two main steps, hydroxylation and ring fission control the degradation (see Gibson, 1978 for a review). Compounds with alkyl substituents may undergo oxidation of either the aromatic nucleus or the alkyl sidechain leading to the catechol or protocatechuic acid routes as indicated in Fig.13.The degradation of linear alkylbenzenes (LAB) or their sulphonated counterparts (LABs), which constitute the most widely used anionic surfactant, takes place by an analogous mechanism (Swisher, 1987).

Two different metabolic routes have been identified for the ring-fission step (Fig. 14). The ortho-fission pathway involves cleavage of the ring between the hydroxylated carbon atoms, whereas in the meta-fission it is the adjacent bond that is cleaved. It has become apparent that the ortho pathway is used almost exclusively for the metabolism of the catechol itself and that reactions of the meta-pathway are most tolerant for both catechol and substituted congeners (Hopper, 1978).

Polycyclic aromatic hydrocarbons behave in a similar way. Some accepted biodegradation pathways in Pseudomonas sp are exemplified in Fig. 15. Two interesting aspects arise from these schemes which have been focused recently. The first is the effect that methyl substitution may have in the degradation rate of different isomers. As it is known, petroleum contains, predominantly, alkylsubstituted aromatics and it will be important to assess these rates in order to understand better the degradation of petroleum in the marine environment. Very recently, Bayona et al. (1986) have shown the different reactivity of methyl-substituted naphthalenes, phenanthrenes, pyrenes and chrysenes according to the position of the substituents. In general, the β- or 2-methyl isomers were more easily degraded. The explanation of this feature, if the oxidation of the α, β-positions unoccupied in the molecule was enhanced or, by the contrary, a co-oxidation of the methyl group took place, is to be investigated.

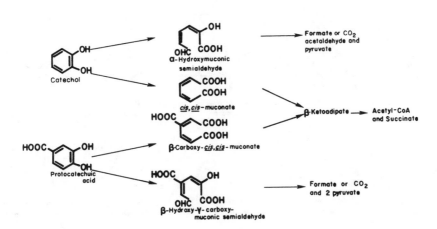

FIGURE 14. Ring-fission products formed from hydroxylated aromatic compounds (Gibson, 1978).

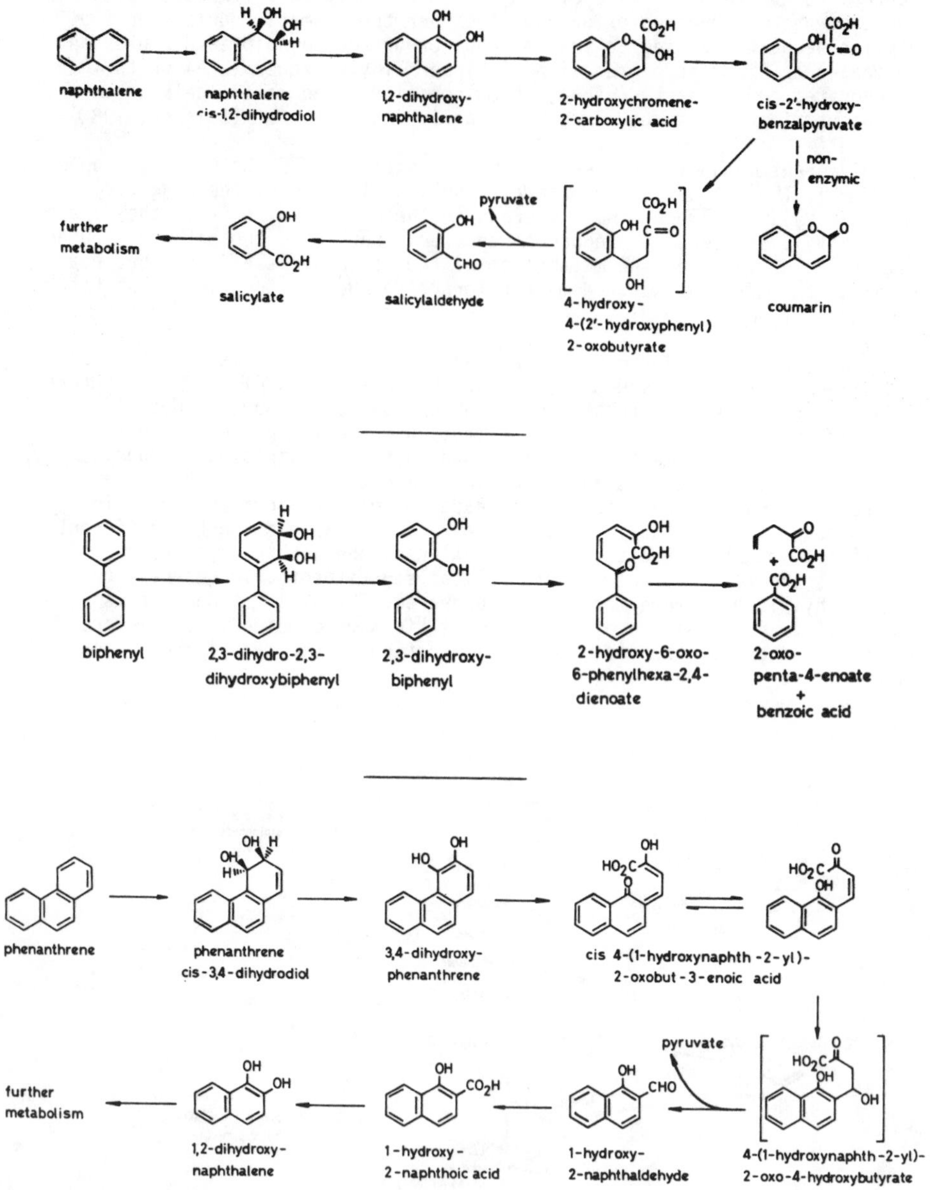

FIGURE 15. The metabolism of naphthalene, biphenyl and phenanthrene in pseudomonads.

120

The second aspect relates with the degradation of the highly aromatized petroleum residues. Albaigés and Cuberes (1980) observed that the molecular weight of the asphalthenic fraction of seawater exposed crudes decreased dramatically together with the aromaticity and proposed that a degradation of the previously photooxidized benzylic positions of the alkyl chains of the polycondensed asphaltenic cluster could explain these modifications.

More recently, Rontani et al. (1985 and 1987) have also shown a positive interaction between photochemical and microbiological degradation of different aromatic hydrocarbons (e.g. anthracene and alkylbenzenes) in the sense that the photooxidation of certain positions of the molecule allows an accelerated elimination of the substrate by the use of the resulting products by bacteria. In this sense, for example, the oxigenated products shown in Fig. 8 for the photooxidation of alkylbenzenes can be easily assimilated by Alcaligenes sp. Another illustrative example is shown in Fig. 16, where a more biologically refractory molecule can be degraded at a higher rate.

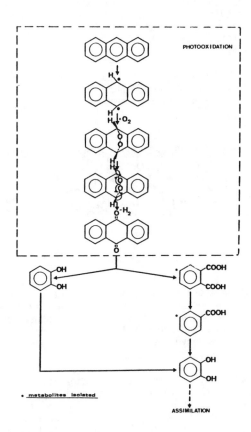

FIGURE 16. Interaction between photooxidation and bacterial degradation of anthracene.

Particular interest has been devoted to the understanding of the degradation of aromatic compounds bearing halogen substituents, such as PCBs, DDT, chlorophenols and chlorobenzoic acids, which are widely distributed in the marine environment and are known to be highly persistent.

It has become clear that the presence of electron-withdrawing substituents (chlorine, nitro, sulphonate) in the aromatic ring hinders the degradation of the compounds in accordance with the assumption that dioxygenation is an electrophilic reaction. However, a number of reports show that these compounds may also undergo microbial degradation.

PCBs degrading bacteria recovered from estuarine and marine environments include seven genera: Pseudomonas, Vibrio, Aeromonas, Micrococcus, Acinetobacter, Bacillus and Streptomyces (Sayler et al., 1978)

The results obtained by Furukawa et al. (1986) indicate that degradation rate decreases as chlorine substitution increases and that microorganisms preferentially metabolize the unsubstituted rings or at least require an unsubstituted 2,3-position in order to initiate the degradation of chlorinated biphenyls. It has also been observed that some congeners that do not degrade easily when present alone in the culture medium do so when present in a mixture, clearly indicating that co-metabolism could be involved in the degradation process (Baxter et al., 1975).

Like other aromatic hydrocarbons PCBs are considered to be metabolized via a deoxygenase catalyzed reaction. Metabolic products include chlorobenzoic acids, hydroxyderivatives and meta-cleavage compounds:

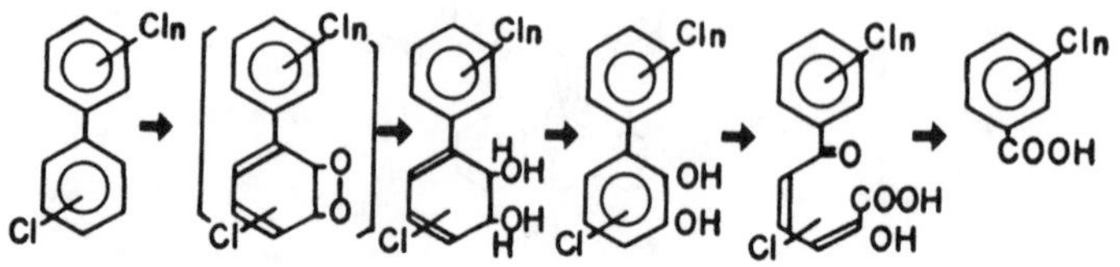

Although hydroxylation and ring cleavage are the main mechanisms in biodegradation of PCBs, other products have also been identified. Incubation of Aroclor 1242 with a bacterial culture from a harbor produced a number of alkenes and alkylbenzenes (Kaiser and Wong, 1974). The degradation in presence of nitrate produced nitrohydroxyderivatives that were more toxic than the precursors (Sylvestre et al., 1982).

These microorganisms usually do not have the capacity to dehalogenate either the parent compounds of the chlorinated degradation products in order to avoid the synthesis of recalcitrant metabolites, as in the case of some chlorobenzoic acids. There are a few examples in which the replacement of the chlorine substituent by hydrogen or a hydroxyl group occurs. However, another mechanism to avoid this synthesis has been developed by certain strains which are able to utilize 3- and 4- chlorobenzoates by synthetizing catechol 1,2-dioxygenase instead of the 2,3-dioxygenase   (Neilson et al., 1985):

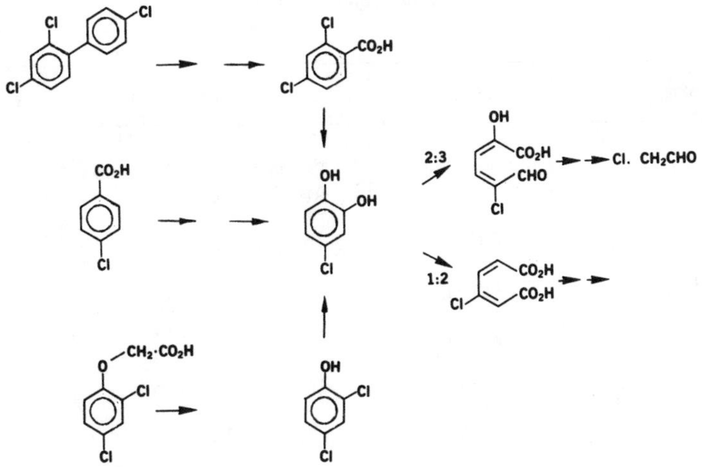

Marine microorganisms are involved also in the transformation and mobilisation of metals and metalloids, particularly heavy metals, by microbially-mediated changes in valence state. The capacity to oxidize and/or reduce particular metals is so widespread among prokariotes that is unlikely that specific enzymatic systems are often involved. Microbial transformations of this type have been extensively reviewed (Mitchell, 1978; Grant and Long, 1981) and are summarized in Table V.

Particularly significant has been the methylation of metals, namely of mercury, lead and tin, because the production of highly toxic end products. It appears that methylation involves the transfer of either methylcarbonium ion (CH_3^+) from S-adenosyl methionine or methyl homocysteine, or the transfer of methyl carbanion ion (CH_3^-) from methylcobalamin. Prokariotes including archaebacterial methanogens have been implicated in methylations of this kind.

This transformation is probably a detoxification process since the metabolites are volatile and thus escape from the vicinity of the microorganism. There is evidence that maximum methylation rates takes place at oxidative anaerobic sediment interfaces. Interesting to note that the ratio of methyl to total mercury in various polluted sediments seems to vary between 0.1% and 0.5%. If mercury was really being removed from sediments by methylation, then in many locations no more mercury would remain. However, breakdown processes of methylmercury exist in the aquatic environment, that will return mercury to the sediment. Then, there is a dynamic equilibrium between methylation and demethylation in sediments.

A potentially important cycling process for organomercury compounds and metals in general is shown by the role of hydrogen sulfide in the mobilization of methyl mercury and sulfides:

$$2CH_3Hg^+ + H_2S \longrightarrow (CH_3)_2Hg + HgS$$

TABLE V

Microbial transformations of environmentally significant metals and metalloids
(after Grant and Long, 1981)

Transformation	Metal	Microorganism
Oxidation	As(III)	Pseudomonas, Actinobacter, Alcaligenes
	Cu(I)	Thiobacillus
Reduction	As(V)	Chlorella
	Hg(II)	Pseudomonas
	Se(IV)	Corynebacterium
Methylation	As(V)	Aspergillus
	Cd(II)	Pseudomonas
	Se(IV)	id.
	Sn(II)	id.
	Pb(IV)	id.
	Hg(II)	Bacillus, Clostridium, Methanogens

In the real environment, hydrophobic substrates may also establish a number of interaction mechanisms with the marine humic substances, which may result in a reduction in their available concentration for microbiological attack. The binding between the two substances is usually by ionic or covalent bonding or it may be the result of absorption by van der Waals attractions, hydrogen bonding, charge transfer and hydrophobic bonding. Binding, particularly in sediments, may be catalysed by microbial enzymes. Degradation of certain pollutants results in reactive intermediates which may bind more easily to humic substances than the parent chemicals.

As far as the ecosystem is concerned, the significance of any metabolic activity cannot be assessed only from the standpoint of the organism producing the metabolite. There are some situations in which a metabolite, though nontoxic to the organism producing it, is highly toxic to other components of the system. As indicated above, microbial C-methylation is one of the detoxification systems developed by an organism in response to exposure to Hg^{+2}. This results in the synthesis of the extremely toxic methylmercury for the higher forms of marine life.

The formation of nitrophenol or nonylphenols during the respective degradation of the pesticide parathion or the widely used alkylphenolethoxylated detergents are also examples of microbial production of toxic components in the natural ecosystem. Finally, the natural occurrence of O-methylation of halogenated phenolic compounds present in the effluents of paper mills has been documented. The resulting chloroguaiacols have additional deleterious effects on fishes.

Aside from the importance of controlling the input of pollutants into the marine environment, an understanding of the transfer and transformation processes through the different marine compartments and that of the ultimate fate of a pollutant, i.e., whether it becomes degraded or immobilized in the sediments, are of considerable interest and also necessary for modelling the situation and for mass balance calculations.

References

Albaigés J, P.Albrecht (1979). Fingerprinting marine pollutant hydrocarbons by COM-GC-MS. Internat.J.Environ.Anal.Chem., 6, 171-190.

Albaigés J (1980). The fate and source identification of petroleum tars in the marine environment. Coll.Internat.CNRS No. 293, pp 233-247. CNRS, Paris.

Albaigés J, MR Cuberes (1980). On the degradation of petroleum residues in the marine environment. Chemosphere, 9, 539-545.

Anderson C, K Hochgeschwender, H Weidermann and R Wilmes (1987). Studies of the oxidative photoinduced degradation of silicones in the aquatic environment. Chemosphere, 16, 2567-2577.

Atlas RM (1984). Petroleum microbiology. 692 pp. MacMillan Pub. Co. New York.

Baughman GL, LA Burns (1980). Transport and transformation of chemicals: A perspective. In: The Handbook of Environmental Chemistry (ed. O.Hutzinger) Vol. 2A, pp 1-18. Springer-Verlag Berlin.

Baxter RA, PE Gilbert, RA Lidgett, JH Mainprize, HA Vodden (1975). The degradation of PCBs by microorganisms. Sci.Total Environ., 4, 53

Bayona JM, J Albaigés, AM Solanas, R Parés, P Garrigues and M Ewald (1986). Selective aerobic degradation of methyl-substituted polycyclic aromatic hydrocarbons in petroleum by pure microbial cultures. Intern.J.Environ.Anal.Chem., 23, 289-303.

Bolin B, ET Degens, S Kempe, P Ketner (1979). The global carbon cycle. 491 pp J.Wiley. Chichester.

Bollag GM (1979). Transformation of xenobiotics by microbial activity. In Microbial degradation of pollutants in marine environments (Ed. A.W.Bourquin and P.E.Prichard). pp. 19-22. Gulf Breeze

Burwood R, GC Speers (1974). Some chemical and physical aspectes of the fate of crude oil in the marine environment. Adv.Org.Geochem., 1973, 1005-1027.

Crosby DG, AS Wong (1973). J.Agric.Food Chem., 21, 1052

Ehrhardt M, F.Bouchertall, HP Hopf (1982). Aromatic ketones concentrated from Baltic Sea water. Mar.Chem., 11, 449-461

Ehrhardt M (1984). Marine Gelbstoff. In: The Handbook of Environmental Chemistry (ed. O.Hutzinger) Vol. 1C, pp. 63-77. Springer Verlag.Berlin.

Ehrhardt M, G Petrick (1984). On the photooxidation of alkylbenzenes in seawater. Mar.Chem., 15, 47-58.

Furukawa K (1986). Modification of PCBs by bacteria and other microorganisms. In PCBs and the Environment (Ed. JS Waid) Vol. 2, p. 89, CRC Press

Gibson DT (1978). Microbial transformations of aromatic pollutants. In Aquatic Pollutants. Transformation and biological effects (Ed. O.Hutzinger et al.) p. 187. Pergamon Press. Oxford

Grant WD, PE Long (1981). Environmental Microbiology. Blackie. Glasgow.

Grant WD, PE Long (1985). Environmental Microbiology. In The Handbook of Environmental Chemistry (Ed. O.Hutzinger). Vol. 1D, pp. 127-237. Springer-Verlag. Berlin.

Gschwend PM, OC Zafiriou, RFC Mantoura, RP Schwarzenbach (1982). Volatile organic compounds at a coastal site. Environ.Sci.Technol., 16, 31-38

Gundlach ER, PD Boehm, M Marchand, RM Atlas, DW Ward, DA Wolfe (1983). The fate of Amoco Cadiz oil. Science, 221, 122-129.

Haag WR, and J Hoigné (1986). Singlet oxygen in surface waters. w. Photochemical formation and steady-state concentrations in various types of waters. Environ.Sci.Technol., 20, 341-348.

Harvey GR, DA Borau, LA Chesal, JM Tokar (1983). Mar. Chem., 12, 119.

Hopper DJ (1978). Microbial degradation of aromatic hydrocarbons. In Developments in biodegradation of hydrocarbons (Ed. RJ Watkinson) pp 85-112. Applied Sci.Pub.London.

Hutzinger O, AAM Roof (1981). Polychlorinated byphenyls and related halogenated compounds. In Analytical Techniques in Environmental Chemistry (Ed. J.Albaigés) p. 167-184. Pergamon Press.

Jordan RE, JR Payne (1980). Fate and weathering of petroleum spills in the marine environment. 174 pp. Ann Arbor Sci. Ann Arbor.

Kahn SU, M Schnitzer (1978). J.Environ.Sci.Health, B13, 299.

Kaiser KLE, PTS Wong (1974). Bacterial degradation of PCBs. I. Identification
of some metabolic products from Aroclor 1242. Bull.Environ.Contam.
Toxicol., 11, 291.

Malins DC (1977). Effects of petroleum on Arctic and Subarctic marine environments and organisms. Academic Press N.Y.

Mantoura RFC (1981). Organo-metallic interactions in natural waters. In Marine
Organic Chemistry (Ed. EK Duursma, R Dawson) p. 179. Elsevier.
Amsterdam.

Mitchell R (1978). Water Pollution Microbiology. John Wiley.

Moilanen KW (1975). Environ.Sci.Res., 6, 45.

Momzikoff A, R Santus, M Giraud (1983). A study of the photosensitizing properties of seawater. Mar. Chem., 12, 1-14.

Neilson AH, AS Allard, M Remberger (1985). Biodegradation and transformation of
recalcitrant compounds. In The Handbook of Environmental
Chemistry (Ed. O.Hutzinger) Vol. 2C, pp 29-86. Springer-Verlag.
Berlin
Parlar M, M Mansour (1982). GC determination of several cyclodiene insecticides
in the presence of PCBs by photoisomerization reactions. In
Analytical Techniques in Environmental Chemistry. 2. (Ed. J
Albaigés), pp 241-247. Pergamon Press. Oxford.

Patton JS, MW Rigler, PD Boehm, DL Fiest (1981). Ixtoc 1 oil spill: flaking
of surface mousse in the Gulf of Mexico.
Pilpel N (1968). Fate of oil in the sea. Endeavour, 27, 11-13.

Raymond RL, VW Jamison, JO Hudson (1970). Hydrocarbon cooxidation in microbial
systems. Lipids, 6, 453-457.

Richards DJ   and WK Shieh (1986). Biological fate of organic priority pollutants in the aquatic environment. Water Res., 20, 1077-1090.

Risebrough RW, WM Jarman, AM Springer, W Walker, W Hunt (1986). A metabolic derivation of DDE from Kelthane. Environ.Toxicol.Chem., 5, 13-19.

Rontani JF, P Bonin and G Giusti (1987). Mechanistic study of interactions
between photooxidation and biodegradation of n-nonylbenzene in
sea water. Mar.Chem., 22, 1-12

Rontani JF, E Rambeloarisoa, JC Bertrand and G Giusti (1985). Favourable
interaction between photooxidation and bacterial degradation of
anthracene in sea water. Chemosphere, 14, 1909-1912

Roof AAM (1982). Aquatic photochemistry. In The Handbook of Environmental
Chemistry (Ed. O.Hutainger) Vol. 2B pp. 43-72. Springer-Verlag.
Berlin.

Ross RD, DG Crosby (1973). Photolysis of ethylemethiourea. J.Agric.Food Chem.
21, 335.

Ross RD, DG Crosby (1975). The photoxidation of aldrin in water. Chemosphere, 4,
227.

Ruzo LO, MJ Zabik and RD Schmetz (1974). Photochemistry of bioactive compounds. Photochemical processes of PCBs. J.Am.Chem.Soc., 96, 3809.

Sayler GS, R Thomas, RR Colwell (1978). PCB degrading bacteria and PCB in estuarine and marine environment. Estuarine Coastal Mar.Sci. 6, 553.

Scully FE and J Hoigné (1987). Rate constants for reactions of singlet oxygen with phenols and other compounds in water. Chemosphere, 16, 681-694.

Slater JH (1978). In: The oil industry and microbial ecosystems (Ed. KWA Chater, HJ Somerville) pp 137-154, Hayden and Son, London

Smith JH (1978). Environmental pathways of selected chemicals in freshwater systems. EPA-600/7-78-074, Cincinatti.

Stuermer DH, GR Harvey (1978). Structural studies on marine humus: a new reduction sequence of carbon skeleton determination. Mar.Chem., 6 55-70.

Sundstrom G, LO Ruzo (1978). Photochemical transformation of pollutants in water. In: Aquatic Pollutants. Transformation and biological effects (Ed. O.Hutzinger et al.) p. 205. Pergamon Press. Oxford.

Suzuki J, T Hagino and S Suzuki (1987a). Formation of 1-nitropyrene by photolysis of pyrene in water containing nitrite ion. Chemosphere, 16, 859-867.

Suzuki J, T Sato, A Ito, S Suzuki (1987b). Photochemical reaction of biphenyl in water containing nitrite and nitrate ion. Chemosphere, 16, 1289-1300.

Swisher RD (1987). Surfactant Biodegradation. Surfactant Sci.Ser. Vol. 1B (Mar-1085 pp. Marcel Dekker. New York.

Sylvestre M, R Masse, F Messier, J Fanteux, JG Bisaillon, R Beaudet (1982). Bacterial nitration of 4-chlorobiphenyl. Appl.Environ.Microbiol. 44, 871.

Sydness LK, SH Hansen and IC Burkow (1985). Factors affecting photooxidation of oil constituents in the marine environment. Chemosphere, 14, 1043-1055.

Thominette F, J Verdu (1984). Photooxidative behaviour of crude oils relative to sea pollution. Mar.Chem., 15, 105-115.

Tjessem K, A Aaberg (1983). Photochemical transformation and degradation of petroleum residues in the marine environment. Chemosphere, 12, 1373-1394.

Wangersky P (1980). Chemical Oceanography. I. The Handbook of Environmental Chemistry (Ed. O.Hutzinger) Vol. 1A, p. 51, Springer-Verlag. Berlin.

Watkinson RJ (1978). Developments in biodegradation of hydrocarbons. 232 pp. Applied Sc. London.

Wong AS, DG Crosby (1979). J.Agric.Food Chem., 29, 125-130.

Zafiriou OC, J Joussot-Dubien, RG Zepp, RG Zika (1984). Photochemistry of natural waters. Environ.Sci.Technol., 18, 358A-371A.

Zepp RG, NL Wolfe, JA Gordon, RC Fincher (1976). Light induced transformations of methoxychlor in aquatic systems. J.Agric.Food Chem., 24, 727

Zepp RG, NL Wolfe, JA Gordon, GL Baughman (1975). Dynamics of 2,4-D esters in surface waters: hydrolysis, photolysis and vaporization. Environ. Sci.Technol., 9, 1144.

Zepp RG (1982). Experimental approaches to environmental photochemistry. In The Handbook of Environmental Chemistry (Ed. O.Hutzinger) Vol. 2B, pp 19-41. Springer-Verlag. Berlin

Zepp RG (1977). Rates of direct photolysis in aquatic environment. Environ. Sci.Technol., 11, 359.

Zepp RG, GL Baughman (1978). Prediction of photochemical transformation of pollutants in the aquatic environment. In Aquatic Pollutants. Transformation and biological effects. (Ed. O. Hutzinger et al.) pp 237-263. Pergamon Press.

Zepp RG (1978). Quantum yields for reaction pollutants in dilute aqueous solution. Environ.Sci.Technol., 12, 327.

Zika RG (1981). Marine organic Photochemistry. In Marine Organic Chemistry (Ed. EK Duursma, R Dawson) p. 299. Elsevier. Amsterdam.

The Biological Effects of Marine Pollutants

B. L. BAYNE
Institute for Marine Environmental Research
Prospect Place
The Hoe
Plymouth PL1, 3DH, England

INTRODUCTION

The very term "pollution" implies a biological response to a man-made disturbance and is to be distinguished from the term "contamination", which refers only to the chemical consequences of environmental impact from industrial and other sources. The need for convincing measures of biological response to contamination has been recognised for many years and has prompted a great deal of research. More recently, there has been discussion on the various merits of the many known responses to contamination for detecting pollution in the natural environment, in attempts to recommend procedures for general use (ICES, 1978; papers in McIntyre and Pearce, 1980; National Research Council of Canada, 1985; Bayne et al., 1985). This contribution will consider some of the biological effects of contamination that are proving appropriate in field studies and in monitoring of environmental impact. In order to achieve a focus in what is a very wide field of endeavour, I will concentrate on research conducted on marine invertebrates and on marine benthic communities. There is no reason to believe that the principles involved in this research are not applicable also to other groups of organisms.

Any effective programme of "biological effects" research, whether designed to understand the fundamental features of toxic damage, or to provide the means for monitoring environmental impact (and these two are clearly related) must consider responses at many levels within the biological "systems heirarchy" (Table 1). Inherent in a systems approach to biology is the concept that responses at one level are driven by processes at lower levels in the heirarchy, and express their main biological consequences at the next higher level. Thus, a change in temperature may result in a change in the rate of oxygen consumption by an organism (a physiological response), which is caused by biochemical changes at the level of individual enzymes and metabolic pathways and which causes, in turn, a modification to the animal's energy balance, with consequences for growth and reproduction.

This concept has proven to be particularly effective in research on ecotoxicology, which embraces a wide spectrum of processes, from those affecting the bioaccumulation of potentially toxic substances to those which reflect damage to communities of organisms and to whole ecosystems. It is this idea of the integration of response, from the biochemical mechanisms of toxicity, through consequent cellular damage to physiological and population reaction, that I hope to illustrate, by drawing on research carried out particularly on marine molluscs and especially on the common marine mussel, Mytilus edulis. An integrated understanding of responses to contaminants does more than provide conviction and

Table 1. Components of a biological "systems heirarchy"

                    Molecular (biochemical) processes

                    Sub-cellular organelles

                    Cells

                    Tissues

                    The whole organism (the physiological level)

                    The population

                    The community of organisms

                    The ecosystem.

integrity to the research data, for knowledge of such effects at different levels of biological complexity also address different aspects of the applied requirements of such research. For example, the closer the effect of interest is to the molecular mechanism that represents the toxic action of a contaminant, the greater the probability that knowledge of that effect will provide specific information on the chemical nature of the contaminating compound. Equally, we might expect that measures of community response (e.g. those concerned with relative species abundances) will be less specific in respect of particular contaminants but will bear more directly on the concerns of legislators and others involved with environmental regulation.

These ideas are expressed in Figure 1, based on criteria of sensitivity to contamination and on the time-scale associated with the first receipt of the contaminant stimulus and the detection of a response. The important point is that measures of response across the whole spectrum of biological complexity (from 'the cell to the ecosystem') are necessary, both for the purposes of monitoring (because they provide different information to the environmental scientist) and for the purposes of scientific rigour (because a framework of cause and effect, from molecular events to population change, would be much more meaningful than a mere collection of unrelated observations of effects). Appropriate research is still a long way from providing a linked series of cause/effect relationships from processes at the site of toxic action to consequences for populations and communities, but the convincing demonstration of such relationships comprises a realistic objective for such research programmes.

A research programme of this kind, which aims to provide an integrated understanding of toxic effects, would clearly benefit from a conceptual 'umbrella' that expresses the fundamental similarities of responses at all levels of the biological heirarchy. Such a paradigm does not yet exist, but ideas expressed by Selye (1950) provide a possible basis. Selye addressed a specific aspect of biological adaptation viz. the responses of mammals to general environmental stressors. He described a "general adaptation syndrome" comprising an "alarm" response, which represents the immediate effects of an environmental stimulus, followed by the "coping" responses, during which various homeostatic mechanisms within the cells and tissues counteract the potentially damaging consequences of the stimulus; finally, with prolonged stress, an "exhaustion" stage sets in which will ultimately lead to the death of the organism. Selye's model cannot simply be extrapolated to the much wider realm of biological effects of contamination in general, although his observations on hormonal

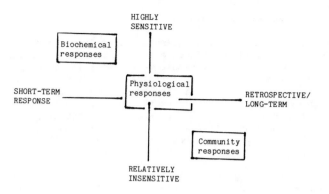

FIGURE 1. Biological responses to contaminants, characterised with respect to sensitivity and to time-scale.

responses, and on the primary and secondary features of the so-called "stress syndrome", have proved applicable, with minor modification, in studies of the effects of environmental change (including pollution) on fish (Mazeaud & Mazeaud, 1981). Nevertheless, the basic ideas expressed by Selye, including the notion of a cascade of biochemical and physiological responses which integrate to provide the organism with a capacity to compensate for an environmental stress, do seem to be widely applicable, and his distinctions between "distress" and "eustress", and between the alarm and coping stages of response, have been applied widely (Rapport et al., 1985).

RESPONSES TO CONTAMINANTS

Biochemical responses

Research into the biochemical effects of contamination may take two, rather different, points of departure: (1) An examination of general biochemical processes which can be shown empirically to be responsive to stress from pollutants; (2) A study of specific biochemical mechanisms by which organisms are known to deal with potentially damaging compounds. The first approach has proved less informative than the second, not least because it ignores the most relevant contribution that biochemical investigations can make, viz. to offer insight into the specific processes of toxic action (Figure 1), whilst suffering the same difficulties of other studies of having first to describe the full "range of normality" (Weber, 1963) for the process being measured, in order that departures from normality can be recognised and ascribed to stress from pollution. This difficulty is illustrated by a study by Livingstone (1984) on

133

mussels from the Shetland Islands, UK.

Livingstone(loc.cit) measured the specific activity (μmoles/min/gram of tissue) of the enzyme phosphofructokinase extracted from the adductor muscle of mussels from two sites, Sullom Voe (animals potentially impacted by hydrocarbons from an oil terminal) and Gluss Voe (selected as a relatively unimpacted reference site). This enzyme, PFK, is involved in the regulation of carbon flux through the glyco lytic pathway which supplies carbon skeletons for both aerobic and anaerobic energy-producing reactions. There were significant differences between the sites, with mussels from Sullom Voe consistently showing higher PFK activities than mussels from Gluss Voe. However, Livingstone (1984) was unable, from these data alone, to suggest which of the two groups of mussels was more affected by stressful environmental conditions. Mussels from Sullom Voe were known from other data to have higher rates of oxygen consumption than mussels from Gluss Voe, a difference consistent in direction with the difference of observed enzyme activities. On the other hand, mussels from the Sullom Voe site were also expec ted to experience longer periods of anaerobic metabolism as a result of living in a generally more disturbed environment; this too may explain the observed differences in PFK activity. As Livingstone (1984) concludes, in such an exampl physiological (i.e. "whole animal") measurements are more likely to elucidate the true condition of the organisms, and the biochemical measurements can con- tribute little to this interpretation.

On the other hand, when the biochemical measurements bear on processes known to be responsive to specific classes of contaminants, both this specificity and its associated sensitivity can provide unique information from which to infer the effects of pollutants. Currently two such biochemical systems in marine mollusc are the objects of much research, the cytochrome P-450-related mono-oxygenase system (involved in the detoxication of organic compounds) and the system of metal-binding, or metallothionein-like, proteins (involved in detoxifying trace metals).

The mono-oxygenase or MFO system. Figure 2 is an illustration of some component of the P-450 mono-oxygenase system. This system is widely distributed in marine organisms; the component enzymes convert lipid-soluble (non-polar) organic compounds (e.g. polyaromatic hydrocarbons) to more water-soluble, and hence excretable, metabolites. Phase I reactions involve oxygenation of the compound; phase II involves conjugation to other molecules prior to excretion. An impor- tant feature of this system is that the activities of the enzymes and the con- centration of the catalyst cytochrome P-450 may be increased if the animal is exposed to certain organic contaminants. This phenomenon is termed induction an it has been proposed as a possible index of the biological impact of compounds such as polycyclic aromatic hydrocarbons in marine organisms.

Table 2 shows some results from a study by Livingstone et al. (1985) on mussels which had been exposed for four months to ca. 30ug of diesel oil per litre in outdoor flow-through tanks at the Solbergstrand mesocosm facility in Norway; the control animals were held under similar conditions in water without added diesel oil. The concentration of P-450 doubled in the exposed mussels; activities of both NADH-dependent and NADPH-dependent cytochrome c reductase also increased significantly in the exposed mussels. Such results are encouraging in the search for diagnostic indications of the effects of hydrocarbons on bivalve molluscs, though further research is necessary for an understanding of the bio- logical consequences of these observed levels of induction.

	Control	Exposed
P-450	46.6 ± 14.3	92.0 ± 10.9
NADH-CYTCRED	95.4 ± 8.1	116.4 ± 4.3
NADPH-CYTCRED	10.3 ± 1.2	13.5 ± 0.9

Table 2. Responses of components of the digestive gland microsomal MFO system of
Mytilus edulis exposed to ca. 30ug diesel oil per litre for four months, compared
with control, unexposed animals. Values are means ± standard error, for sample
sizes between 4 and 7; each sample is the pooled tissue of eight mussels. NADH-
CYTCRED (NADH cytochrome c reductase) and NADPH-CYTCRED (NADPH cytochrome c
reductase) activities are quoted as nanomoles per minute per milligram protein.
Cytochrome P-450 concentrations are quoted as picomoles per gram protein. Data
from Livingstone et al. (1985). All differences between control and exposed
were significant at P<0.05.

MIXED FUNCTION OXYGENASE SYSTEM (MFO)

$$R + O_2 + NADPH_2 = R{-}OH + H_2O + NADP$$

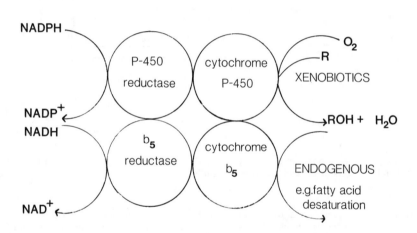

FIGURE 2. Diagram representing the mixed function oxidase system. R: organic
substrate. ROH: hydroxylated product.

Metal binding proteins. Many bivalve molluscs are known to accumulate high concentrations of certain metals without apparant toxic effects; they have therefore
served not only as good monitors of contamination (Goldberg, 1980) but also as
useful objects for the study of the cellular processes that effect detoxication
of metals accumulated beyond normal concentrations. Figure 3 is a simplified
version of a model of regulation of cellular metal levels, modified from George
& Viarengo (1985). Some metals are taken into the cell cytoplasm where they are
bound to metal-binding proteins (or metallothioneins). These proteins may themselves be sequestered into the lysosomes. There are three stages in the genesis
of lysosomes: (1) Primary lysosomes containing digestive enzymes are pinched off

from the Golgi apparatus of the cell; (2) These lysosomes then engulf cellular macromolecules and organelles (e.g. membrane components, pinocytotic vesicles), becoming the so-called secondary lysosomes; (3) The lysosomal enzymes hydrolyse (digest) the contents of the secondary lysosome to their monomeric units, retaining undigestible remnants which then comprise the tertiary lysosome, subsequently to be exocytosed from the cell and excreted.

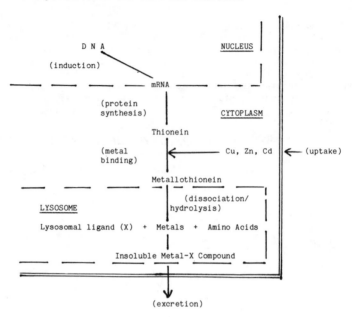

FIGURE 3. Some components of metabolic pathways for metals in animal cells. Modified from George and Viarengo (1985).

There are two processes of particular note here, the role of the metallothionein their interaction with the lysosome. As with the MFO system discussed earlier, animals are known to induce higher than normal levels of metallothionein production in response to high rates of metal intake into the cell, particularly cadmium, copper, silver and mercury (Viarengo et al., 1980; George & Viarengo, 1985). The presence of the metal-binding protein is detected by gel-permeation chromatography, followed by analysis for metal content and Viarengo et al. (1982 amongst others, have demonstrated higher levels of these proteins in mussels fro a metal-polluted area than in mussels from a less contaminated site. Knowledge of the biochemistry of metallothionein, its relationships with other metal-binding proteins and its response to mixtures of different trace metals provides specific information on the exposure history of the organism.

The second process of interest in the present context concerns the events within the lysosomes, as the metal-binding proteins are hydrolysed, with release of the component amino acids to the cell cytoplasm. The metals themselves may then become tightly bound within intra-molecular cross linkages formed by lipid peroxidation, eventually to be excreted from the bivalve kidney, as so-called residual bodies, in particulate form (George, 1983). This process may be accompanied by a general increase in the lipid content of the lysosomes which may be measured cytochemically (Moore, 1987) to provide a relatively simple indication of contaminant exposure and effect.

These biochemical studies on mono-oxygenase systems and on metal-binding proteins have many elements in common when referred to the general aims of 'biological effects' research i.e. the detection of meaningful biological responses to contamination. They both bear on fundamental molecular processes at the sites of toxic action within the cell. In so doing they provide information on the specific contaminants involved. Insight into the appropriate mechanisms also indicates the limits of dose beyond which there will be a breakdown in normal function leading to pathological damage. A good example of this is described by Moore & Viarengo (1987), in a study to be discussed in the next section. Finally, these studies demonstrate the high degree of sensitivity and the consequent "early-warning" properties of biochemical processes, typical perhaps of the alarm stage of biological response, that are so important in properly integrated studies of pollution impact.

Cellular responses

Cellular and sub-cellular responses to contaminants have been reported for a wide range of marine organisms. I will concentrate here on studies carried out by Moore and his colleagues (e.g. Moore et al., 1987) on the sub-cellular organelle, the lysosome, and on some of the consequences of disruption to lysosomal function caused by pollution. Some basic features of the lysosomal system have already been described. Figure 4 illustrates an important aspect of lysosomal function. The lysosome contains hydrolytic enzymes which are normally retained within the lysosomal membrane and which function in a controlled manner by the exchange of substrates and products across the limiting membrane. Lysosomes are also known to act as sites for the concentration and accumulation of many organic compounds as well as metals. When the lysosomes become 'overloaded', in a process not yet fully understood, the integrity of its limiting membrane may be damaged, causing the release of hydrolytic enzymes into the cytoplasm where they may cause damage to the cell. Moore (1976) developed a simple cytochemical technique to measure the integrity of the lysosomal membrane by assessing the "latency" of the lysosomal enzymes; these normally exhibit high latency (signifying an intact membrane) but, as the membrane is damaged, latency reduces, with release of enzyme and a breakdown of normal controls on hydrolytic activity within the cell.

In a series of studies Moore has catalogued the effects of various aspects of environmental stress, including contamination, on the latency of the lysosomal membrane and shown how simple measures of lysosomal latency may be used as sensitive indicators of the health and condition of the organism. Figure 5 is taken from a study by Moore & Viarengo (1987). Mussels were exposed to different concentrations of the polyaromatic hydrocarbon phenanthrene, which causes a sharp destabilisation (decline in labilisation period) of the lysosomal membrane at approximately 150 parts per billion in seawater (Figure 5A). In the same experiment, an indirect measure was made of cellular protein catabolism, indexed as the specific activity of a radio-isotope label incorporated into the cytosolic proteins (Figure 5B). The results showed a marked increase in protein breakdown (a decline in the amount of label associated with the proteins) at the same phenanthrene concentration that caused disruption to the lysosomal membranes. Moore & Viarengo (1987) conclude that membrane disruption, caused by phenanthrene, released hydrolytic enzymes to the cytoplasm and so brought about increased breakdown of cell proteins.

I LYSOSOME IN 'NORMAL' CELL   II LYSOSOME IN 'STRESSED' CELL

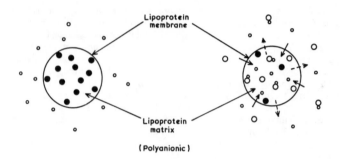

FIGURE 4. Lysosomes in nonstressed cells are largely impermeable to many substrates and the lysosomal hydrolases are mostly inactive or latent. Stress
destabilizes the lysosomal membrane resulting in increased permeability to
substrates, activation of hydrolytic enzymes (reduced latency) and release of
the hydrolases into the cytoplasm, resulting in limited cytolysis. Small
open circles, substrates; filled circles, acid hydrolases (bound or inactive);
big open circles, free or active acid hydrolases. After Moore, in McIntyre
and Pearce (1980).

These studies, when taken together with biochemical results and with some physiological research (to be discussed later), suggest an integrated series of events
linked to contaminant exposure. What then are the consequences to the cell and
to the animal's tissues of this disruption to normal lysosomal function? This
question has been addressed particularly in the digestive gland of mussels, a
tissue rich in lysosomes. Again using long-term exposures to oil in order to
disturb animal function, Lowe et al. (1981) measured the mean thickness of the
digestive cells of mussels, as well as certain features of lysosomal size and
numerical density, and recorded significantly reduced cell thickness in the oil-
exposed mussels, interpreted as the direct result of enhanced hydrolytic activity
caused, in turn, by disturbed lysosomal function.

Cytochemical and cytological observations of this type provide sensitive
measures of pollution effect, with clear consequences for pathological damage to
tissues and organ systems, and with reference to known biochemical responses.
Cytological effects do not, however, share the same specificity observed at the
biochemical level. For example, lysosomal membrane latency is known to be responsive to salinity and temperature changes as well as contaminant exposure.
Nevertheless, they provide a necessary link in any integration of effects, and
their relative ease of measurement makes them strong candidates for monitoring
programmes. Of the aspects of these studies briefly discussed here, the relationship between lysosomal processes and features of protein metabolism point to
certain physiological correlates of the cytological responses that help to carry
the integration forwards to the whole-animal level.

138

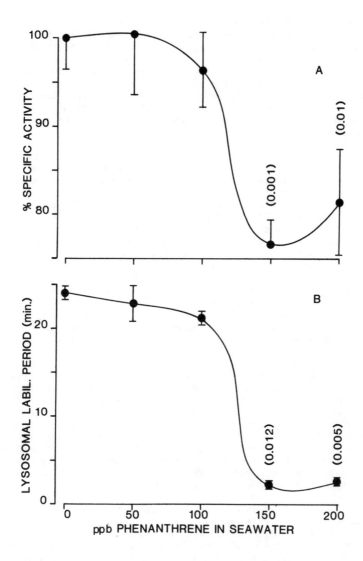

FIGURE 5.   A: The effect of phenanthrene in seawater on the specific activity
of C-labelled cytosolic proteins (as % of control values): each point is the
mean, with SE, of at least 5 replicate experiments.

B: The effect of phenanthrene in seawater on the stability of lysosomes.   Values in brackets show the probability that the results might be the
same as control values.   From Moore and Viarengo (1987).

Physiological responses.

In this category of effects are included those that relate to the physiology and the behaviour of the whole animal including, for example, the processes of feedi respiration, growth and reproduction. As with all other levels of biological organization, there have been very many relevant studies; I will concentrate on those that hold some promise in linking to observed biochemical and cytological effects, and on those that have proved to be effective in detecting contaminant effects on bivalve and other molluscs in the field. In considering these res- ponses we are dealing with the secondary effects of stress and with some of the mechanisms that may help to compensate the organism for primary damage due to th toxicity of contaminants.

Many biochemical responses to pollutants impose a demand for protein synthesis, either directly, as for example by induction in the MFO system and the produc- tion of metal-binding proteins, or in compensating for enhanced protein breakdow caused by lysosomal disruption, or in simply meeting the requirements for repair of cell membranes and other organelles. In order to measure appropriate aspects of protein metabolism, it is necessary to consider the <u>turnover</u> of the body's proteins, that is, to estimate both protein synthesis and protein breakdown, since net protein balance is the difference between these two processes. It has recently become possible, by using the stable isotope of nitrogen (nitrogen-15) to estimate the various processes involved, and to relate rates of protein synthesis to the consequent energy demands that this imposes on the organism (Hawkins, 1985). It is important to realise, in interpreting these results, tha a net increase in protein balance can result from a reduction in protein synthesis if this is accompanied by a decline also in protein breakdown; if, as a result of an environmental stress, both synthesis and breakdown were to increa but the latter were to increase proportionately more than the former, the result would be a decline in net protein content.

Figure 6A is a graph showing a relationship between the rate of protein synthesi and the rate of oxygen consumption for 12 individual mussels at 10°C; it can be estimated that protein synthesis may account for up to 30% of the total routine metabolic demand in these animals. Results of such experiments indicate another important relationship (Figure 6B), namely that the rate of protein synthesis is a negative function of the efficiency of protein synthesis (measured as protein balance as a proportion of gross synthesis). Not surprisingly therefore, individuals that have a high efficiency of protein synthesis also show a high net energy balance and high rate of growth, since they incur lower energy costs than individuals with a lower efficiency.

I have shown these results in some detail because they provide a link between biochemical and cellular responses and the effects of pollution at a 'higher', physiological level of organization. If biochemical or cellular responses lead to a net increase in protein synthesis, as many of them do, there will be a con- comitant increase in net energy demand, leading to less energy available for growth and reproduction. Indeed, studies of energy balance within individuals under various degrees of stress, from contaminants and other environmental con- ditions, have demonstrated considerable potential for measuring the physio- logical condition of animals in the field. These studies draw on the standard energy balance equation, in order to estimate the "scope for growth and repro- duction" (Bayne <u>et al</u>., 1979, 1982).

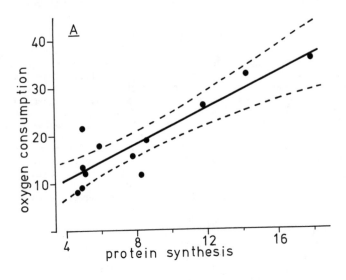

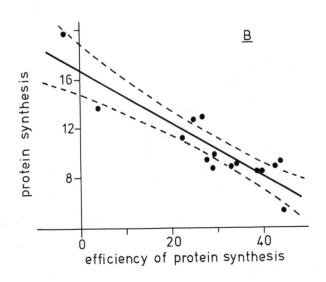

FIGURE 6. A: Rates of oxygen consumption by mussels (μl per hour) related to rates of protein synthesis (mg per day).

B: The relationship between the rate (mg per day) and the efficiency (percent) of protein synthesis in mussels. From Hawkins and Bayne (unpublished).

The scope for growth (SFG) is defined as the energy available to the animal for growth and reproduction after the demands of respiratory metabolism have been met from the absorbed ration. It is calculated as:

$$SFG = A-R-U$$

where A = the rate of absorption of energy from the diet, R = the rate of energy loss through respiration and U = the rate of energy lost as excreta. A is usual determined as I.$\underline{e}$, where I = the rate of ingestion and $\underline{e}$ = the efficiency with which the animal absorbs energy from the ingested ration. When the SFG is positive, this signifies that the animal has surplus energy available for somati growth and for reproduction; when SFG is negative, the organism is having to utilise any reserves of energy that have been laid down in its tissues in order to provide sufficient metabolic fuel for the demands of body maintenance. A zero SFG signifies that the animal is exactly in balance with its food resource i.e. that it is gaining from its ration just enough to maintain a constant body size. A prolonged period of negative SFG would result eventually in the death of the animal. These ideas are reviewed by Bayne & Newell (1983).

Figure 7 summarises many studies undertaken by Widdows and colleagues to describ the relationship between the scope for growth and the concentration, in the wate and in the animal's tissues, of petroleum aromatic hydrocarbons. When plotted in this way, on a logarithmic axis for tissue hydrocarbon concentration, there is a tight linear relationship which predicts that SFG will be negative if hydro carbons were to exceed 100ug/gram wet mass of tissue (measured in terms of two- and three-ringed hydrocarbons, and at a mean ration level of 0.5mg particulate organic matter per litre). The reliability of this relationship was recently

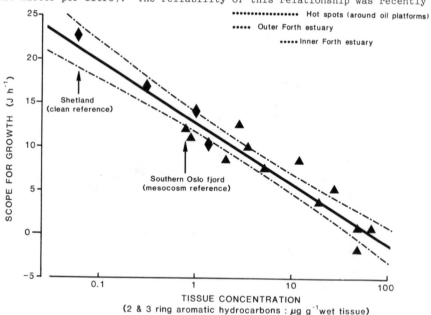

FIGURE 7. Relation between scope for growth (Joules per hour) and the concentration of two- and three-ringed aromatic hydrocarbons (ug per gram wet mass) in the tissues of Mytilus edulis. Horizontal bars represent estimates of tissue concentration based on observed concentrations in the water. Based on analyses by J. Widdows, from Moore et al. (1987).

evaluated by Widdows & Johnson (1987) in a careful study in Norway; their results supported the predictive nature of earlier studies, demonstrated that the SFG is sensitive to environmentally realistic levels of contamination and also showed how, through a detailed understanding of the chemical form of the contaminant and its relative toxicity (based on chemical principles and on laboratory experiments), physiological data can provide a certain specificity in unravelling the pollutants causing damage in field situations where the animals are exposed to complex mixtures of compounds.

We can summarise the integrated nature of these studies to this point in a simple diagram:

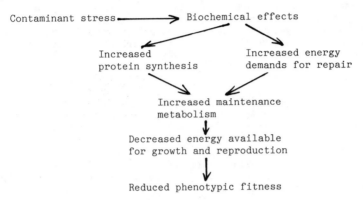

The phenotypic fitness of an organism may be defined as the sum of those physiological processes (traits) which affect the prospects for successful reproduction. Many relevant features of contaminant effects on molluscs have been examined as possible attributes of fitness in this context, some as direct effects, others as indirect, operating via the energy balance or scope for growth. For example, Lowe & Pipe (1985) recorded changes in reproductive tissue in mussels subjected to long-term exposures to hydrocarbons at low concentrations (Table 3). By using quantitative stereological techniques they were able to demonstrate a decline, due to contamination, in the numbers of gametes and of the main energy storage cells (the vesicular cells) in reproductive tissue, and an increase in the number of degenerating oocytes.

	Gametes	Vesicular Cells	Degenerating Oocytes
Control	0.77 ± 0.27	0.53 ± 0.07	0.05 ± 0.01
30ug/1	0.44 ± 0.16	0.38 ± 0.07	0.35 ± 0.13
130ug/1	0.30 ± 0.11	0.14 ± 0.02	0.31 ± 0.09
Recovery	0.66 ± 0.18	0.54 ± 0.04	0.01 ± 0.01

Table 3. Effects of hydrocarbon exposure, and a recovery period of 53 days, on reproductive tissue in the mussel. Values are means ± standard error, measured as volumes (cubic millimetres) for samples of 10 individuals. Data from Lowe & Pipe (1985). All values for exposed animals were significantly different ($P < 0.05$) from controls; recovery was complete in all conditions.

These data indicate a direct impairement of reproduction due to hydrocarbons and suggest that reproductive efficiency would be reduced not only by the loss o gametes, but also by reduction in the energy reserves available for further gametogenesis. Other effects that have been documented for mussels under environmental stress confirm a reduction in fecundity and include a decline in the mean size and lipid content of eggs produced under stress, with impairement of subsequent rates of growth of the larvae (Bayne et al., 1982). These experiments suggested that relatively brief periods of reduced scope for growth would have no effect on reproductive output, due to the use of nutrient reserves for the maintenance of gamete production. Should the scope for growth be reduced for prolonged periods of time, however, gametes already produced may be resorbed to provide nutrient for body maintenance, with overall decline in fecundity and reproductive success. These results suggest that we may integrate further, from observed physiological effects of pollution to likely population consequences, though at this level of integration more serious problems occur.

Population responses

In the biological heirarchy under consideration the attributes of the population 'level' include recruitment, mortality, size and age structure, biomass and population production. Many of these attributes are affected directly by responses at the physiological (whole organism) level, since they are made up from the simple sum of effects on the individuals that comprise the population. For example, a reduced scope for growth amongst individuals will, when summed for the population as a whole, be reflected as a reduction in population production. Figure 8 shows results from a study by Bayne & Worrall (1980) on two populations

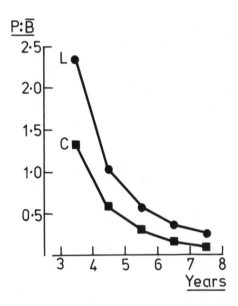

FIGURE 8. Age-related production: biomass ratios in two populations of mussels: C, Cattewater (a stressed site); L, Lynher (a relatively uncontaminated site). From Bayne and Worrall (1980).

of mussels, one of which (Cattewater) was subjected to thermal stress at the cooling-water outfall of an electricity generating station. The individual mussels at this site had a supressed scope for growth relative to the other site and this was reflected, at the population level, as reduced production and smaller production:biomass and production:respiration ratios.

Other population attributes are equally the reflections of effects on the individuals; the age/size distribution within a population will depend on rates of growth of individuals; individuals stressed by pollutants may be more sensitive to predation, parasitism and disease; mortality may be higher as individuals already responding to contamination face further environmental stress. Such population effects have been observed in both field studies (Gilfillan & Vandermeulen, 1978) and in studies carried out using experimental mesocosms (Grassle et al., 1981; Widdows et al., 1985). It is clear both from theoretical considerations and as a result of experimental observations that adverse effects on individuals ultimately manifest themselves at the population level.

However, when considering populations of marine invertebrates, many of which reproduce by means of a dispersive larval phase, a major difficulty in linking effects on individuals to the processes maintaining population size (i.e. recruitment) is apparent, for we can not be sure that larvae released from a population are the same larvae that, three to four weeks later, will recruit back into the 'parent' population. There is, in other words, an apparent uncoupling between the adults of a population and their larvae, at least as regards the processes of recruitment. Marked effects on the reproductive potential of individuals may be observed, as discussed in the previous section, but it is currently not possible with assurance to suggest that reduced reproductive efficiency amongst individuals will be linked with reduced recruitment, since the recruits into the population may originate from other sites, many kilometres away, and may not be subjected to the same contaminant stresses.

This difficulty bears on a major area of current research effort in marine ecology, viz. the factors controlling recruitment in marine populations and the relationships between recruits settling to the benthos from the water column and the various density-dependent and density-independent factors operating within the adult population. Until progress in fundamental understanding of these processes is made, the linkage and integration that we seek, between cellular, physiological and population effects of pollution, will be incomplete. However, this is not to argue that population consequences of impaired physiological function cannot be made explicit for, as argued above, many population-level attributes are the direct expression of individual effects, summed across the age and size spectrum of the population. Measured changes in the scope for growth, for example, will have unequivocal implications for population production, biomass and for the survival of age classes. Also, as our understanding of the genetics of invertebrate populations develops, and as we learn to link physiological with genetic variability within these populations so we will be able increasingly to interpret physiological responses in evolutionary terms and possibly to predict the long-term consequences to populations chronically exposed to contamination.

Community responses

Individual organisms differ in their sensitivity to contaminants. Along a gradient of pollution, therefore, some species will resist the consequent stress, some will succomb and other, mobile, species may migrate away from the contamination altogether. The net result will be a change in the pattern of species that constitute the community. Attempts to detect community-level effects of pollution start from this observation. However, faunal communities are subjected to many very complex forces, both natural and anthropogenic, with considerable variability over time and space scales and reflecting many varied 'internal' inter-relationships (complex trophic interactions, parasitism etc.), quite apart from 'external' forcing by pollution, for example, making the problems of objective description of community structure formidable indeed. It is inevitable, therefore, that considerable research effort is deployed to develop descriptive procedures that are rigorous enough to detect the effects of disturbance by pollution in spite of the very large variability inherent in the structure of natural communities.

Broadly speaking, such procedures can be divided into three categories:

(1) Descriptive statistics. The attempt here is to describe the number of species detected per sample, the abundance of individuals and of biomass, and then to summarise this information in the form of measures of diversity and species richness.

(2) Patterns of species abundance and biomass. Huston (1979) described a general hypothesis of species diversity which states that, under conditions of infrequent disturbance, competition between species will result in competitive displacement, whereby a few competitively superior species come to dominate the community, and species diversity is relatively low. When subjected to disturbance by pollution, competitive equilibrium is prevented from occurring, and species diversity increases. At higher levels of disturbance, species are eliminated from the community and diversity declines. Various measures of ranked abundance and diversity have been proposed as measures of community effects, within this broad hypothesis, and one such procedure is discussed below.

(3) Multivariate analysis. These procedures attempt to group co-occurring species into multivariate clusters, and then to describe the degree of similarity between species groups and various features of the environment, including contaminant loads.

It would be inappropriate here to attempt a general discussion of these various techniques (see Gray et al., 1987; Clarke & Green, 1987). I will instead draw attention to a new method, proposed by Warwick (1986), which addresses a fundamental problem for community-level effects measurements, namely the requirement for reference samples which are needed to control for temporal and spatial variability in order to be able to interpret changes observed within the samples of concern as due to pollution. Warwick emphasises the difficulties, both conceptual and logistical, inherent in the normal requirement for such controls, and he proposes a technique which is based, instead, on "internal controls" in which different structural properties of the same faunal assemblage are expected to respond differently to the effects of pollution.

Warwick's (1986) idea is illustrated in Figure 9. He postulates that under stable, unpolluted conditions, the biomass of the community will be dominated by one or a few large species, each represented by rather few individuals; the distribution of numbers of individuals among species will be more even than biomass, and the resulting "k-dominance curves" will show biomass diversity plotting above numbers diversity along its entire length. In a grossly polluted situation, however, the community becomes increasingly dominated numerically by one or a few very small species, and few large species are present. In such a case, numbers diversity will plot above biomass diversity. In moderately polluted situations an intermediate situation might prevail. The reader is referred to Warwick's paper (and to Gray et al., 1987) for further discussion and evaluation of this procedure.

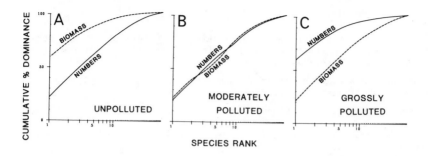

FIGURE 9.   Hypothetical k-dominance curves for species biomass and numbers, showing unpolluted, moderately polluted and grossly polluted conditions. From Warwick (1986).

Techniques such as these illustrate a simple point, namely that communities are comprised of species populations, and changes in community structure reflect the responses of these populations to contamination.   The integration from population- to community-level effects is self-evident.

We may now summarise the integrated nature of effects, from those marking the health of individual organisms to those indicative of community change, in the following way, bearing in mind that much research remains to be done to progress from an anecdotal to a fully analytical linkage of the kind suggested here;

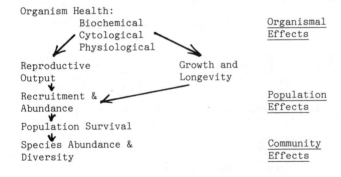

The "Ecosystem Distress Syndrome": Further integration and speculation.

Many ecologists have considered properties characteristic of the final level in our heirarchy, the ecosystem, with a view to recognising symptoms indicative of damage. Some of these symptoms are:

Reductions in species diversity; elimination of longer-lived, larger species in favour of short-lived opportunistic species.

Changes in the ratios of production:biomass and production:respiration.

Departure from tightly regulated nutrient cycles towards looser coupling of nutrient pathways.

Departure from population stability towards instability, accompanied by marked changes in reproductive cycles of component species.

In a critical review of such studies, Rapport et al. (1985) adapted Selye's model of an organism's response to stress for application to ecosystems, whilst recognising that the analogy of "ecosystem as organism" has to be treated with caution They recognised three stages in ecosystem response to stresses of various kinds, which resemble the alarm, coping and exhaustion stages discussed earlier. To quote a few examples:

(1) Alarm Reactions. Abnormal fluctuations in sensitive populations, with changes in the distribution of sensitive species; disruption to the reproductive behaviour of certain species. Appearance of biochemical, pathological and physiological abnormalities.

(2) Coping Reactions. Feedback mechanisms which tend to counteract or deactivate the stress. Replacement of sensitive species with functionally similar but more resistant species. The development of external biogeochemical feedback loops which serve to maintain nutrient flux. Activation of detoxication pathways, for example by increased microbial activity. Increase in community-level respiration and in respiration:biomass ratio.

(3) Exhaustion. Represents the failure of homeostatic mechanisms leading to ecosystem breakdown.

Rapport et al. (1985) pose the interesting suggestion that the ideas underlying Selye's (1950, 1974) model for a syndrome of stress effects may provide coherence to the wider field of biological effects studies in general. Certainly there are similarities that may be useful in guiding future research. For example, a partial list of symptoms characteristic of the "alarm" and "coping" stages in the response of marine mussels to contaminant stress is as follows:

Stage	Symptoms
Alarm	Altered lysosomal properties
	Increased autolysis
	Enhanced protein catabolism
Coping	Increased protein synthesis
	Increased respiration
	Altered feeding behaviour
	Modified growth and reproduction.

Certain similarities then emerge between symptoms at the organismal and the ecosystem levels:

148

(1) Alarm or early-warning signals occur at a lower level in the systems heirarchy than the symptoms of coping. For organisms the alarm symptoms are biochemical and cellular; for ecosystems they occur at the population level.

(2) There is increased respiration, associated with increased maintenance costs, during the coping stage in both individuals and ecosystems.

(3) Also during the coping stage there is an increase in processes of detoxication, involving metabolic systems in organisms and microbial activity in ecosystems. When these systems are overloaded damage occurs, and insight into the processes of detoxication, both for organisms and ecosystems, should provide some specificity to the symptoms of stress.

(4) Functional properties are preserved during the coping responses and this may involve structural modification. At the organism level, for example, rates of digestion and absorption may be maintained in spite of cellular damage to the digestive system, by modification to rates of feeding; within ecosystems rates of nutrient flux may be maintained by changes in species' abundance.

There is a considerable element of speculation in this discussion, if only because no studies have been undertaken specifically to seek out any generalities that may exist over the whole field of biological effects research. Such a study would involve considerable effort, spread over the many biological sub-disciplines involved, and covering wide time and space scales in order to be relevant to molecular events associated with toxicity on the one hand, and changes in ecosystem structure and function on the other. Nevertheless, examples of chronic anthropogenic stress on natural systems are now legion and can be viewed as providing 'natural' experiments, as a basis for some far-reaching and potentially exciting research effort.

REFERENCES

Bayne, B.L. and Worrall, C.M., Growth and production of mussels, Mytilus edulis from two populations, Mar. Ecol. Progr. Ser., vol. 3, pp. 317-328, 1980.

Bayne, B.L. and Newell, R.C., in The Mollusca, vol. 4, Physiology, part 1, eds. A.S.M. Saleuddin and K.M. Wilbur, Academic Press, New York, 1983.

Bayne, B.L., Moore, M.N., Widdows, J., Livingstone, D.R. and Salkeld, P., Measurements of the responses of individuals to environmental stress and pollution: studies with bivalve molluscs, Phil. Trans. Roy. Soc. Lond., B. vol. 286, pp. 562-581, 1979.

Bayne, B.L., Widdows, J., Moore, M.N., Salkeld, P., Worrall, C.M. and Donkin, P., Some ecological consequences of the physiological and biochemical effects of petroleum compounds on marine molluscs, Phil. Trans. Roy. Soc. Lond., B. vol. 297, pp. 219-239, 1982.

Bayne, B.L., Dixon, D.R., Ivanovici, A., Livingstone, D.R., Lowe, D.M., Moore, M.N., Stebbing, A.R.D. and Widdows, J., The effects of stress and pollution on marine animals, pp. 384, Praeger Scientific, New York, 1985.

Clarke, K.R. and Green, R.H. in The biological effects of pollutants in the sea: the results of a practical workshop, eds. B.L. Bayne, K.R. Clarke and J.S. Gray, Inter-Science, Hamburg, 1987 (in press).

George, S.G., Heavy metal detoxication in Mytilus kidney: an in vitro study of Cd and Zn-binding to isolated tertiary lysosomes, Comp. Biochem. Physiol., vol. 76C, pp. 59-66, 1983.

George, S.G. and Viarengo, A., in Marine pollution and physiology; recent advances, eds. F.J. Vernberg, F.P. Thurberg, A. Calabrese and W. Vernberg, Univ. S. Carolina Press, USA, 1985.

Gillfillan, E.S. and Vandermeulen, J.H., Alterations in growth and physiology in chronically oiled soft-shell clams, Mya arenaria, exposed to Bunker C from Chedabucto Bay, Nova Scotia, 1970-1976, J. Fish. Res. Bd. Can., vol. 35, pp. 630-636, 1978.

Goldberg, E.D. (ed.) The International Mussel Watch, 248 pp., National Academy of Science, Washington, USA, 1980.

Grassle, J.F., Elmgren, R. and Grassle J.P., Response of benthic communities in MERL experimental ecosystems to low level, chronic additions of No. 2 fuel oil, Mar. Envir. Res., vol. 4, pp. 279-297, 1981.

Gray, J.S., Ashan, M., Carr, M., Green, R.H., Pearson, T.H., Rosenberg, R. and Warwick, R., in The biological effects of pollution in the sea: the results of a practical workshop, eds. B.L. Bayne, K.R. Clarke and J.S. Gray, Inter-Science, Hamburg, 1987 (in press).

Hawkins, A.J.S., Relationships between the synthesis and breakdown of protein, dietary absorption and turnover of nitrogen and carbon in the blue mussel, Mytilus edulis L., Oecologia, vol. 66, pp. 42-49, 1985.

Huston, M., A general hypothesis of species diversity, Am. Nat., vol. 113, pp. 81-101, 1979.

ICES, Report of the ICES advisory committee on marine pollution, Cooperative Research Report, No. 132, 1978.

Livingstone, D.R. in Toxins, drugs and pollutants in marine animals, ed. J. Bolis, Springer-Verlag, Berlin, 1984.

Livingstone, D.R., Moore, M.N., Lowe, D.M., Nasci, C. and Farrar, S.V., Responses of the cytochrome P-450 monooxygenase system to diesel oil in the common mussel, Mytilus edulis L. and the periwinkle Littorina littorea L., Aquat. Toxicol., vol. 7, pp. 79-91, 1985.

Lowe, D.M. and Pipe, R.K., Cellular responses in the mussel Mytilus edulis following exposure to diesel oil emulsions: reproductive and nutrient storage cells, Mar. Env. Res. vol. 17, pp. 234-237, 1985.

Lowe, D.M., Moore, M.N. and Clarke, K.R., Effects of oil on digestive cells in mussels: quantitative alterations in cellular and lysosomal structure, Aquat. Toxicol., vol. 1, pp. 213-226, 1981.

Mazeaud, M.M. and Mazeaud, F., in Stress and Fish, ed. A.D. Pickering, Academic Press, Toronto, 1981.

McIntyre, A.D. and Pearce, J.B. (eds.), Biological effects of marine pollution and problems of monitoring, Rapp. P.-V. Reun. Cons. Int. Explor. Mer., vol. 179, 1980.

Moore, M.N., Cytochemical demonstration of latency of lysosomal hydrolases in digestive cells of the common mussel Mytilus edulis, and changes induced by therman stress, Cell Tissue Res., vol. 175, pp. 279-287, 1976.

Moore, M.N., in The biological effects of pollution in the sea: the results of a practical workshop, eds. B.L. Bayne, K.R. Clarke and J.S. Gray, Inter-Science, Hamburg, 1987 (in press).

Moore, M.N. and Viarengo, A., Lysosomal membrane fragility and catabolism of cytosolic proteins: evidence for a direct relationship, Experientia, In press, 1987.

Moore, M.N., Livingstone, D.R., Widdows, J., Lowe, D.M. and Pipe, R.K. Molecular, cellular and physiological effects of oil-derived hydrocarbons on molluscs and their use in impact assessment, Phil. Trans. Roy. Soc. Lond. B., 1987 (in press).

National Research Council Canada, The role of biochemical indicators in the assessment of ecosystem health - their development and validation, Publications NRCC/CNRC, Ottawa, Canada, 1985.

Rapport, D.J., Regier, H.A. and Hutchinson, T.C., Ecosystem behaviour under stress, Am. Nat., vol. 125, pp. 617-640, 1985.

Selye, H., Stress and the general adaptation syndrome, Br. med. J., vol. 1, pp. 1383-1392, 1950.

Seyle, H., Stress without distress, Lippincott, New York, 1974.

Viarengo, A., Pertica, M., Mancinelli, G., Capelli, R. and Orunesu, M., Rapid induction of copper-binding proteins in the gills of metal-exposed mussels, Comp. Biochem. Physiol., vol. 67C, pp. 215-218, 1980.

Viarengo, A., Pertica, M., Mancinelli, G., Palinero, S., Zanicelic, G. and Orunesu, M., Evaluation of general and specific stress indices in mussels collected from populations subjected to different levels of heavy metal pollution, Mar. Env. Res., vol. 6, pp. 235-243. 1982.

Warwick, R.M., A new method for detecting pollution effects on marine macrobenthic communities, Mar. Biol., vol. 92, pp. 557-562, 1986.

Weber, G., in Advances in enzyme regulation, ed. G. Weber, Pergamon, Oxford, 1963.

Widdows, J. and Johnson, D., in The biological effects of pollution in the sea: the results of a practical workshop, eds. B.L. Bayne, R.K. Clarke and J.S. Gray, Inter-Science, Hamburg, 1987 (in press).

Widdows, J., Donkin, P. and Evans, S., Recovery of Mytilus edulis L. from chronic oil exposure, Mar. Env. Res., vol. 17, pp. 250-253, 1985.

Monitoring and Surveillance Systems

ROBERT W. RISEBROUGH
Institute of Marine Sciences
University of California
Santa Cruz, California 95064, USA

The international scientific community studying the nature and the magnitude of the chemical changes that are currently occurring in the marine environment makes frequent use of the word "monitoring". In English, the meaning of this word is somewhat narrower than the meanings of the words into which it is translated in other languages. This more restricted useage has contributed to the confusion about what is properly considered "monitoring", as distinguished from "surveillance" or "research". The essence of monitoring includes a concept of repeated measurement, usually over time but frequently also over space, of an environmental parameter. Munn (1973) has provided a formal definition:

> "the process of repetitive observing for defined purposes of one or more elements or indicators of the environment according to pre-arranged schedules in space and time, and using comparable methodologies for environmental sensing and data collection"

One of the best examples of environmental monitoring has been the measurements of atmospheric levels of carbon dioxide, undertaken over a period of years at the same localities, which have shown a continuing global increase in these levels resulting from the combustion of fossil fuels (Keeling et al., 1976a; 1976b; Freyer, 1979). An example in a local area is the program undertaken in the New York Bight after bottom waters became anoxic in 1976, killing fish and benthic organisms (Swanson and Sindermann, 1979). Repeated measurments over both time and space of dissolved oxygen and other environmental parameters have been undertaken, coinciding with the moving further offshore of the dump site for sewage sludge from the New York City region (Suszkowski and Santoro, 1986; Levine, 1986).

Most programs dealing with contaminants in coastal and marine environments are more properly considered as "surveillance", that is, watching out for trouble, as in the general meaning of this word in both English and French. Such programs employ whatever combination of sampling and measurements that might best characterize the distribution or effects of contaminants, whether or not these are repeated over time or space. Not included in the meanings in English of either "monitoring" or "surveillance",

153

however, is the concept of "control", implying an active participation to reduce damage, that is included in the meanings of the German and Russian equivalents.

The best monitoring and surveillance programs increase our understanding of the processes whereby contaminants, as well as natural elements and compounds, move through and perhaps influence ecosystems, - even though they are not primarily research programs. The scientists participating in the International Mussel Watch Workshop held in Barcelona in 1978 proposed the following objectives of monitoring and surveillance programs (National Academy of Sciences, 1980):

- to advance the state of knowledge and understanding of environmental processes;

- to support the processes of environmental regulation, standard setting, and enforcement;

- to determine and assess the level of contamination and warn of potentially dangerous conditions;

- to develop methods and instrumentation; and

- to train scientists.

Unless monitoring and surveillance programs are addressed to well-defined, focussed questions it is likely that the results will be of little greater value than collections of random numbers. Some discussion of priorities among programs is therefore in order. Programs might then be considered from two approaches, - addressing the contaminants of concern, and the most appropriate media for their measurement.

PRIORITY PROBLEMS

Our conceptions of what are the most important problems in marine pollution have changed considerably since the Stockholm meeting in 1972 that lead to the establishment of the United Nations Environment Programme. At that time the global community was becoming aware of major changes in the chemical environment of the world's oceans. DDT had been detected in Antarctica (Sladen et al., 1966; George and Frear, 1966) and had found to be a contaminant in the marine environments at levels that threatened higher members of food webs (Risebrough, 1969). Polychlorinated biphenyls (PCBs) were suddenly recognized as significant global pollutants, particularly in aquatic environments (Jensen et al., 1969; Koeman et al., 1969; Risebrough et al., 1968). Mercury accumulated in fish and shellfish had poisoned people in Japan (Irukayama, 1966), and was being reported in marine fish such as tuna and swordfish at levels considered dangerous for human health. The global marine environment appeared to be truly threatened.

The monitoring, surveillance, and research programs undertaken in the intervening 15 years that have examined the chemical changes occurring in the oceans have brought both perspective and clarification. Chemical changes in the atmosphere, however, appear to have greater immediate and longer-term importance, since they bring the possibilities of climate modification and increases in sealevel (Bolin et al., 1979) and are already associated with damages to forest ecosystems (Tomlinson, 1983; Wang et al., 1986). International concern for the maintenance of the ozone layer in the upper atmosphere has lead in 1987 to the signing of an international convention restricting the atmospheric input of chlorofluoromethanes. In comparison, the threats to the oceans appear to be of a lower magnitude.

Genuine concern for the health of the oceans, however, persists, prompted both by the kinds and by the magnitude of chemical change (Goldberg, 1983). Dominant chlorinated hydrocarbons in surface waters are toxaphene and the hexachlorocyclohexanes; the use of these and of other chlorinated hydrocarbons continues in many countries. Generally, however, the thesis has been accepted that uses of persistent, toxic compounds that accumulate in food webs must be phased out. This particular threat to the oceans, while still present, appears to be diminishing. The high levels of mercury in the tuna and swordfish from the open oceans documented in the early 1970s have since been attributed largely to natural sources. Mercury levels in a deep sea fish were shown, from the analysis of museum specimens, not to have changed over the past century (Barber et al., 1984), although fluxes of mercury into the marine environment have increased as a result of man's activities, as have fluxes of lead and of other elements (Nriagu, 1979). Because of this, and since there are continuing inputs of petroleum and of a wide variety of other substances into the oceans beyond the coastal zones, continuing surveillance programs are called for. The immediate threats to the open oceans perceived in the early 1970s do not, however, exist at this time, in part because there has been a global shift in attitudes about the use of toxic, mobile, persistent substances such as the PCBs.

In a companion lecture (Risebrough, this volume) the distinction between the oceans and the coastal zones has been made. The coastal zones receive almost all of the wastes discharged through wastewater treatment systems, from rivers, directly from factories and refineries, or from offshore dumping of waste chemicals. The coastal zones, rather than the truly marine systems of the oceans, will be the focus of the monitoring and surveillance programs considered in this lecture.

PRIORITY CONTAMINANTS

Monitoring and surveillance programs, such as the U.S. Mussel Watch Program undertaken in 1976-78, have traditionally considered four principal classes of contaminants: 1) petroleum and combustion products; 2) chlorinated hydrocarbons in the medium-molecular weight range; 3) heavy metals and several trace elements; and 4) radioactive isotopes (Goldberg et al., 1978; 1983; Farrington et al., 1983). Because of very different analytical methodologies used for organics, metals, and radioactive isotopes, such analyses are frequently undertaken in different laboratories.

In the companion lecture (Risebrough, this volume), another category, plastics and related floating debris, was added to those substances that are of concern in the marine environment. In addition, there are three other classes of organic compounds which are of concern, or are of potential concern, in coastal environments: 1) volatiles; 2) neutral compounds not detected, or not routinely detected, by procedures using electron capture gas chromatography; and 3) polar compounds. Examples of each are in the List of 129 Unambiguous Priority Pollutants of the U.S. Environmental Protection Agency (Keith and Telliard, 1979; Table 1).

The volatiles are principally of concern in groundwaters and drinking waters (Mackay et al., 1985); in coastal areas they are unlikely to exert significant effects except perhaps in the immediate vicinity of the sites of wastewater discharges or of chronic input of petroleum. From surface waters they readily partition into the atmosphere (Roberts and Dandliker, 1983; Wakeham et al., 1983), where they are degraded in a variety of oxidative or photochemical processes.

Among the "other neutrals", chlorinated benzenes have been reported in organisms and sediments in the vicinity of a major ocean outfall (Young et al., 1980) and in the Great Lakes (Oliver and Nichol, 1982). There are, however, comparatively few data on the environmental distributions of these and other neutral compounds in the Priority Pollutant List.

Polar compounds are not reported in many analyses of environmental samples. They are not extracted with the procedures used for the familiar environmental pollutants such as the PCBs which are largely non-polar, poorly soluble in water but highly soluble in lipids. Extraction techniques that would recover polar compounds generally apply to only limited amounts of water, such that only higher concentrations are detected. Yet PCBs and other chlorinated hydrocarbons such as the DDT compounds are metabolized or degraded to polar compounds (Sundstrom et al., 1975). Similarly, aromatic hydrocarbons are metabolized or degraded to more polar derivatives which are not presently detected or characterized (Hinga and Pilson, 1987).

Even though polar compounds are accumulated by organisms to a lesser extent than are the non-polar compounds, they must be considered as potential toxicants in coastal waters. All organisms are exposed to them, amd there is a potential for interference with biochemical and physiological processes.

The List of Priority Pollutants was compiled by the U.S. Environmental Protection Agency in response to a suit brought by a coalition of environmental groups against the Agency; the Federal Water Pollution Control Law had required that EPA provide effluent limitations and guidelines for a number of principal pollutants. It was not the original intent that this list be written in stone, but rather that compounds would be dropped from it or others added on the basis of monitoring experience. This has not yet happened, and in many programs that include the analysis of environmental contaminants it is required that these compounds be analyzed by methodologies specified by EPA. Some compounds clearly should be

Table 1. The 129 Unambiguous Priority Pollutants of the U.S.
Environmental Protection Agency.

Volatiles - 33

Acrolein
Acrylonitrile
Benzene
Bis(chloromethyl)ether
Bis(2-chloroethyl)ether
Bis(2-chloroethoxy)methane
Bromoform
Carbon tetrachloride
Chlorobenzene
Chlorodibromomethane
Chloroethane
Chloroform
Dichlorobromomethane
Dichlorofluoromethane
Ethyl benzene
Methyl bromide
Methyl chloride
Methylene chloride
Tetrachloroethylene
Toluene
Trichloroethylene
Trichlorofluoromethane
Vinyl chloride
1,1-Dichloroethane
1,1-Dichloroethylene
1,1,1-Trichloroethane
1,1,2-Trichloroethane
1,1,2,2-Tetrachloroethane
1,2-Dichloroethane
1,2-Dichloropropane
1,2-Trans-dichloroethylene
1,3,-Dichloropropene
2-Chloroethyl vinyl ether

Pesticides and PCBs - 28

Aldrin
alpha-HCH
beta-HCH
gamma-HCH
delta-HCH
Chlordane
Dieldrin
Endosulfan sulfate
Endosulfan, beta
Endosulfan, alpha
Endrin
Endrin aldehyde
Heptachlor epoxide
Heptachlor
Hexachlorocyclopentadiene

Pesticides and PCBs - 28

Hexachlorobutadiene
Hexachlorobenzene
PCB 1221
PCB 1232
PCB 1242
PCB 1248
PCB 1254
PCB 1260
PCB 1016
Toxaphene
4,4'-DDD (p,p'-DDD)
4,4'-DDE (p,p'-DDE)
4,4'-DDT (p,p'-DDT)

Other Neutrals -10

Bis-2-chloroisopropyl ether
Hexachloroethane
1,2-Dichlorobenzene
1,2,4-Trichlorobenzene
1,3-Dichlorobenzene
1,4-Dichlorobenzene
2-Chloronaphthalene
4-Bromophenyl phenyl ether
4-Chlorophenyl phenyl ether
Isophorone

Phenols - 11

p-Chloro-m-cresol
Pentachlorophenol
Phenol
2-Chlorophenol
2-Nitrophenol
2,4-Dichlorophenol
2,4-Dimethylphenol
2,4-Dinitrophenol
2,4,6-Trichlorophenol
4-Nitrophenol
4,6-Dinitro-o-cresol

N-Containers - 4

1,2-Diphenyl hydrazine
Nitrobenzene
2,4-Dinitrotoluene
2,6-Dinitrotoluene

Asbestos

Total cyanides

Table 1, page 2.

<table>
<tr><td>Hetero-Carcinogens - 6</td><td>Phthalates - 6</td></tr>
<tr><td>
N-nitroso-di-n-propyl amine
N-nitroso-diphenyl amine
N-nitroso-dimethyl amine
2,3,7,8-Tetrachlorodibenzodioxin
Benzidine
3,3'-Dichlorobenzidine
</td><td>
Bis(2-ethyl hexyl) phthalate
Butyl benzyl phthalate
Di-n-butyl phthalate
Diethyl phthalate
Dimethyl phthalate
Di-n-octyl phthalate
</td></tr>
<tr><td>Aromatic hydrocarbons - 16</td><td>Metals - 13</td></tr>
<tr><td>
Acenaphthylene
Acenaphthene
Anthracene
Benz(e)acephenanthrylene
Benzo(k)fluoranthene
Benzo(a)pyrene
Chrysene
Fluoranthene
Fluorene
Indeno(1,2,3-c,d)pyrene
Naphthalene
Phenanthrene
Pyrene
Benzo(g,h,i)perylene\
Benzo(g,h,i)anthracene
1,2,5,6-Dibenzanthracene
</td><td>
Antimony
Arsenic
Beryllium
Cadmium
Chromium
Copper
Lead
Mercury
Nickel
Selenium
Silver
Thallium
Zinc
</td></tr>
</table>

Reprinted with permission from L.H. Keith and W.A. Telliard, 1979. Environmental Science and Technology 13:416-423. Copyright, 1979, American Chemical Society.

dropped from the list. Endrin aldehyde, for example, is hardly ever reported. Aldrin readily converts to dieldrin in the environment and would not therefore occur except immediately after application. Similarly, heptachlor converts to heptachlor epoxide. The Aroclors listed are all commercial products of mixtures of polychlorinated biphenyls, with characteristic compositions. They are never found as such in the environment in almost all usual situations, and reporting environmental PCBs in this format is not scientifically defensible. Moreover, present methodologies permit the measurement of a number of individual PCB compounds and reporting should be done in part at least on that basis. The list of polynuclear aromatics is somewhat arbitrary and should include compounds that would indicate origins, whether from petroleum or from combustion processes, such as the total of the methyl phenanthrenes and/or the totals of the methyl pyrenes and fluoranthenes (Sporstol et al., 1983). Only the most toxic dioxin is included, but other dioxins are also toxic and reporting these would provide a guide to origins, whether from combustion processes or from chlorinated phenol synthesis. Similarly the chlorinated dibenzofurans are highly toxic compounds of environmental significance (Czuczwa and Hites, 1984).

Local programs might therefore take this list as a guide, but in assessing impacts in coastal environments it would seem appropriate to consider first those compounds produced by the local industries, and design monitoring and analytical strategies accordingly.

PROPERTIES OF MARINE CONTAMINANTS RELEVANT TO MONITORING

A widely diverse group of compounds such as those on the Priority Pollutant List possesses a wide range of physical and chemical properties. Those particularly relevant in monitoring and surveillance programs include:

1) <u>Water Solubility</u>. This property profoundly affects the environmental distribution of a contaminant and particularly if it is an organic, the level to which it is bio-accumulated by organisms. The earliest studies of accumulation of chlorinated hydrocarbons found higher levels of accumulation at successive levels of the food chains. Birds eating earthworms had higher levels than the earthworms, and hawks eating these birds had still higher levels. This was also found to be true for mammals and birds in aquatic environments, but in fish and invertebrates the burden of organochlorines was found to be closely related to the lipid content. The suggestion was made that the amounts of lipid might be more important than position in the food web in determining the level of bioaccumulation of chlorinated hydrocarbons (Hamelink <u>et al</u>., 1971).

More recent research has supported this hypothesis. Strong correlations were found between the accumulation by fish of organochlorine compounds and their solubilities in water (Clayton <u>et al</u>., 1977; Chiou <u>et al</u>., 1977; Dexter and Pavlou, 1978; Veith <u>et al</u>., 1979). Specifically it was the accumulation in lipids that was correlated with what is now known as the octanol-water partition coefficient, the logarithm of the ratio of solubilities in n-octanol and in water (Veith <u>et al</u>., 1979; Kenaga, 1980; Mackay, 1982). The ratios in solubilities in octanol and in water are, moreover, closely correlated with the ratios of solubilities in triolein and in water. Triolein is a triglyceride that can be considered a model lipid. These similarities indicate that water solubilities are the major determinants for the coefficients (Chiou, 1985).

The distributions between water and lipid are therefore determined by passive equilibrium partitioning, such that the chemical potentials are equivalent in both phases. This principle also governs the distributions of contaminants in all phases of aquatic systems, including the atmosphere.

2) <u>Associations with Sediments and Particulate Material</u>. The partitioning between water and lipid phases of organic contaminants is furthermore correlated with the partitioning between water and sediments, when concentrations in the latter are expressed on a total carbon basis (Karickhoff <u>et al</u>., 1979; Chiou

159

et al., 1983; MacIntyre et al., 1984). Also, correlations have been found between levels in the sediments and in lipids of fish (Connor, 1984; Breck, 1985).

Solubilities in water of inorganic contaminants do not determine bioaccumulation in a comparable manner, whether from water or from sediments. Although the principle of maintenance of chemical potential still applies, many other factors not reviewed here but including the formation of insoluble compounds, complexing with organic materials, metabolic regulation, and adsorptive phenomena affect both bioavailability and bioaccumulation.

Many contaminants are associated with the particulate phase in seawater and the behavior of the particles strongly influences the patterns of local fluxes (Olsen et al., 1982). We shall provide below examples of the partitioning between particulate and dissolved phases of a number of compounds. Here also the distributions appear to be governed by solubility considerations; the more soluble compounds are predominantly in the "dissolved" phase, whereas the less soluble compounds are associated with the particles.

Association of hydrophobic contaminants with dissolved organics in seawater, such as the humic and fulvic acids, may increase their relative amounts in the "dissolved" phase (Boehm and Quinn, 1973; Leversee et al., 1983). Again, the physical chemical factor governing the distributions is the maintenance of chemical potential throughout the system.

Volatility. Volatility, or vapor pressure, is evidently another property that fundamentally affects the distribution of a contaminant in the ecosystem. We have noted above that the "volatile" compounds readily partition out of surface waters to enter the atmosphere where the majority are soon destroyed by oxidative or photochemical processes. One exeption is the chlorofluorocarbons which have greater persistence, giving rise to reactions in the upper atmosphere which result in the depletion of ozone.

A number of compounds with low volatilities, including most of the synthetic chlorinated hydrocarbons such as the PCBs and a number of pesticides, nevertheless are dispersed primarily through the atmosphere (Risebrough et al., 1968; Risebrough, 1969; Risebrough et al., 1976; Bidleman and Olney, 1974; Bidleman et al., 1976; Bidleman and Christensen, 1979; Atlas and Giam, 1981). Equilibrium concentrations in the atmosphere are rarely reached, because of incorporation into food webs and other organic matter; evaporation therefore proceeds from soils and other sites of application (Spencer, 1975; Spencer and Cliath, 1972; Spencer et al., 1974). Rapid disappearance of DDT from soils in Kenya (Sleicher and Hopcraft, 1984) and from soils of other countries where DDT and other chlorinated hydrocarbon insecticides are still used account for their continuing ubiquitous presence in the global environment. In surface waters of the Eastern Pacific toxaphene and the alpha isomer of hexachlorocyclohexane are currently the dominant chlorinated hydrocarbons of medium molecular weight (de Lappe et al., 1983; B.W. de Lappe and R.W. Risebrough, unpublished).

160

Earlier studies of transport of organic contaminants to the sea tended to use the concept of "fallout", an essentially irreversible process based on observations of the fate of radioactive isotopes created in atomic explosions. Equilibrium, however, is a two-way process. We can expect therefore lichens in the arctic tundra, and remote sea surfaces in the central Pacific, to be net sources of these compounds to the atmosphere if atmospheric concentrations fall. We require therefore a model of continuing global circulation through the atmosphere, with reversible partitioning into organic systems.

The implication for coastal systems is that persistent compounds of this kind discharged with waste waters will tend to leave the local area through volatilization into the atmosphere. The transfers between the surface waters and the atmosphere are best understood using the fugacity concept.

4) Fugacity. The concept of fugacity has been extensively developed by Professor Mackay of the University of Toronto and his students (Mackay, 1979; 1980; Mackay and Wolkoff, 1973; Mackay and Leinonen, 1975; Mackay and Peterson, 1981; Mackay et al., 1983). It is best described as an "escaping tendency", and has units of pressure. In the atmosphere it is simply the vapor pressure. In the aqueous phase it is the product of the concentration and the Henry's law constant. It can therefore be considered an extension of the expression of Henry's law in equation form

$$P = HC$$

where P is the partial pressure of the gas, C is the concentration of the dissolved gas, and H is the Henry's law constant. As either the partial pressure in the atmosphere or the concentration in water decreases, this law predicts a net movement either into or out of the water. Many of the compounds of interest, such as the PCBs, have both low vapor pressures and low solubilities in water, such that it is not immediately obvious whether coastal and marine waters should be expected to be sources or sinks of these compounds. A combination of theoretical studies, laboratory experiments, and field measurements is currently underway to gain a better understanding of the movements of these compounds through the global environment.

CHOICE OF THE MEDIUM

It is frequently asked whether it is more appropriate to measure contaminant levels in organisms, and if so which species, or whether sediments or water might be more appropriate. The answer evidently depends on the problem to be addressed and requires, moreover, the development of a conceptual model of the physical, biological, and social systems affected. If wastes are to be discharged for example from a coastal factory it would be necessary to study the local circulation patterns, the biological communities, sensitivities to the contaminants that would be discharged, and ways to reduce or eliminate any threat to the integrity of the local coastal ecosystem.

In this section we consider several applications of the uses of organisms, sediments, and coastal waters to describe the kind and magnitude of chemical change.

Organisms as Indicators. In the companion lecture (Risebrough, this volume) we noted that reproductive failures of several species of fish-eating birds in coastal areas of North America first called attention to a deleterious kind of chemical change in coastal environments. Birth defects of fish-eating birds have also provided an "early-warning" of the presence of teratogenic chemicals in the food web (Hays and Risebrough, 1972; Gilbertson et al., 1976). Mindful that the fish consumed by these birds are also eaten by people, many monitoring and surveillance programs include the analysis of edible species.

In the late 1960s, extraordinary things were happening in the marine environment of southern California. All of the Brown Pelicans, Pelecanus occidentalis, and Double-crested Cormorants, Phalacrocorax auritus, were laying thin-shelled eggs which crushed under the weight of the incubating adults (Risebrough et al., 1971; Risebrough, 1972; Gress et al., 1973). About one quarter of the California Sea Lion, Zalophus californianus, pups were being born prematurely (DeLong et al., 1973). Exceptionally high levels of DDT compounds, in the order of 1000 ppm, but also of PCBs, were found in the body fat or egg lipid. This exceptionally high level of contamination in a marine environment raised more than a few questions, and perhaps the two most important priorities were to determine the sources, and to set up a baseline to establish concentration trends over time.

The data from the birds and sea lions did not permit much to be said about sources, and it was not feasible to obtain samples from them on a regular basis. Mussels, Mytilus californianus, were therefore selected as an "indicator species". They were abundant on the coasts of both mainland and islands, and adequate material could therefore be obtained for analysis wherever and whenever they were needed. The first survey, in 1971, showed a strong gradient of concentrations of DDE, the principal DDT compound, away from the area of Los Angeles where a DDT manufacturer was located (Figure 1), indicating that this was the major source of DDT compounds. The sources of PCBs were shown to be more diffuse. A repeat survey in 1974 indicated that concentrations of both groups of compounds were going down (Risebrough et al., 1980), following the ending of discharge of DDT wastes and reductions of PCB inputs into the environment.

Contamination problems in southern California persisted in the mid-1980s and mussels were again collected to compare contaminant concentrations with those in the early 1970s. Methodologies have improved considerably since then, and mussels collected in the early 1970s and since preserved in frozen storage were retrieved for analysis. Equivalent results were obtained in 1986 for DDE and PCB values when mussels obtained in the early 1970s were re-analyzed. Compounds not measured 15 years ago such as those of the chlordane group were detected and quantified. This approach determined that at Santa Monica Pier in Los Angeles, levels of DDT compounds, PCBs, and chlordane compounds had dropped by factors of 8, 5, and 7-10, respectively between 1971 and 1986. Throughout

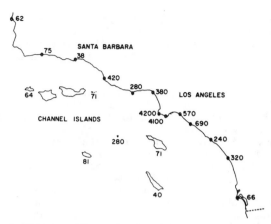

Figure 1.   Total DDT residues in <u>Mytilus californianus</u> in Southern
California in 1971.   Nanograms/gram (parts per billion)
of the wet weight.   Data of B.W. de Lappe,   R.W.
Risebrough, and D.R. Young.

the area, however, DDT levels after a sharp drop in the early
1970s, have shown no further decrease since about 1977.   This was
attributed to continuing input of DDT compounds from a reservoir
in the coastal sediments, derived from past discharges of the DDT
factory,   estimated to be in the order of 200 tons (Young <u>et al</u>.,
1976; Risebrough, 1987).

In these studies mussels served as a good "indicator organism" for
measuring   changes   in   levels   of   the   several   chlorinated
hydrocarbons over both space and time.   The concept of "indicator
organisms" has been reviewed by Phillips (1980) and by Anderlini
(in press).   Some organisms are indicators of certain kinds of
pollution   or   contamination   merely   by   being   present.   Thus the
presence of the bacterium <u>Escherichia coli</u> is an indication of
untreated sewage and the probable presence of pathogenic viruses.
High   concentrations   of   the   polychaete   <u>Capitella capitata</u>   are
associated with sediments containing wastes with a high organic
loading.   Certain   blue-green   algae   are   associated   with
eutrophication in coastal waters.   Usually, however, an indicator
organism is selected because it accumulates contaminants from
ambient waters, and concentrations in the organism are assumed to
reflect those in the ambient ecosystem, increasing or decreasing
in a linear relationship.   The ideal indicator organism would
therefore possess the following properties (Phillips, 1980;
Anderlini, in press):

1) It does not regulate concentrations of a compound or element
that is a contaminant.   Many species regulate concentrations of
metals, particularly those that participate in physiological
processes, and have mechanisms for sequestering the toxic metals
that   do   not   participate   in   physiological   processes.   Many
taxonomic groups possess an inducible enzyme system that responds
to certain kinds of organic contaminants by making them more
water-soluble such that they are excreted or partition into the

163

ambient waters. This capacity is poorly developed in molluscs, making them more suitable as an indicator for compounds such as the polynuclear aromatics. This capacity generally increases from molluscs to crustacea to fish to birds and mammals.

2) The concentrations in tissues are a linear function of concentrations in the ambient environment. In practice this condition is rarely achieved, since it does not consider reservoirs. Thus concentrations of DDE in the lipids of a mussel are related to the levels in ambient waters, but the exact relationship may depend upon the total burden of particulate matter and of dissolved organics. The levels in the water system are in turn related to those in the sediments, but do not reflect the total burden of the sediments. From the levels of DDE recorded in coastal mussels in the Los Angeles area it is not therefore possible to conclude that there is a reservoir of some 200 tons of DDE in the coastal sediments that is a continuing source to the local food webs (Young et al., 1976; Risebrough, 1987).

3) It is sedentary, preferably in areas where it can be readily collected, such as the intertidal zone.

4) It is abundant in the study area.

5) It is of reasonable size.

6) It tolerates brackish water.

A number of secondary factors may influence accumulation. They include lipid content, seasonal variations, size, age, weight, sex, sampling parameters, pollutant interactions, and altered behavior. Designs of effective monitoring programs must therefore take these into account (Phillips, 1980).

Mussels and other filter-feeding bivalves fulfill these conditions better than other groups such as fish, and for this reason they have been chosen for a number of "Mussel Watch" programs around the world. They are deficient, however in responding to three major needs in coastal monitoring and surveillance programs:

1) Identification of uncharacterized compounds. Many electron capture chromatograms of mussel extracts contain peaks that are unidentified and which represent contaminants of potential concern. Mass spectrometry is the technique of choice and necessity for their identification, but because mussels frequently contain a complex mixture of petroleum and biogenic compounds, a "clean" spectrum that would permit identification can not be obtained. Species higher in the food web that do not accumulate these compounds because of their capacity to degrade them are much more suitable for such identifications.

2) Poor accumulation of more polar compounds. Only the least polar compounds among the organic contaminants are accumulated by mussels to the extent that they can be detected and quantified. For those of higher polarilty, examination of the water system is necessary.

3) Documentation of the historical record. Archived collections,
such as those used in the Los Angeles studies, are rarely
available to establish trends over time. Sediments frequently
provide such an opportunity.

Sediments.

Sediments provide a major advantage over intertidal organisms such
as mussels since samples may be obtained over two dimensions
rather than one in documenting the present extent of
contamination, and at many sites over three dimensions in
documenting the history of past contamination. The distribution
of DDT compounds in surface sediments in the Los Angeles region
(Young et al., 1976) provides a good example. The gradient of
concentrations from the discharge point followed the water
circulation patterns in the area. The literature contains many
other examples in which sediments have been used to document the
spatial extent of contamination from a point source discharge.

Many metals are bound in insoluble complexes in sediments such
that they are not accumulated by organisms; polynuclear aromatics
associated with inorganic carbon particles may also not be
"bioavailable" to the benthic food webs. Contaminant burdens in
sediments do not necessarily totally reflect therefore the
potential accumulation by organisms.

In many areas sediments accumulate over time, and unless there is
frequent resuspension or continued mixing by burrowing organisms
the successive layers can be dated by several available techniques
and analyzed to provide a historical record of local
contamination. Figure 2 shows the historical deposition of DDE
and PCB in dated sediments of the Santa Barbara Basin of Southern
California. DDE first appeared in detectable quantities in
sediments deposited after 1952 and thereafter increased; PCBs
first appeared in sediments deposited around 1945 (Hom et al.,
1974). In the Western Baltic, the sedimentary record shows that
DDT compounds appeared around 1945, the PCBs around 1940 (Muller
et al., 1980).

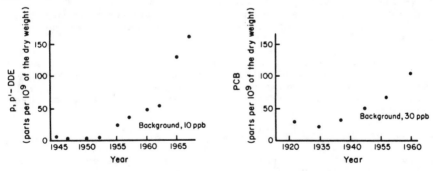

Figure 2. Deposition of DDE and PCB in dated sediments of the
          Santa Barbara Basin in Southern California. From Hom et
          al., 1974. Copyright 1974 by the American Association
          for the Advancement of Science.

The oceanic basins of southern California also have provided records of the fluxes of lead into the local marine environment. Rates of accumulation began to increase in the 1940s; analyses of the isotopic compositions indicated that the combustion of lead additives in gasoline was the principal source (Chow et al., 1973). In the western Baltic, the increase was less rapid, but beginning about 1880 (Muller et al., 1980).

Continuing analyses of sediment cores will document both the extent and the magnitude of chemical change in environmental contaminant levels that result from changes in use and environmental inputs.

Water Systems.

The analysis of water has not yet become routine in coastal monitoring and surveillance programs, largely because the low concentrations of almost all contaminants lead to problems of both contamination and recovery. Increasingly, however, as sampling methodologies improve, water contamination problems will be addressed by the sampling and analysis of water rather than of an "indicator".

Almost all of the first attempts to analyze seawater for metals and organics yielded values that are now considered too high because of contamination at some stage of the sampling process. Prior to about 1975, almost all of the determinations of lead and zinc in seawater were too high because of contamination during collection (Patterson et al., 1976; Bruland et al., 1978). Extreme precautions must be taken to prevent contamination in determination of cadmium (Martin et al., 1979). A review of the development of methodologies for the sampling of organics in seawater (de Lappe et al., 1980) pointed out that a typical concentration of 1 ng/liter of PCBs in seawater, a level reported by a number of investigators, would require more PCB in the world's oceans than has been produced. A typical value of 1 microgram/liter of petroleum hydrocarbons in the top 100 meters would represent about 40 million tonnes, several times higher than the estimated total yearly input into the oceans of petroleum. These considerations alone indicate that the true values must be substantially lower, except in more contaminated areas near major sources.

The methodologies for sampling metals in seawater are not reviewed here, because they are outside of the author's experience and expertise. They can currently be carried out, however, only by relatively few laboratories in the world which have adequate "clean rooms".

The use of in situ systems for the sampling of organics in seawater provides fewer opportunities for contamination than methodologies which require storage in containers. Moreover, sample volumes are not limited, an important consideration when concentrations are in the range of picograms per liter. Three systems have been developed, one which uses liquid-liquid extraction, and two others which use either polyurethane foam or XAD resin as solid sorbents.

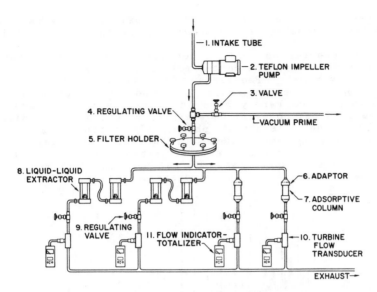

Figure 3. A shipboard trace organic filtration-extraction system.
Any combination of sorbents of liquid-liquid extractions
from the "dissolved"phase is possible. From de Lappe et
al., 1983. Reprinted with permission of the Canadian
Journal of Fisheries and Aquatic Sciences.

All systems use a pre-filter, usually a glass fiber filter, for
retention of a separate particulate phase (Figure 3; de Lappe et
al., 1983). Water then passes to a liquid-liquid extractor or to
one or more solid sorbents. The process is best carried out in a
ship-board clean room which receives seawater through a dedicated
intake system.

The liquid-liquid systems (Ahnoff and Josefsson, 1974) have lower
flow rates and organic solvents are usually more difficult to
manage and to maintain free from contamination than are the solid
sorbents. Although they are more efficient in extraction of some
contaminants than is the polyurethane foam (de Lappe et al.,
1983), they are not now frequently used.

The   XAD   macroreticular   resins   are   styrene-divinylbenzene
copolymers and among these XAD-2 is probably the most frequently
used for the extraction of organics from seawater (Harvey et al.,
1973; Duinker, 1984). Exhaustive cleaning is frequently required
to obtain an adequate blank. Storage, particularly by freezing,
may open new surfaces of the Amberlite beads, resulting in higher
blanks. Some preparations may give irregular recoveries (Picer
and Picer, 1980). Nevertheless the XADs are being increasingly
used for extraction of organics from the dissolved phase.

Polyurethane foams (Braun et al., 1985) are being increasingly
used for the sampling of organics in air (Billings and Bidleman,
1983 and cited references). Some foams have proved difficult to
clean-up such that satisfactory blanks may be difficult to obtain;

satisfactory blanks have been obtained, however, (de Lappe et al., 1983) and the reasons for the disparities are not yet apparent.

Inadequate recoveries as well as contamination may contribute to the errors in reporting the concentrations of organic contaminants in seawater systems. Recoveries from the particulate phase, however, appear to be good, since compounds characteristic of the pariculate phase are rarely seen in extracts of the "dissolved" phase. Few studies, however, have so far reported on estimates of recovery from the "dissolved" phase. De Lappe et al. (1983) mounted two polyurethane columns in series. When levels of a compound in the first were several times higher than in the second, recoveries were assumed to be quantitative. Good recoveries were obtained for polynuclear aromatics, PCBs, and DDT

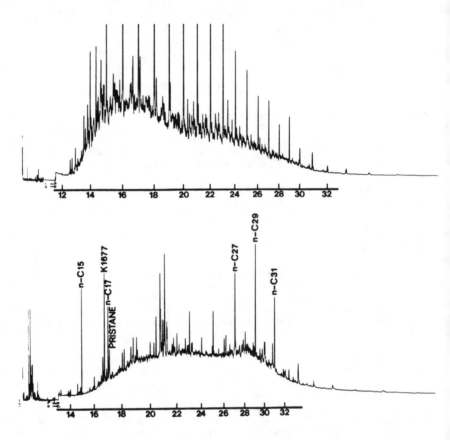

Figure 4. Chromatograms obtained with the Flame Ionization Detector of saturated fractions of the dissolved (top) and particulate (bottom) phases of seawater sampled in San Francisco Bay, August 1980. Numbers represent the emergence times of the n-alkanes. From de Lappe et al., 1983. Reprinted with permission of the Canadian Journal of Fisheries and Aquatic Sciences.

168

compounds, but recoveries for the lower-carbon alkanes and the hexachlorocyclohexanes were lower. Further work is necessary to determine recovery efficiencies and "break-through" volumes for both solid absorbents.

Two examples, both provided from the same sample of water of San Francisco Bay, indicate some of the differences between the particulate and dissolved phases. In the saturate fraction (Figure 4), lower carbon alkanes predominate in the dissolved phase. The particulates contain the odd-carbon higher alkanes, n-C27, n-C29, and n-C31, that derive from terrestrial plants, n-C15 and n-C17, as well as a cluster of hydrocarbons, that are characteristic of marine phytoplankton, pristane which is a marker

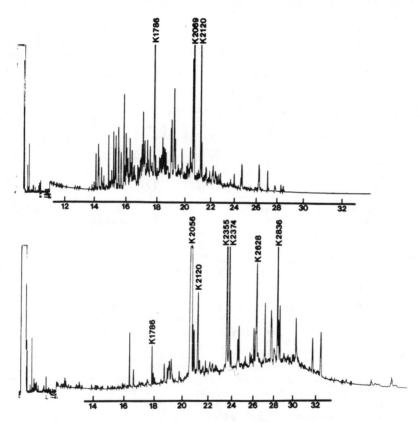

Figure 5.   Chromatograms obtained with the Flame Ionization Detector of aromatic fractions of the dissolved (top) and particulate (bottom) phases of seawater sampled in San Francisco Bay, August 1980. Numbers represent the emergence times of the n-alkanes. K1786: phenanthrene; K2069: fluoranthene; K2120: pyrene; K2056: heneicosa-hexaene. From de Lappe et al., 1983. Reprinted with permission of the Canadian Journal of Fisheries and Aquatic Sciences.

for zooplankton, and an unresolved complex mixture which by GC/MS analyses was shown to contain compounds characteristic of petroleum (de Lappe et al., 1983). In the aromatic fractions, the lower molecular weight aromatics through pyrene predominate in the dissolved phase; the particles contain the higher-molecular-weight aromatic compounds that are derived from combustion processes and which are also present in petroleum. Also present is heneicosa-hexaene, a compound found in marine phytoplankton (Figure 5).

These sampling and analytical data provide only a glimpse of the complexity, but also of the beauty, of the processes that occur in coastal ecosystems that are now being changed through man's activities. Neither the complexity, nor the beauty, are apparent in most reports of coastal monitoring and surveillance programs.

LITERATURE CITED

Ahnoff, M., and B. Josefsson. 1974. Sample apparatus for on-site continuous liquid extraction of organic compounds from natural waters. Anal. Chem. 46:658-663.

Atlas, E., and C.S. Giam. 1981. Global transport of organic pollutants: ambient concentrations in the remote marine atmosphere. Science (Washington, D.C.) 211:163-165.

Anderlini, V.C. Bioindicators. In Pollution and the Protection of Water Quality, R.W. Risebrough (ed.). Hemisphere, New York. (in the press).

Barber, R.T., P.J. Whaling, and D.M. Cohen. 1984. Mercury in recent and century-old deep-sea fish. Environ. Sci. Technol. 18:552-555.

Bidleman, T.F., and C.E. Olney. 1974. Chlorinated hydrocarbons in the Sargasso Sea atmosphere and surface water. Science (Washington, D.C.) 183:516-518.

Bidleman, T.F., C.P. Rice, and C.E. Olney. 1976. High molecular weight chlorinated hydrocarbons in the air and sea:rates and mechanisms of air/sea transfer. Pages 323-352 in H.L. Windom and R.A. Duce, eds., Marine Pollutant Transfer. Lexington Books. D.C. Heath and Company. Lexington, Mass. 391 pp.

Bidleman, T.F., and E.J. Christensen. 1979. Atmospheric removal process for high molecular weight organochlorines. J. Geophys. Res. 84:7857-7862.

Billings, W.N., and T.F. Bidleman. 1983. High volume collection of chlorinated hydrocarbons in urban air using three solid adsorbents. Atmos. Environ. 17:383-391.

Boehm, P.D., and J.G. Quinn. 1973. Solubilization of hydrocarbons by the dissolved organic matter in sea water. Geochim. Cosmochim. Acta 37:2459-2477.

Braun, T., J.D. Navratil, and A.B. Farag. 1985. Polyurethane Foam Sorbents in Separation Science. CRC Press. Boca Raton, Florida. 219 pp.

Bruland, K.W., G.A. Knauer, and J.H. Martin. 1978. Zinc in north-east Pacific water. Nature (London) 271:741-743.

Chiou, C.T., V.H. Freed, D.W. Schmedding, and R.L. Kohnert. 1977. Partition coefficient and bioaccumulation of selected organic chemicals. Environ. Sci. Technol. 11:475-478.

Chiou, C.T., P.E. Porter, and D.W. Schmedding. 1983. Partition equilibria of nonionic organic comopunds between soil organic matter and water. Environ. Sci. Technol. 17:227-231.

Chiou, C.T. 1985. Partition coefficients of organic compounds in lipid-water systems and correlations with fish bioconcentration factors. Environ. Sci. Technol. 19:57-62.

Chow, T.J., K.W. Bruland, K. Bertine, A. Soutar, M. Koide, and E.D. Goldberg. 1973. Lead pollution: records in Southern California coastal sediments. Science 181:551-552.

Clayton, J.R., S.P. Pavlou, and N.E. Breitner. 1977. Polychlorinated biphenyls in coastal marine zooplankton: Bioaccumulation by equilibrium partitioning. Environ. Sci. Technol. 11:676-682.

Connor, M.S. 1984. Fish/sediment concentration ratios for organic compounds. Environ. Sci. Technol. 18:31-35.

Czuczwa, J., and R.A. Hites. 1984. Environmental fate of combustion-generated polychlorinated dioxins and furans. Environ. Sci. Technol. 18:440-450.

DeLong, R.L., W.G. Gilmartin, and J.G. Simpson. 1973. Premature births in California sea lions: association with high organochlorine pollutant residue levels. Science 181:1168-1169.

Dexter, R.N., and S.P. Pavlou. 1978. Mass solubility and aqueous activity coefficients of stable organic chemicals in the marine environment: polychlorinated biphenyls. Mar. Chem. 6: 41-53.

Duinker, J.C. 1984. The determination of polychlorinated biphenyls in open ocean waters. IOC Tech. Ser. 26. Unesco, Paris.

Farrington, J.W., E.D. Goldberg, R.W. Risebrough, J.H. Martin and V.T. Bowen. 1983. U.S. "Mussel Watch" 1976-1978: An overview of the trace metal, DDE, PCB, hydrocarbon and artificial radionuclide data. Environ. Sci. Technol. 17:490-496.

Freyer, H.-D. 1979. Variations in the Atmospheric CO_2 Content. Pages 79-99 in B. Bolin, E.T. Degens, S. Kempe, and P. Ketner, eds. The Global Carbon Cycle. SCOPE 13. John Wiley and Sons, Chichester and New York. 491 pp.

George, J.L., and D.E.H. Frear. 1966. Pesticides in the Antarctic. J. Appl. Ecol. 3(suppl.)155-167.

Gilbertson, M.,R.D. Morris, and R.A. Hunter 1976. Abnormal chicks and PCB residue levels in eggs of colonial birds on the lower Great Lakes (1971-73). Auk 93:434-442.

Goldberg, E.D. 1983. Can the oceans be protected? Can. J. Fish. Aquat. Sci. 40(Suppl. 2):349-353.

Goldberg, E.D., V.T. Bowen, J.W. Farrington, G. Harvey, J.H. Martin, P.L. Parker, R.W. Risebrough, W. Robertson, E. Schneider and E. Gamble.1978. The mussel Watch. Environmental Conservation 5: 1-25.

Goldberg, E.D., M. Koide, V. Hodge, A.R. Flegal, and J. Martin. 1983. U.S. Mussel Watch: 1977-1978. Results on trace metals and radionuclides. Est. Coast. Shelf Sci. 16:69-93.

Gress, F., R.W. Risebrough, D.W. Anderson, L.F. Kiff, and J.R. Jehl. 1973. Reproductive failures of double-crested cormorants in southern California and Baja California. Wilson Bulletin 85: 197-208.

Hamelink, J.L., R.C. Waybrant, and R.C. Ball. 1971. A proposal: Exchange equilibria control the degree chlorinated hydrocarbons are biologically magnified in lentic environments. Trans. Amer. Fish. Soc. 100:207-214.

Harvey, G.R., W.G. Steinhauer, and J.M. Teal. 1973. Polychlorobiphenyls in North Atlantic Ocean water. Science 180:643-644.

Hays, H., and R.W. Risebrough. 1972. Pollutant concentrations in abnormal young terns from Long Island Sound. Auk 89:19-35.

Hinga, K.R., and M.E.Q. Pilson. 1987. Persistence of benz(a)anthracene degradation products in an enclosed marine ecosystem. Environ. Sci. Technol. 21:648-653.

Hom, W., R.W. Risebrough, D.R. Young, and A. Soutar. 1974. Deposition of DDE and PCB in dated sediments of the Santa Barbara Basin. Science 184: 1197-1199.

Irukayama, K. 1966. The pollution of Minimata Bay and Minimata disease. Adv. Water Pollut. Res. 3:153-180.

Jensen, S., A.G. Johnels, M. Olsson, and G. Otterlind. 1969. DDT and PCB in marine animals from Swedish waters. Nature (London) 224:247-250.

Karickoff, S.W., D.S. Brown, and T.A. Scott. 1979. Sorption of
    hydrophobic pollutants on natural sediments. Water Res.
    13:421-428.

Keeling, C.D., R.B. Bacastow, A.E. Bainbridge, C.A. Ekdahl, P.R.
    Guenther, L.S. Waterman, and J.F.S. Chin. 1976a. Atmospheric
    carbon dioxide variations at Mauna Loa Observatory, Hawaii.
    Tellus 28:538-551.

Keeling, C.D., J.A. Adams, C.A. Ekdahl, and P.R. Guenther. 1976b.
    Atmospheric carbon dioxide variations at the South Pole.
    Tellus 28:552-564.

Keith, L.H., and W.A. Telliard. 1979. Priority Pollutants.
    I - a perspective view. Environ. Sci. Technol. 13:416-423.

Kenaga, E.E. 1980. Correlation of bioconcentration factors of
    chemicals in aquatic and terrestrial organisms with their
    physical and chemical properties. Environ. Sci. Technol.
    14:553-556.

Koeman, J.H., M.C. ten Noever de Brauw, and R.H. de Vos. 1969.
    Chlorinated biphenyls in fish, mussels and birds from the
    river Rhine and the Netherlands coastal area. Nature
    (London) 221:1126-1128.

Leversee, G.J., P.F. Landrum, J.P. Giesy, and T. Fannin. 1983.
    Humic acids reduce bioaccumulation of some polycyclic
    aromatic hydrocarbons. Can. J. Fish. Aquat. Sci. 40(Suppl.
    2):63-69.

Levine, E.A. 1986. Program for monitoring sewage dump site in the
    New York Bight. Pages 760-763 in Oceans '86 Conference
    Record. IEEE Service Center, Piscataway, New Jersey, and
    Marine Technology Society, Washington, D.C.

Nriagu, J.O. 1979. Global inventory of natural and anthropogenic
    emissions of trace metals to the atmosphere. Nature (London)
    279:409-411.

Mackay, D. 1979. Finding fugacity feasible. Environ. Sci.
    Technol. 13: 1218-1223.

Mackay, D. 1982. Correlation of bioconcentration factors. Environ.
    Sci. Technol. 16:274-278.

Mackay, D., and P.J. Leinonen. 1975. Rate of evaporation of
    low-solubility contaminants from water bodies to atmosphere.
    Environ. Sci. Technol. 9:1178-1180.

Mackay, D., and S. Paterson. 1981. Calculating fugacity.
    Environ. Sci. Technol. 15:1006-1014.

Mackay, D., W.Y. Shiu, and E. Chau. 1983. Calculation of
    diffusion resistances controlling volatilization rates of
    organic contaminants from water. Can. J. Fish. Aquat. Sci.
    40 (Suppl.2):295-303

Mackay, D.M., P.V. Roberts, and J.A. Cherry. 1985. Transport of organic contaminants in groundwater. Environ. Sci. Technol. 19:384-392.

MacIntyre, W.G., C.L. Smith, and Chiou, C.T., P.E. Porter, and T.D. Shoup. 1984. Comment on "partition equilibria of nonionic organic compounds between soil organic matter and water". Environ. Sci. Technol. 18:295-297.

Martin, J.H., K.W. Bruland, and W.W. Broenkow. 1976. Cadmium transport in the California Current. Pages 159-184 in Marine Pollutant Transfer, H.L. Windom and R.A. Duce, eds. D.C. Heath and Co. Lexington, Massachusetts.

Muller, G., J. Dominik, R. Reuther, R. Malisch, E. Schulte, L. Acker, and G. Irion. 1980. Sedimentary record of environmental pollution in the Western Baltic Sea. Naturwissenschaften 67:595-600.

Munn, R.E. 1973. Global Environmmental Monitoring System (GEMS). SCOPE 3. Scientific Committee on Problems of the Environment. c/o Royal Society, London. 130 pp.

National Academy of Sciences. 1980. The International Mussel Watch, E.D. Goldberg, ed. National Academy of Sciences, Washington, D.C. 248 pp.

Oliver, B.G., and K.D. Nicol. 1982. Chlorobenzenes in sediments, water, and selected fish from Lakes Superior, Huron, Erie, and Ontario. Environ. Sci. Technol. 16:532-536.

Olsen, C.R., N.H. Cutshall, and I.L. Larsen. 1982. Pollutant-particle associations and dynamics in coastal marine environments: a review. Mar. Chem. 11:501-533.

Patterson, C.D., D. Settle, B. Schaule, and M. Burnett. 1976. Transport of pollutant lead to the oceans and within ocean ecosystems. Pages 23-38 in Marine Pollutant Transfer, H.L. Windom and R.A. Duce, eds. D.C. Heath and Co. Lexington, Massachusetts.

Phillips, D.J.H. 1980. Quantitative Aquatic Biological Indicators. Applied Sciences Publishers, Ltd. London. 488 pp.

Picer, N., and M. Picer. 1980. Evaluation of macroreticular resins for the determination of low concentrations of chlorinated hydrocarbons in sea water and tap water. J. Chromatogr. 193:357-369.

Risebrough, R.W. 1969. Chlorinated hydrocarbons in marine eco-systems. Pages 5-23 in M. W. Miller and G. G. Berg, eds., Chemical Fallout. Charles C. Thomas, Springfield, Ill.

Risebrough, R.W. 1972. Effects of environmental pollutants upon animals other than man. Pages 443-463 in L. Le Cam, J. Neyman and E. L. Scott, eds., Proceedings Sixth Berkeley Symposium Mathematical Statistics and Probability. Univ. Calif. Press, Berkeley and Los Angeles. 599 pp.

Risebrough, R.W. 1987. Distribution of organic contaminants in coastal areas of Los Angeles and the Southern California Bight. Los Angeles Regional Water Quality Control Board. Los Angeles. 114 pp.

Risebrough, R.W., R.J. Huggett, J.J. Griffin, and E.D. Goldberg. 1968. Pesticides: transatlantic movements in the northeast trades. Science 159: 1233-1236.

Risebrough, R.W., P. Reiche, D.B. Peakall, S.G. Herman, and M.N. Kirven. 1968. Polychlorinated biphenyls in the global ecosystem. Nature (London) 220:1098-1102.

Risebrough, R.W., F.C. Sibley, and M.N. Kirven. 1971. Reproductive failure of the brown pelican on Anacapa Island in 1969. American Birds 25(1): 8-9.

Risebrough, R.W., B.W. de Lappe, and W. Walker II. 1976. Transfer of higher-molecular weight chlorinated hydrocarbons to the marine environment. Pages 261-321 in Marine Pollutant Transefer, H. L. Windom and R. A. Duce, eds. D. C. Heath and Company, Lexington, Massachusetts. 391 pp.

Risebrough. R.W., B.W. de Lappe, E.F. Letterman, J.L. Lane, M. Firestone-Gillis, A.M. Springer and W. Walker II. 1980. California State Mussel Watch, Volume III. Organic pollutants in mussels, Mytilus californianus and M. edulis. California State Water Resources Control Board, Water Quality Monitoring Report 79-22. Sacramento. 108 pp + seven appendices.

Roberts, P.V., and P.G. Dandliker. 1983. Mass transfer of volatile organic contaminants from aqueous solution to the atmosphere during surface aeration. Environ. Sci. Technol. 17:484-489.

Sladen, W.J.L., C.M. Menzie, and W.L. Reichel. 1966. DDT residues in Adelie Penguins and a Crabeater Seal from Antarctica. Nature 210:670-673.

Sleicher, C.A., and J. Hopcraft. 1984. Persistence of pesticides in surface soil and relation to sublimation. Environ. Sci. Technol. 18:514-518.

Spencer, W.F. 1975. Movement of DDT and its derivatives into the atmosphere. Residue Rev. 59:91-117.

Spencer, W.F., and M.M. Cliath. 1972. Volatility of DDT and related compounds. J. Agr. Food Chem. 20:645-649.

Spencer, W.F., M.M. Cliath, W.J. Farmer, and R.A. Shepherd. 1974. Volatility of DDT residues in soil as affected by flooding and organic matter applications. J. Environ. Quality 3:126-129.

Sporstol, S., N. Gjos, R.G. Lichtenthaler, K.O. Gustavsen, K. Urdal, and F. Oreld. 1983. Source identificatifon of aromatic hydrocarbons in sediments using GC/MS. Environ. Sci. Technol. 17:282-286.

Sundstrom, G., B. Jansson, and S. Jensen. 1975. Structure of phenolic metabolites of p,p'-DDE in rat, wild seal and guillemot. Nature (London) 255:627-628.

Suszkowski, D.J., and E.D. Santoro. 1986. Marine monitoring in the New York Bight. Pages 754-759 in Oceans '86 Conference Record. IEEE Service Center, Piscataway, New Jersey, and Marine Technology Society, Washington, D.C.

Swanson, R.L., and C.J. Sindermann. 1979. Oxygen depletion and associated benthic mortalities in New York Bight. 1976. US Dept. of Commerce, NOAA Professional Paper No. 11. 345 pp.

Tomlinson II, G.H. 1983. Air polluants and forest decline. Environ. Sci. Technol. 17:246A-256A.

Veith, G.D., D.L. DeFoe, and B.V. Bergstedt. 1979. Measuring and estimating the bioconcentration factor of chemicals in fish. J. Fish Res. Board Can. 36:1040-1048.

Wakeham, S.G., A.C. Davis, and J.L. Karas. 1983. Mesocosm experiments to determine the fate and persistence of volatile organic compounds in coastal seawater. Environ. Sci. Technol. 17:611-617.

Wang, D., F.H. Bormann, and D.F. Karnosky. 1986. Regional tree growth reductions due to ambient ozone:evidence from field experiments. Environ. Sci. Technol. 20:1122-1125.

Young, D.R., D.J. McDermott, and T.C. Heesen. 1976. DDT in sediments and organisms around southern California outfalls. J. Water Pollut. Control Fed. 48:1919-1928.

Young, D.R., T.C. Heesen, and R.W. Gossett. 1980. Chlorinated benzenes in southern California municipal wastewaters and submarine discharge zones. Pages 471-486 in Water Chlorination: Environmental Impact and Health Effects, Vol. 3. R.L. Jolley, W.A. Brungs, and R.C. Cummings, eds. Ann Arbor Science, Ann Arbor, Michigan.

Sources and Transport of Oil Pollutants in the Arabian Gulf

MOHAMED I. EL SAMRA
Marine Sciences Department
University of Qatar

ABSTRACT

The paper presents data on dissolved/dispersed hydrocarbons obtained through five cruises carried out in the Arabian Gulf from Kuwait at the north to the Strait of Hormuz at the south. The data indicate that among sources of oil pollution, the oil-fields and tanker routes are the main contributors in the Gulf area. The highest values (40-500 ppb) were found near to the oil-fields, while intermediate values (~ 22 ppb) were recorded at stations located in the tanker routes. The coastal areas, on the other hand, showed the lowest values of dissolved/dispersed hydrocarbons (< 10 ppb). The current regime of the Arabian Gulf is discussed as a transport mechanism of oil pollutants as well as a natural protection process for different areas.

INTRODUCTION

The compiled data, in literature, reveal extraordinary lack of extensive observations on oil pollution in the Arabian Gulf. The great concern over the problem has stimulated research in two major branches:

1) measurements of in situ levels of hydrocarbons in water, sediments and biota. These measurements are limited in frequency, regularity and synopticity. However, the recent accumulation of data could give, more or less, the true status of oil pollution levels in some coastal areas in the Gulf.

2) statistical analysis of oceanographic variables (e.g. current direction) and/or data obtained from vessel observations of oil slicks. Vessel reports can be negative or repetitive in many cases. Oostdam (1980) calculated a mean percentage of 16.9% of positive reports in the Gulf region. However, statistical analysis could give a general predictive estimation of the magnitude and fate of oil in the area. It could also be possible to establish some computerized models for oil movements within the Gulf, such as the GULFSLIK program for the Arabian Gulf reported by Lehr et al. (1982).

This paper presents in situ measurements of petroleum hydrocarbons in the seawater of the Arabian Gulf, from Kuwait to the Strait of Hormuz (Fig. 1), during the period 1984-1986. According to the data given and previous one, the major sources of oil pollution in the area are identified and the mechanisms of pollutant transport are discussed.

A brief review of the literature

Among the first published studies on oil pollution in the Gulf region, Oostdam and Anderlini (1978) calculated the annual average of oil spilled in front of Kuwait (183 m³), from vessel reports in that area. Oostdam (1980) used vessel reports and estimated a volume of 16 x 10 m for the oil of all slicks in the Gulf. It was concluded that slick dispersal patterns agreed with surface current circulation, Anderlini and Al-Harmi (1979) stated that, in Kuwait beaches, most tar (maximum 2 Kg/m) was accumulated on the windward side of sand cusps or obstructions. Furthermore, according to Burns et al. (1982), the average quantities of tar in the beaches at the Strait of Hormuz ranged between 647 and 2325 g/m.

Burns et al. (1982) reported gas chromatographic analysis of sediments, oysters, mussels and fish tissue in the Gulf of Oman and the Omani coast at the Strait of Hormuz on the Arabian Gulf side. Accordingly, 0.8 ug/g wet weight were found in sediments at the Strait of Hormuz and 2.5-19 ug/g in those at the coasts of the Gulf of Oman. Resolved fractions of hydrocarbons in oysters collected at Hormuz amounted 0.8 ug/g wet tissue, while unresolved fractions reached to 6.0 ug/g wet weight. Anderlini et al. (1981), IAEA (1985) and Badawy et al. (1985) gave mean values of 111, 257 and 111 ug/g dry weight for oysters collected from Kuwait, Bahrain and Oman, respectively.

Among the most extensive works on marine pollution in coastal waters of the Arabian Gulf, was that conducted by Fowler during a period of 18 months. In this work, he stated that the highest recorded loads of tar (14-858 g/m) were found in Bahrain beaches, while, less tar (4-233 g/m) was present on the UAE beaches and the lowest loads for the entire 18 month period were reported at the beaches of Abu Dhabi. In the same paper, Fowler (1985) used fluorescence analysis to give values for total petroleum hydrocarbon levels in seawater of Bahrain (up to 5.7 ug/l) and UAE (up to 3.75 ug/l) and stated that current levels of petroleum hydrocarbons in the Gulf were not exceptionally high (20 ug/l). Also, according to Fowler (1985), sediments and oysters from Bahrain showed higher contents of hydrocarbons (36.36 ug/g dry wt for sediments at the eastern coast of Bahrain) compared to values lower that 1.0 ug/g for the UAE coast.

Literathy et al. (1985) gave a mean value of 5.8 ug/g for the clay-silt fraction of bottom sediments very nearshore to Kuwait, while Zerba et al. (1985) found hydrocarbon concentrations exceeding 100 ug/g for bottom sediments at the offshore area south of Kuwait (near to the offshore drilling and production area at the border between Kuwait and Saudi Arabia). Precisely, El Samra et al. (1986) found the highest value of total hydrocarbons (100-500 ppb) in the surface water of this offshore oil production area. Zerba et al. (1985) and Douabul et al. (1984) gave values around 13 ug/g for sediments collected at Shatt At-Arab area at the very north of the Arabian Gulf, while Grimalt et al (1985) identified n-alkanes by GC-MS technique for sediments collected at the same area and found values ranging between 3.3 and 18.8 ug/g dry sediments.

Finally, two reports (UNEP, 1984; Anonymous, 1986) provide some estimates of the amounts of oil contributed by different sources in the Arabian Gulf and give models for oil transport mechanisms through the region. These results will be discussed together with the present data given in the text.

MATERIALS AND METHODS

Present observations were made from R.V. Mukhatabar Al-Bihar, the Qatar University research vessel. Five cruises were made; the first in October-November, 1984 covering the area of the northwestern part of the Gulf, the second and the third in September-October, 1985 and 1986 covering the area east of Qatar to the Strait of Hormuz. In addition, two other cruises were made in the Qatarian sea-area during the years 1984 and 1985.

The sampling plan adopted provided a wide coverage of the area. Surface samples (1 m depth) of seawater were collected from 44 locations distributed in the Gulf (Fig. 1). Some of these locations were chosen to collect water samples from different depths including 1m, 5m, 10m and bottom. Sampling, experimental procedures, and analysis were made following the methods described by Parsons et al. (1985). Accordingly, PVC Niskin bottles (2 litre) were used for sample collection. Extraction of dissolved/dispersed petroleum hydrocarbons using dichloromethane was carried out on board. Analysis of the extracts was made by UV-fluorescence, using a Turner 430 spectrofluorometer and Kuwait crude oil as standard.

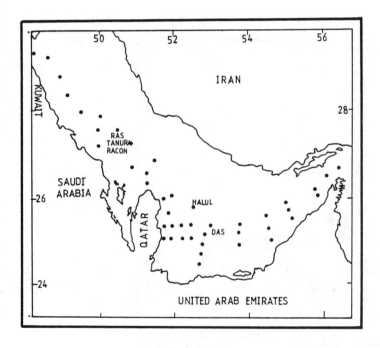

FIGURE 1. Sampling locations in the Arabian Gulf.

179

RESULTS AND DISCUSSION

Present observations on the levels of dissolved/dispersed hydrocarbons in the Arabian Gulf (Fig. 2) and those of previous work (Fowler, 1985, in particular) indicate that oil concentrations in the coastal waters of the Arab countries are, generally, lower than 20 ppb. The coastal water in front of UAE did not exceed 17 ppb, while that in front of the southern part of Qatar did not exceed 11 ppb. Values of 4.3 and 3 ppb were measured in the coastal waters of Saudi Arabia and Kuwait, respectively. A mean value of 6 ppb is calculated from the present data for the coastal waters of the Arabian Gulf.

On the other hand, the open sea waters show values higher than 20 ppb, up to 25.5 ppb, with a mean value of 22 ppb. This area (80 kilometres from the shore-line) was considered by El Samra and El Deeb (1987) as the oil tankers routes in the Arabian Gulf.

The highest values of petroleum hydrocarbons are found in the area affected directly by offshore oil-fields. Values of 40.9 ppb were measured near the biggest oil-field zone of UAE at Das Island, while a value of 68.6 ppb was measured near the oil-fields belonging to Qatar at Halul Island. However, the highest values of oil pollution (199-546.4 ppb) were measured at the area of Ras Tanura, in front of the Saudi Arabia coast, where the largest centre of oil production and refinery in the world is found. This high value is supported by that given by Zerba et al. (1985) for sediments at the very near area.

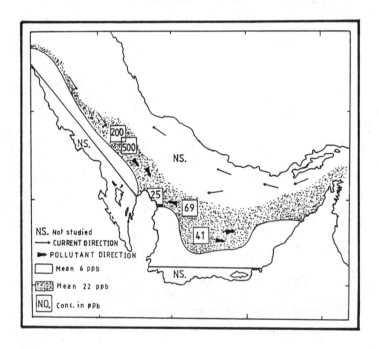

FIGURE 2. Levels of dissolved/dispersed hydrocarbons in the Arabian Gulf waters.

180

In an attempt to estimate the budget of petroleum hydrocarbons in the Arabian Gulf, the following data were used: the total area of the western half of the Arabian Gulf, which is 115.000 Km ; the mean concentration of hydrocarbons measured in the different areas, namely, coastal, open-sea and oil-fields waters; and, finally, the mean depth of the water column, that according to El Samra and El Deeb (1987) has no significant changes in its pollution levels from surface down to 10m depth.

The total area of coastal waters could be roughly considered as 30% of the total area of investigation. Thus, the estimated amount of dissolved/dispersed hydrocarbons in this area is 2070 tonnes. On the other hand, the total area of the open sea water is taken as 60% of the total and thus, the estimated amount of oil pollutants in it could be 30360 tonnes, approximately. The oil content of the water at the areas where oil terminals are found ranged between 40 ppb and 546 ppb with a mean value of 200 ppb. Considering that these fixed-point sources of oil pollution could affect about 10% of the study area, an standing stock of 23000 tonnes of oil is calculated for the areas very near to offshore oil-fields.

Golob and Brus (1984) estimated the amount of oil entering the Gulf region from offshore platforms and pipelines to be 32162 tonnes, whereas the oil spilled from tanker transportation, coastal refineries and coastal non-refinery wastes was estimated in 82032, 1347 and 4911 tonnes, respectively. These values are very high compared to the present in-situ measurements, but, this may be due to the fact that the estimation made by Golob and Brus (1984) was based on data of 1979 when oil production, exportation and transportation were much more higher than in the present days. In any case, present values reflect the in-situ situation of the oil pollution in the western half of the Arabian Gulf.

Mechanisms of pollutant transport

The current system prevailing in the Arabian Gulf is one of the main dynamic factors that affect transport of oil slicks and oil contaminated water streams. The resultant current movement goes along the axis of the Arabian Gulf (Hunter, 1984; Galt et al., 1984, among others). Thus, pollutants could be advected by the current in a direction parallel to the coast-line with the net effect that pollutants do not come very close to shore. The rip current prevailing in the Gulf coasts are offshore-directed, thus, causing advection of oil from the nearshore zone to the offshore region. Because of its geographical location, Qatar peninsula is impacted directly by the main current circulation. Therefore, the northern tip of Qatar and the eastern coast of Bahrain are among the primary threatened areas for coastal impacts of oil. On the other hand, the southern coast of Qatar and the coast of UAE, are shadwed and protected from direct impact of oil movement. The low concentrations obtained at these areas in the present and previous works support this idea.

CONCLUSION

The sea-area of the Arabian Gulf is divided into three main zones according to the sources and transport mechanism of oil pollutants. The zone of coastal water showed lower values of oil concentrations compared to the zone of tanker routes. The effect of offshore oil-fields as a focal point source of oil pollution is clearly noticed by the very high concentrations obtained there. Exchange of waters between these three zones is affected mainly by the current systems prevailing in the Gulf. In addition, due to evaporation, the specific gravity of surface oil increases and the residual oil may become denser than seawater, thus increasing the possibility of sinking. Oil, then, may be dispersed in the water column and outflow into the Gulf of Oman, taking into account that water of high density (bottom) is directed from the Arabian Gulf into the Gulf of Oman through the sill-less Strait of Hormuz (Hunter, 1984).

REFERENCES

Anderlini, V.C. and Al-Harmi, L. (1979). A survey of tar pollution on beaches of Kuwait Institute for Scientific Research, Marine Pollution Programme Report EES-11.

Anderlini, V.C., Al-Harmi, L., DeLappe, B.W., Risebrough, R.W., Walker, W., Simoneit, B.R.T. and Newton, A.S. (1981). Distribution of hydrocarbons in the oyster, Pinctada margaritifera, along the coast of Kuwait. Mar Pollut. Bull. 12, 57-62.

Anonymous (1986). Marine Environment and Pollution. In Proceedings of First Arabian Gulf Conference of Environment and Pollution, Kuwait, 7-9 February 1982 (Riad, H., Clayton, D. and Behnehani, M., ed.). Univ. of Kuwait, 348 pp.

Badawy, M.I. Al Harthy, F.T. and Al-Kayoumi, A.A. (1985). Petroleum and chlorinated hydrocarbons in the coastal waters of Oman. ROPME Symposium on Regional Marine Monitoring and Research Programmes, UAE Univ. 8-11, Dec. 1985.

Burns, K.A., Villeneuve, J.P., Anderlini, V.C. and Fowler, S.W. (1982). Survey of tar, hydrocarbon and metal pollution in the coastal waters of Oman. Mar. Pollut. Bull., 13, 240-247.

Douabul, A.A.Z., Al-Saad, H.T. and Darmoian, S.A. (1984). Distribution of petroleum residues in surficial sediments from Shatt Al-Arab river and the north-west region of the Arabian Gulf. Mar. Pollut. Bull., 15, 198-200.

El Samra, M.I., Emara, H.I. and Shunbo, E. (1986). Dissolved petroleum hydrocarbon in the Northwestern Arabian Gulf. Mar. Pollut. Bull., 17, 65-68.

El Samra, M.I. and El Deeb, K.Z. (1987). Horizontal and vertical distribution of oil pollution in the Arabian Gulf and the Gulf of Oman. Mar. Pollut. Bull. (in press).

Fowler, S.W. (1985). Coastal baseline studies of pollutants in Bahrain, UAE and Oman. ROPME Synposium on Region. Mar. Monit. and Res. Prog., UAE Univ. 8-11 December 1985.

Galt, J.A., Payton, D.L., Torgremson, G.M. and Glen, W. (1984). Applications of trajectory analysis for Nowruz oil spill. UNESCO Rep. in Mar. Sci., 28, 55-66.

Golob, R. and Brus, E. (1984). Statistical analysis of oil pollution in the KAP Region and the implications of selected world-wide oil spills to the region. UNEP Regional Seas Reports and Studies, 44, 7-34.

Grimalt, J., Albaigés, J., Al-Saad, H.T. and Douabul, A.A.Z. (1985). n-Alkane distributions in surface sediments from the Arabian Gulf. Naturwissenschaftem, 72, 35-37.

Hunter, J.R. (1984). A review of the residual circulation and mixing processes in the KAP region, with reference to applicable modelling studies. UNESCO Rep. in Mar. Sci. 28, 47-45.

IAEA (1985). Pollution sampling and analytical programme for Bahrain, UAE and Oman. Final Report, 1985.

Lehr, W.J., Belen, M., Cekirge, H.M. and Gunay, N. (1986). GULFSLIK- an oil spill modelling program for the Arabian Gulf. In: Preceedings of the First Arabian Gulf Conference on Environment and Pollution, Kuwait, 7-9 February 1982, 263-276.

Literathy, P., Nasser Ali, L., Zarba, M.A. and Ali, M.A. (1985). Problems of monitoring bottom sediment and heavy metals in the marine environment. ROPME Symposium on Regional Marine Monitoring and Research Programmes. UAE Univ., 8-11 December, 1985.

Oostdam, B.L. (1980). Oil pollution in the Persian Gulf and approaches 1978. Mar. Pollut. Bull. 11, 138-144.

Oostdam, B.L. and Anderlini, V. (1978). Oil spills and tar pollution along the coast of Kuwait. MPP Rept., Kuwait Inst. Sci. Res., Kuwait, 54 pp.

Parsons, T.R., Maita, Y. and Lalli, C.M. (1985). Determination of petroleum hydrocarbons. In: A Manual of Chemical and Biological Methods for Seawater Analysis, 56-59. Pergamon Press, Oxford.

UNEP (1984). Combating oil pollution in the Kuwait Action Plan Region. UNEP Regional Seas Reports and Studies, 44, 397 pp.

Zerba, M.A., Mohammad, O.S., Anderlini, V.C., Literathy, P. and Shunbo, F. (1985). Petroleum residues in surface sediments of Kuwait. Mar. Pollut. Bull. 16, 209-211.

■
Trace Metals and Petroleum Hydrocarbons in Syrian Coastal Waters and Selected Biota Species

F. ABOSAMRA, R. NAHHAS, N. ZABALAWI, S. BABA, G. TALJO and F. KASSOUMEH
Scientific Studies and Research Centre
Damascus, Syria

ABSTRACT

Total aliphatic hydrocarbons (TAH) and trace metals (Pb, Cd, Cu and Zn) were determined in sea-water at 12 sites along the syrian coast. Hydrocarbon concentrations ranged from 100 ng/l to 1460 ng/l. The values for Pb ranged from 10 to 160 ug/l, Cd from 5-50 ug/l, Cu from 5-30 ug/l and Zn from 10-360 ug/l.

Three fish, a shrimp and a bivalve species were selected for determining TAH and the above trace elements in addition to Hg. All species exhibited significant concentrations of TAH (1.62-113.51 ug/g wet w/t) with Pinctada martensi and Euthynnus alleteratus exhibiting the highest values. The same species showed also higher values of Pb and Cu, but the highest Hg and Cd values were observed in Sardina pilchardus.

INTRODUCTION

The syrian coast -from the Turkish border North to the Lebanese border South- does not exceed 181 Km (Fig. 1). Yet, it is densely populated as four cities and many small towns and villages are located in the narrow plains of the coastal belt that lies immediately at the feet of a series of hills and mountains. The area attracts many tourism during summer. Resort places in the mountains are frequented, but the beaches still have the biggest attraction. On the other hand, agriculture is an all-year-round practice as this is the main function of the coastal-plains.

Two of the cities, Latakia and Tartous, are major ports. Banias is the site of an oil refinery. Industries and varied activities are expanding in the area. Most of these, use the sea or the river systems to dump their sewage and refuse. Sewage wastes from cities are, so far, dumped directly into the sea near the beach without any treatment, although the situation is now being changed. In addition, ship maintenance works including painting exist at least North of Latakia and in Latakia harbour. Table I shows the different industries and their possible polluting products. Local practices are not the only potential source of pollution. The predominant current system (Fig. 1) indicates that pollution from external sources is also possible.

The lack of knowledge about concentrations of pollutants in our country, and the awareness of the need to protect our coastal environment have prompted this research. Furthermore, the fact that marine pollution is not a local but a regional problem, and that the eastern part of the Mediterranean is the less studied, encouraged us to start an extensive pollution monitoring survey of the area, including petroleum hydrocarbons and trace metals.

Sampling sites (A-L) were selected taking into account the variety of pollutant inputs that may occur. Station A was included as a reference. The examined compartments were coastal waters and biota. This consisted in three fish, a shrimp and a bivalve species (Table II). The organisms were taken from the coastal waters in the vicinity of Latakia.

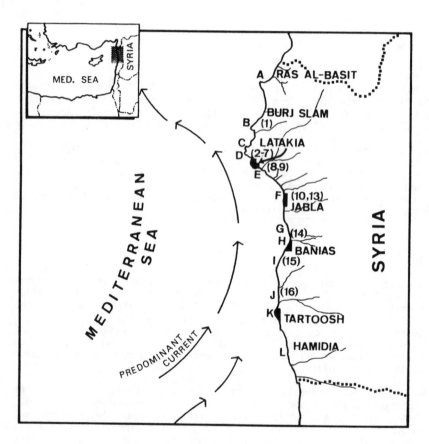

FIGURE 1. The Syrian coast with sampling stations (letters) and industrial activities (numbers listed in Table I)

TABLE I. Industrial activities along the syrian coast and their possible pollutant inputs.

Nos. (in Fig.1)	Industry	Possible polluting products
1	Cement factory	Metals
2	Textiles	Organics
3	Refreshments (including production of tins)	Metals
4	Electric motors	Metals
5	Aluminium works	Metals
6	Mill (cereals)-Batteries	Metals-acids
7	Tobacco	
8	Wood processing	Organics
9	Rubbish treatment	Metals
10	Weaving and Textiles	Organics
11	Bakery	
12	Food conserves	Organics
13	Tobacco	
14	Oil refinery	Oil
15	Power station	Oil
16	Cement factory	Metals

TABLE II. Species selected for analysis

Species	Habitat	Main food	Fishing procedure
Sardina pilchardus	Pelagic	Phyto & Zoo plankton	Gill nets
Mullus barbatus	Bottom	Bottom invertebrates	Bottom trowl
Euthynnus alleteratus	Pelagic	Fish, mollusks	Floating line
Penaeus kerathurus	Sandy-mud bottom	Zooplankton	Bottom trowl
Pinctada martensi	Rocks	Filter-feeding	Direct

MATERIALS AND METHODS

Sampling

Sea water was collected for metal analyses in plastic containers, acidified to pH 3.5 and tranferred to the lab at 4°C.

Sampling for hydrocarbon analyses was accomplished as described elsewhere (Abosamra, 1986). In summary, sea-water is passed through a tube containing a polyurethane foam (PUF) piece which is fitted tightly inside. After about 200 l of water have passed through, the PUF plug is taken for extraction in the laboratory.

Analyses

Trace metals.- 100 ml of sea-water, acidified to pH 3.5, by adding (HCl 0.01 N) was extracted using sodium diethyldithiocarbamate trihydrate. After mixing, the complex was extracted with methyl-isobutyl ketone. After separation and drying of the organic layer, metal concentrations were determined by AAS (Grasshof, 1983).

Organisms were prepared using instruments made of plastic materials. Only the muscle was removed from the side of the fish in the vicinity of the dorsal fin. The edible flesh of the shrimp was removed in one piece from segments II to V of the abdomen. The edible soft parts of the bivalve were removed alltogether.

A known weight of homogenized wet tissue material was digested in concentrated nitric acid overnight. The mixture was then heated on a hot plate until the volume was reduced to 1/5 of the original. The same previous quantity of nitric acid was added again and the operation was repeated until the mixture became clear.

The solution was diluted and filtered while still hot into a 50 ml volumetric flask. Double distilled water was used to bring the volume to 50 ml. The solution was analysed by flameless AAS (IAEA, 1980).

Petroleum aliphatic hydrocarbons.- The PUF plugs were extracted using distilled dichloromethane, then the extract was dried over sodium sulfate and the aliphatic hydrocarbons isolated by column chromatography on Sephadex 200 (20 g), eluting by distilled dichloromethane. After gentle evaporation under nitrogen the extract was redissolved in distilled hexane and analysed by GC.

The muscular tissues of the organisms were obtained as for metals but using s.s. instruments. The method outlined by Warner (1976) was adopted in treating this material. After digesting the homogenate tissue in NaOH at 90°C for 2 hours (with shaking at half time) the solution was cooled at room temperature and extracted with diethylether. Then, it was centrifuged for 10 min at 3000 r.p.m. The ether layer was removed and the operation repeated. The combined extracts were dried over anhydrous magnesium sulfate (1g) and fractionated in the Sephadex column.

The aliphatic hydrocarbon fractions (eluted with 100-155 ml of dichloromethane) were analyzed by capillary column GC under the following conditions:
column: SE-30 (30 m x 0.25 mm i.d.). Detector and injector temperatures: 280 and 240°C, respectively. Oven temperature: 70° C for 1 min and programming to

250° at 5°C/min. Carrier gas: helium. Quantitations of resolved (TAH) and unresolved components (UCM) was carried out using a C_{10}-C_{20} n-alkane standard mixture.

RESULTS AND DISCUSSION

Sea water.

Aliphatic hydrocarbons.- Concentrations of TAH at each sampling site and during five different periods are presented in Fig. 2. It is worth mentioning that the n-alkane profiles and the concurrence of the isoprenoid hydrocarbons pristane and phytabe in most of the samples point to a petrogenic origin of TAH. Generally speaking, variations among different sites do not show any seasonal trend, most values being under 200 ng/l, which can be considered as moderately low levels compared with other Mediterranean areas (Goutx and Saliot, 1980; Burns and Villeneuve, 1983; Albaigés et al., 1984).

Concentrations in July ranged between 100 ng/l at sites D, G and H to approximately 1460 ng/l at site E. Site E is located near the mouth of river AL-KABIR (Fig. 1) and is also exposed to the south because the predominent wind direction during this month and the sea current is SW.

There is a similar situation in January, where a concentration of 1800 ng/l is reached at site K (Tartous city) and 500 ng/l at site L (South of K). In this period the surface near-shore currents are affected by the predominant westerly winds, tending to push materials towards the shore. This is known to happen near Tripoli where small gyre systems counter the direction of the predominant currents. This possibly explains the slight rise in concentration at site L, compared to J. If this were the case, then the high values of site K must be due to the sewage system of Tartous.

March is always characterized by storms and active climatic conditions contributing to considerable mixing. This in turn, might explain the uniformity of concentrations at the various sites which indicates dispersion of the material along the syrian coast. As a matter of fact this is a situation that characterizes, although to a lesser degree, the Eastern Mediterranean surface waters.

It is known that the unresolved complex mixture of hydrocarbons (UCM) in the chromatograms indicate a petrogenic origin. From the values shown in Table III appears that such materials are associated with samples collected in January and March, when mixing is pronounced in the area. High concentrations are found in Latakia harbour where discharges of hydrocarbons are considerable. The two highest concentrations of TAH (sites K and E), however, are not reflected in the UCM pattern possibly indicating a different input source. This agrees with the above explanation regarding Tartous area, but also means that this could be the case in Latakia (site E) and that the origin of these hydrocarbons is merely the river and not necessarily southern sources.

Despite the high concentrations in some sites and periods of TAH and UCM, the general situation corresponds to a rather low levels of pollution, compared with other findings in the region. For example, Oren (1986) reports a value of 15.63 ug/l of total petroleum hydrocarbons in the Bay of Haifa, whereas the near-shore values range from 2.5-19.38 ug/l. The range in Turkish waters is from 8.2-24.4 ug/l (Wahby and El-Deeb, 1981). Demetropoulos and Loizides (1986) report a range of 0.0-6.9 ug/l along the southern coast of Cyprus. The Greek

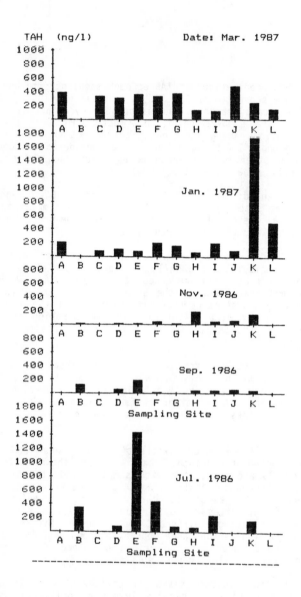

FIGURE 2. Seasonal variation of TAH concentrations (in ng/l) in syrian coastal seawaters. Letters refer to sites in Fig. 1.

islands and harbours exhibit a range of 0.6-28.2 ug/l (Mimicos, 1981), and in Alexandria and the near-shore region the range is 0.7-41.4 ug/l (Aboul-Dahab and Halim, 1981; Wahby and El-Deeb, 1981).

TABLE III. Concentrations of UCM of hydrocarbons at the various sites during the sampling periods (ug/l).

Site	July 86	Sep 86	Nov 86	Jan 87	March 87
A	-	-	-	1.73	1.92
B	< 0.05	< 0.05	0.16	-	-
C	-	-	-	1.28	1.36
D	≤ 0.05	0.48	< 0.05	4.0	6.6
E	≤ 0.05	0.08	< 0.05	6.6	7.0
F	< 0.05	< 0.05	0.24	-	3.8
G	-	< 0.05	< 0.05	4.0	1.65
H	-	< 0.05	-	0.24	2.8
I	0.05	< 0.05	< 0.05	1.44	1.52
J	< 0.05	< 0.05	< 0.05	1.44	< 0.05
K	< 0.05	< 0.05	< 0.05	< 0.05	0.27
L	-	-	-	-	1.84

Trace metals.- Four trace metals were considered in this investigation, namely Pb, Cd, Cu and Zn. The first two are dangerous to the marine environment, whereas the second two are necessary for marine life as they are part of the natural cycles.

Figures 3-6 show the concentrations found at the different sites and seasons.

Concentrations (in ug/l) ranged as follows:

10-160 for Pb, 5-50 for Cd, 5-30 for Cu and 10-360 for Zn.

The concentrations of Pb are much higher than those reported by Breder et al (1981) in the waters of the Magra estuary (Italy) (0.111 ug/l).

The concentrations of Cd in our waters are also much higher than those found in Magra estuary where they are only of 0.020 ug/l (Berdder et al., 1981). On the other hand, Cu and Zn are less than, respectively, the 50-130 ug/l and 60-450 ug/l reported in Alexandria harbour (El-Sayed et al., 1981). However, Scoullos (1981) reported 18 ug/l of Zn in the Bay of Elefsis. These general comparisons indicate that the syrian coast exhibits rather high levels of Pb and Cd, compared to close reference areas.

Pb concentrations (Fig. 3) were similar, with minor exceptions (Sep. 1986), in all sites and seasons. A very high value was recorded at site D in March 1987. This is a site of intense urban and industrial activity. Usually, higher levels of Pb, Zn, Cu and Cr have been observed near sewage outfalls and river mouths. El-Sayed et al. (1981) attributed metal contamination of the Alexandria harbour to sewage and waste disposal.

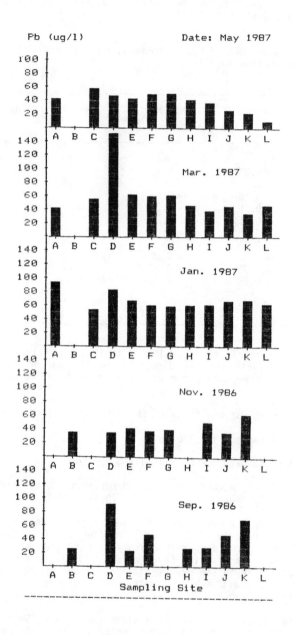

FIGURE 3. Seasonal variation of Pb concentrations (in ug/l) in syrian coastal seawaters. Letters refer to sites in Fig. 1.

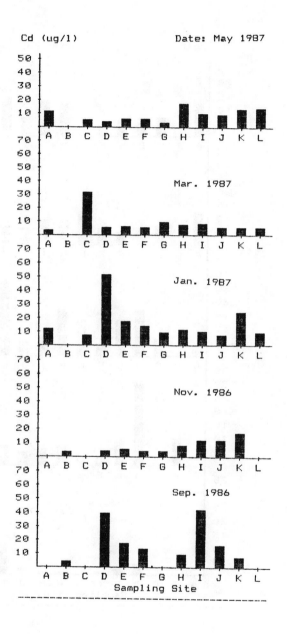

FIGURE 4. As Fig. 3, but for Cd.

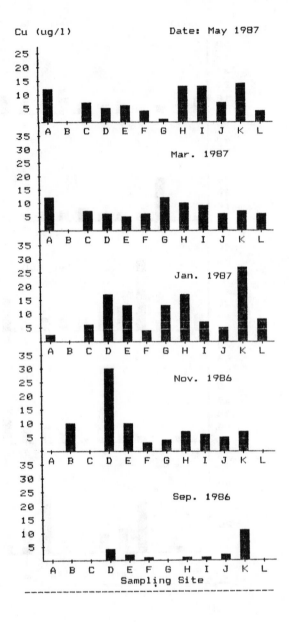

FIGURE 5. As Fig. 3, but for Cu.

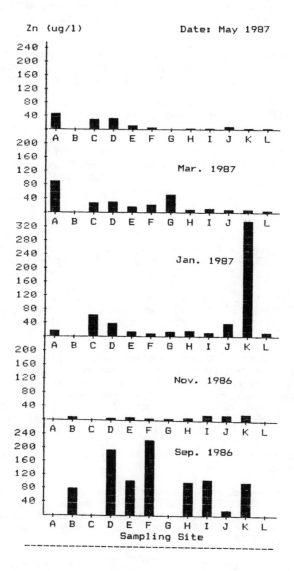

FIGURE 6. As Fig. 3, but for Zn.

The situation for Cd (Fig. 4) is similar. Again site D showed higher concentrations in different seasons. Cu and Zn were more variable (Figures 5 and 6, respectively). Here, site K shares site D with high values. Again K is a site of major sewage outfall that serves the port city of Tartous and it is that a major source of Zn pollution is urban waste-waters. For example, Scoullos (1981) found that the concentrations of this metal in the bay of Elefsis (vicinity of Athens) was four times higher than in other regions of the Aegean sea.

The ship maintenance activity in Latakia harbour also contribute to pollution by metals. El-Sayed et al. (1981) estimated that 253 Kg of copper were released annually from the surface of boats painted with antifouling products.

The presence of Pb, Cd and Cu at site L may be the result of the long range transport of pollutants by the predominant currents from the south, for example Tripoli which is a major urban and industrial area.

In conclusion it can be pointed out that levels of metals in coastal waters are greatly influenced by local inputs and environmental conditions. The monitoring of the area has been able to identify two main hot spots.

Organisms.

Aliphatic hydrocarbons.- All studied species showed the presence of aliphatic hydrocarbons and the existence of an unresolved complex mixture, emphasizing the petrogenic nature of the pollutants. The results are shown in Table IV.

Some individuals of E. alleteratus showed exceptionally high concentrations compared to the rest of the same species and the two other fish species. Pinctada martensi had the highest average concentration and one individual reached over 113 ug/g.

The hydrophobic nature of these pollutants might explain some of the results, although it is known that hydrocarbons are not magnified through the food web. The bivalve lives on rocks and is also associated with the upper sea layers, i.e. these species are more exposed to petroleum hydrocarbons than are M. barbatus or P. kerathurus which are bottom feeders. S. pilchardus is a pelagic fish so that it does not reach the concentrations shown by the other species.

Trace Metals.- The results are also shown in Table IV. The highest concentrations were those of copper in the shrimp and the bivalve, and of lead in E. alleteratus.

Cu concentrations ranged in P. kerathurus from 1.990 - 3.660 ug/g wet wt. This is much less than the 4.725 - 15.365 ug/g found in the same species in the Aegeann by Uysal (1981). However, Cu concentrations in our bivalve (1.406 - 13.105 ug/g) were more than those reported for another bivalve Mytilus galloprovincialis (1.57 - 2.95 ug/g).

Zn concentrations in the shrimp (1.315 ug/g) are also much less than that reported by Uysal (1981) for the same species (12.267 ug/g). Both, Pb and Hg concentrations, were similar in this study and that of Uysal.

TABLE IV. Pollutants concentrations in selected organisms (TAH: total aliphatic hydrocarbons; UCM: unresolved complex mixture; values between brackets: standard deviation from the mean).

Species	Average Standard length (cm)	Average Standard weight (g)	U.C.M. μg g^{-1} Range	U.C.M. μg g^{-1} Average	TAH μg g^{-1} Range	TAH μg g^{-1} Average	Hg Range	Hg Average	Pb Range	Pb Average	Cd Range	Cd Average	Cu Range	Cu Average	Zn Range	Zn Average
							Trace Metals (μg g^{-1} Fresh weight)									
Euthynnus alleteratus	32.9 (1.02)	529.6 (50.6)	0.0-0.2	0.081 (0.1)	22.23-77.33	39.014 (4.50)	0.013-0.716	0.392 (0.319)	0.111-13.123	3.989 (5.367)	0.0-0.003	0.001 (0.001)	0.0-0.717	0.252 (0.293)	0.031-0.216	0.121 (0.072)
Mullus barbatus	12.6 (0.6)	40.8 (6.8)	0.012-0.83	0.093 (0.08)	1.62-11.44	5.844 (3.779)	0.055-0.348	0.162 (0.129)	0.364-1.023	0.589 (0.259)	0.0-0.126	0.075 (0.059)	0.0-0.299	0.146 (0.160)	0.267-0.406	0.360 (0.056)
Sardina pilchardus	14.4 (1.1)	37.5 (8.8)	0.04-0.24	0.115 (0.09)	2.3-11.5	6.788 (4.095)	0.014-1.720	0.614 (0.959)	0.162-1.870	0.841 (0.906)	0.027-1.994	0.684 (1.134)	0.320-1.796	0.909 (0.782)	0.067-1.344	0.602 (0.663)
	14.6 (0.5)	39.1 (4.2)					0.097-0.485	0.301 (0.118)	0.032-0.636	0.319 (0.190)	0.110-0.139	0.124 (0.011)	0.199-0.391	0.310 (0.103)	0.003-0.015	0.009 (0.004)
Penaeus Kerathurus	15.4 (0.7)	20.5 (2.9)	0.006-0.43	0.229 (0.151)	1.80-7.47	4.07 (2.22)	0.021-0.186	0.084 (0.076)	0.108-1.423	0.692 (0.536)	0.023-0.776	0.288 (0.326)	1.990-3.660	2.861 (0.650)	0.076-2.956	1.315 (1.240)
Pinctada martensi	6.2 (0.5)	22.6 (4.3)	0.09-0.52	0.368 (0.194)	4.99-113.51	45.05 (59.57)	0.082-0.358	0.150 (0.118)	0.907-2.458	1.563 (0.580)	0.274-1.305	0.693 (0.377)	1.406-13.102	3.916 (5.138)	0.386-0.607	0.581 (0.184)

197

Hg concentrations in M. barbatus in this study ranged from 0.055 - 0.348 ug/g. This is slightly less than Uysal's finding for the same species (0.086 - 0.437 ug/g), the respective averages being 0.162 and 0.262. Other authors reported values ranging from 0.045 to 2.20 ug/g in Saronikos gulf (Voutsimov and Taliadouri, 1981) and in Alexandria (El-Sokkary, 1981).

Some of these authors report Pb concentrations (in the same species) ranging from 0.305 to 2.316 ug/g, whereas our range was 0.364 - 1.203 ug/g. A similar situation stands for Cd, that showed a range (0 - 0.128 ug/g) also much lower than the others reported (0.057 - 0.545 ug/g).

The fish samples were taken from the vicinity of Latakia. We have seen that syrian coastal waters contain noticeable concentrations of the various metals studied at all sites and seasons, particularly at site D. If we consider metal concentrations in sea-water (Figures 3-6) we find that there is some correlation with the values in fishes collected along different seasons. At this respect, the Euthynnus sample was collected in January and contained the highest concentration of Pb and the lower of Zn with respect to other fishes.

Although it is difficult at the present stage of the study to draw conclusions about the biodynamics of the pollutants in the marine environment, it has demonstrated to be an adequate mean of establishing the quality of these coastal waters and of identifying focal sources of pollution. Moreover, it has contributed to the knowledge of the levels of pollution in the Eastern Mediterranean, an area where data is particularly lacking.

Acknowledgements.- The technical assistance of O. Assi, D. Naser, M. Nouri and E. Der Torossian is highly appreciated. The authors are thankful to Dr. A. Habal and Prof. A.H. Mansour for providing the necessary facilities and means of carrying out this research.

REFERENCES

Abousamra, F. (1986). Sampling and analysis of petroleum hydrocarbons in sediments by HRGC. Science week. Supreme Council of Science. Damascus. Syria.

Aboul - Dahab, O. and Halim, Y. (1981). Oil pollution of the marine environment in the area of Alexandria. In: Workshop on pollution of the Mediterranean, pp. 201-208, I.C.S.E.M. - U.N.E.P., Monaco.

Albaigés, J., Grimalt, J., Bayona, J.M., Risebrough, R., deLappe, B. and Walker, W. (1984). Dissolved, particulate and sedimentary hydrocarbons in a deltaic environment. Org. Geochem., 6, 237-248.

Breder, R., Wurnberg, H.W. and Stoeppler, M. (1981). Toxic trace metal levels in water and sediments from the estuaries of the southern Ligurian and northern Tyrrhenian coasts. A comparative study. In: Workshop on pollution of the Mediterranean, pp. 285-292, I.C.S.E.M. - U.N.E.P., Monaco.

Burns, K.A. and Villeneuve, J.S. (1983). Biogeochemical processes affecting the distribution and vertical transport of hydrocarbon residues in the coastal mediterranean. Geochim. Cosmochim. Acta., 47, 995-1006.

Demetroupoulos, A. and Loizides, L. (1986). In: baseline studies and monitoring of oil and petroleum hydrocarbons in marine waters. U.N.E.P./I.O.C./W.M.O. Reports series No. 1, U.N.E.P., Athens.

Duckworth D.F., and Perry, S.G. (1986). Characterization of spilled oil samples- Purpose, Sampling, Analysis and Interpretation. J. Wiley & Sons, Chichester.

El-Sayed, M. kh., El-Sayed, M.A. and Moussa, A.A. (1981). Anthropogenic material in sediment from the eastern harbour of Alexandria, Egypt. In: Workshop on pollution of Mediterranean, pp. 493-502, I.C.S.E.M. - U.N.E.P., Monaco.

El-Sokkary, A. (1981). Mercury accumulation in fish from Mediterranean coastal area of Alexandria- Egypt. In: Workshop on pollution of the Mediterranean. I.C.S.M. - U.N.E.P., Monaco.

Goutx, M. and Saliot, A. (1980). Relationship between dissolved and particulate fatty acids and hydrocarbons, chlorophyll A and zooplankton biomass in Villefranche Bay. Mediterranean Sea, Mar. Chem., 8, 229-318.

Grimanis, A.P., Zafiropoulos, D., Papadopoulo, C. and Vassilaki-Grimani, M. (1981). Trace elements in the flesh of different fish species from three gulf of Greece. In: Workshop on pollution of the Mediterranean, p. 407-412, I.C.S.E.M. - U.N.E.P., Monaco.

K. Grasshoff, M. Ehrhardt, and K. Kremling (1983). Methods of Sea-Water Analysis. Verlag Chemie, Weinheim.

I.A.E.A. (1980). Elemental Analysis of Biological Materials. Technical Reports Series No. 197 - Vienna.

Mimicos, N. (1981). Pollution by petroleum hydrocarbons along several Greek island coasts and harbours. In: Workshop on pollution of the Mediterranean, pp. 489-492, I.C.S.E.M. - U.N.E.P., Monaco.

Oren, O.H. (1986). In: Baseline studies and monitoring of oil and petroleum hydrocarbons in marine waters (U.N.E.P./I.O.C./W.M.O.). Reports series No. 1, pp. 243-249, U.N.E.P., Athens.

Scoullos, M. (1981). Dissolved and particulate zinc in polluted Mediterranean bay. In: Workshop on pollution of the Mediterranean, pp. 483-488, I.C.S.E.M. - U.N.E.P., Monaco.

U.N.E.S.C.O. - I.O.C. (1982) The determination of petroleum hydrocarbons in sediments. M.G. No 13. Unesco. Paris.

Uysal, M. (1981). Levels of trace elements in some food chain organisms from the Aegean coasts. In: Workshop on Pollution of the Mediterranean, pp. 503-511, I.C.S.E.M. - U.N.E.P., Monaco.

Voutsinou, F., Taliadouri, (1981). Trace metals in marine organisms from the Sarinikos Gulf (Greece). In: Workshop on pollution of the Mediterranean, pp. 275-279, I.C.S.E.M. - U.N.E.P., Monaco.

Wahby, S.D. and El-Deeb, K.Z. (1981). A study of the state of pollution by petroleum hydrocarbons along the coast of Alexandria. In: Workshop on pollution of the Mediterranean, pp. 257-262, I.C.S.E.M. - U.N.E.P., Monaco.

Warner, G.F. (1976). Determination of aliphatic and aromatic hydrocarbons in marine organisms. Anal. Chem., 48, 578-583.

Instrumental Techniques for the Analysis of Trace Pollutants

Joan O. Grimalt
Department of Environmental Chemistry
CID (CSIC)
Jordi Girona, 18-26
08034-Barcelona, Spain

1. INTRODUCTION.

The analytical techniques may be generally characterized by six S-criteria: Specificity, Sensitivity, Speed, Sampling, Simplicity, and $ (McLafferty, 1984). These concepts are also of major concern when they are applied to the analysis of trace pollutants in environmental samples.

Trace pollutant analysis currently involves the identification and measurement of analytes present as minute constituents in complex matrices of unknown composition. Therefore, high **sensitivity** and **specificity** are needed. These matrices include a wide diversity of materials such as sediments, biological tissues, water, aerosols, etc., representing either the environmental compartments where the pollutants may be distributed or the organisms where they may be accumulated. Each of these materials states a different analytical problem evidencing the importance of **sampling** (a concept referring here to the arrangements needed to make the sample available to the instrument). Variability and inhomogeneity are, in turn, common features of the environmental compartments or biological tissues which determine that large series of samples must be analyzed in order to obtain statistically representative data. In consequence, high instrumental **speed** and **simplicity** of operation are required. Finally, the study of trace components in large sets of complex samples is expensive and $ becomes a limiting aspect (as it does in most human activities).

The understanding of the environment, and especially of the aspects concerning chemical pollution, is therefore strongly dependent on the availability of suitable analytical instrumentation. In fact, environmental samples constitute a regular field of application for analytical chemistry in which the more advanced instrumental set-ups are tested. At the same time, the environmental studies represent a permanent challenge for analytical chemists stimulating the development of improved technology with higher S standards.

In this sense, the analytical instrumentation has undergone an impressive development over the last two decades. For example, a part per million ($1/10^6$) was considered an extraordinary sensitivity about twenty years ago. Today, sensitivities in the order of parts per billion ($1/10^9$) are currently available and some routine analysis of trace pollutants involve determinations in the order of parts per trillion ($1/10^{12}$). Some general characteristics of the main instrumental techniques presently used for trace pollutant analysis are summarized in Table 1. The analytical performances of some of these techniques are complementary so that they have been coupled giving rise to a new type of technology: the hyphenated instrumentation.

However, despite all these important improvements many environmental analytical problems still remain unresolved. For example, the measurement of labile organic and inorganic trace components in the atmosphere, the analysis of polar combustion products, the determination of organometallic pollutants in aquatic systems, the structural identification of non-volatile toxic macromolecules in water and sediments, and many others. All them evidencing the lag of the state-of-the-art of the currently available techniques with respect to the actually required level of understanding of many environmental problems. This is perhaps even better illustrated when chemical interpretation of bio-assay tests is intended. Thus, the identification and quantitation of polar metabolites of toxic chemicals in organic tissues or the determination of cause-to-effect relationships based on the analysis of genotoxic components in samples of positive Ames test mutagenicity often become troublesome and of difficult solution. In consequence, the study of trace pollutants requires a good knowledge in analytical chemistry for the application of the best available technique to each specific problem.

In the present chapter, a review of the analytical instrumentation of interest for the identification and measurement of trace pollutants is performed. The discussion is restricted to the techniques commercially available making special emphasis into their usefulness and applications for the analysis of trace components in environmental samples.

Table 1. Comparison of some instrumental analytical techniques for trace pollutant analysis.

Technique	Field of application	Specificity	Det. Lim.	Speed	Cost
Gas chromatography	volatile or volatilizable organics	excellent	10^{-12}g	low	moderate
Liquid chromatography	organic components	good	10^{-10}g	low	moderate
Mass spectrometry	mainly organics	good	10^{-14}g	fast	high
Spectro-fluorimetry	inorganic and organic components	good	10^{-11}g	fast	moderate
Atomic spectroscopy	transition metals and semi-metals	excellent	10^{-11}g^a 10^{-13}g^b	fast	moderate
Anodic stripping voltammetry	selected trace metals	good	10^{-11}g	moderate	low

a) flame. b) furnace.

2. GAS CHROMATOGRAPHY (GC).

GC is a well-developed technique for the analysis of volatile or volatilizable (i.e. by chemical derivatization) substances. The advances in technology of wall-coated open tubular columns (capillary columns) have played a substantial role in the usefulness of GC for environmental trace organic analysis. Their improved resolution as well as their suitability for the combination with mass-spectrometric techniques constitute their two major advantages in comparison with packed columns. Furthermore, lower temperatures are necessary in capillary GC for the elution of chromatographed samples. This temperature reduction can be quite substantial leading to fewer decomposition problems and shorter analysis times. On the other hand, the widespread substitution of glass wall columns by those prepared from flexible fused silica and the increasing availability of ad hoc GC instrumentation have made their successful use much easier.

2.1 Column technology.

The chromatographic resolution between two compounds depends on the number of theoretical plates (N) of the column, the selectivity of the chromatographic system (α) and the solute capacity ratio (K). The resolution (R) may be related to these variables by means of the equation

$$R = \frac{\alpha-1}{\alpha} \cdot \frac{K}{K+1} \cdot \frac{\sqrt{N}}{4} \quad (1)$$

Capillary columns owe their high resolution power (R) to the large number of theoretical plates (N) that can be achieved during their preparation which, in turn, is a consequence of their length. Since the stationary phase is coating the internal surface and not filling the column, their permeability is high allowing the easy use of columns of 10-60 m with only a moderate drop in pressure. At the same time, the diffusion coefficient in the gas phase is high enough for avoiding sensible band broadening despite the high amount of mobile phase with respect to the stationary phase in the system. Capillary columns for most trace organic analytical work need not to be longer than 30 m with internal diameters around 0.2-0.3 mm. Typical columns should have around 3000 theoretical plates per meter.

Besides resolution, an important aspect for the analysis of trace pollutants is thermal stability. This is achieved by bonding some functional groups of the liquid phase to the column wall or by in situ crosslinking low or medium viscosity coatings to form a non extractable polymer. In fact, both processes are produced together in the current methods for stationary phase fixation (Grob et al., 1981). With this system the upper temperature limit for fused silica columns is that corresponding to their outer polyimide coating ($<$ 370°C). However, higher temperatures can be reached when cladding with an aluminum layer. Bonded methyl silicone and methyl 65% phenyl silicone fused silica columns being able to be heated, respectively, to 440°C and 370°C are now commercially available.

2.2 Injection systems.

In capillary gas chromatography the injection systems are vital for obtaining an adequate peak shape, good resolution, reproducibility and low quantitation errors. In trace organic analysis a correct injection mode is even more critical. Two main types of injection are generally selected for dilute solutions: splitless and on-column, the former being more commonly used because of its earlier development.

Splitless injection is performed by flash vaporization of the mixture within the injection chamber followed by condensation in the first coils of the capillary column. The sample solvent must be much more volatile than the solutes and, at the same time, it must have a boiling point higher ($\sim 20^{\circ}$C) than the oven temperature at the time of injection. The condensation of the solvent at the begining of the column (the so-called solvent effect) is essential for band narrowing in this injection mode.

On column injection allows the introduction of the liquid sample directly into the column inlet. With this system the sample loss effects during vaporization and transfer from the vaporizer to the column are eliminated. Therefore, quantitative precision and accuracy for very dilute samples, thermally labile compounds and organics of wide range of volatility are improved with respect to the splitless technique. A secondary cooling circuit for the column inlet is needed in order to assure that the sample remains in liquid phase before oven programming. On column injection has also several drawbacks such as the lower sample volume that can be quantitatively injected without peak splitting and the need for more sample clean-up due to the higher risk of column damage.

Further aspects on these and other injection systems are discussed in Sandra (1985) and Grob (1986).

2.3 Detectors.

Because of the high separation efficiency of the capillary columns, the detectors must have faster response times and smaller extra-column band spreading than those currently used for packed GC. Detectors having response time constants of 50 ms or less are adequate. Band spreading can be minimized using make-up gas at the end of the column. Other characteristics such as sensitivity, stability, detection limit, linearity and reactivity are also important. Many specific detectors developed for packed columns have also been adapted to the open tubular columns.

The **flame ionization detector** (FID) is the most commonly used for its high sensitivity to organic carbon, linear range (10^7) and baseline stability. The **electron capture detector** (ECD) is generally used for the analysis of compounds that have high electron affinities such as pesticides, insecticides and their metabolites. Several modes of operation are possible: coulometric, constant-frequency and constant-current. In the latter two modes, the responses are dependent on sample concentration and electron affinity. Therefore, accurate calibration using standards paralleling the composition of the sample is required for quantitative analysis. The **flame photometric detector** (FPD) is mainly used for the selective detection of sulphur and phosphorous containing components. It has widespread application in the analysis of air, water and coal hydrogenation products. It is constructed in two versions, single-burner and dual-flame, the latter showing less interference from water or hydrocarbons. The **thermionic detector** is specific and highly sensitive for components containing nitrogen and phosphorous (NPD). It uses an alkali metal salt which is placed near the burner of a hydrogen-air flame (alkali flame ionization detector, first generation) or above the detector burner tip (flameless thermionic detector, second generation). The latter version is easier to operate and of longer lifetime providing better response and baseline stability. The nitrogen-phosphorous detector is of widespread use in capillary gas chromatography for the analysis of trace pesticide components.

More information on selective GC detectors is given in Dressler (1986).

Furter discussion on theory and practical applications of open tubular column gas chromatography is available in Ahuja (1984), Shoenmakers (1986), Tranchant (1982) and Lee et al (1984).

3. LIQUID CHROMATOGRAPHY (LC).

Very few instrumental techniques parallel the enormous expansion of LC during the last 20 years. Modern LC instrumentation, while expensive, is suitable for most analytical work. However, unless non-volatile or thermally labile components must be analyzed, gas chromatography is still the technique of choice. Two main aspects must be considerably improved in LC to attain a comparable performance as that of capillary GC: column efficiency and detection, both (and especially the latter) being of high relevance for trace organic analysis.

3.1 Column technology.

Column efficiency may be expressed in terms of the van Deempter equation:

$$H = A + B/u + C \cdot u \qquad (2)$$

where H: column distance equivalent to a theoretical plate.
u: average mobile phase velocity.
A: term related with eddy diffusion.
B: term related with longitudinal diffusion.
C: term related with mass transfer.

This equation may be used for the description of both GC and LC separation processes. The constants A, B and C are related with the characteristics of the column and the mobile phase. Several formulae can be written accordingly (see for instance Poole and Schuette (1984)).

The properties of the mobile phase are especially reflected in the C term because its contribution to the mass transfer resistance is in inverse proportion with the diffusion coefficient of the eluting mixture. Therefore, LC efficiency is considerably limited by the low diffusion coefficients of the liquids whereas this aspect does not represent any major problem in GC.

A way to overcome this limitation is by reduction of the amount of mobile phase in the LC system which involves a decrease in column dimensions. Three main types of small diameter columns have been developed columns with internal diameters of 0.1 to 1 mm (microbore columns), packed columns of smaller diameter (packed capillaries) and wall coated columns with diameters below 0.1 mm (open-tubular capillary columns) (Jorgenson, 1984). The low volumes involved in these dimensions allow to use exotic mobile phases such as perdeuterated or chiral solvents. However, in these cases extra-column band broadening becomes a major problem preventing that the traditional LC equipments can be used. These difficulties are especially important for the injection system, involving strong limitations in terms of sample volume or amount. An additional drawback is the long time usually needed to achieve high separation efficiencies (in the order of hours).

Reduction of packing particle size (i.e. 3 um) has been successful for the increase of the separation speed in columns of regular diameter (10 cm x 4.6 mm i.d.). This approach does also allow to attain some increase in efficiency but the high operational pressures required represent a strong limitation for column length. Therefore the application of these small particle-size packings has been essentially devoted toward analytical speed increase.

The most usual LC columns (30 cm x 4 mm i.d.) are packed with 10 um monomeric octadecylsilane adsorbents yielding efficiencies of about 5000 theoretical plates. Samples are usually run under solvent programming where the solvent polarity is currently increased following the eluotropic series. Typical reversed-phase separations are carried out with the system of methanol-water or acetonitrile-water. In this type of separations the chemical components of increasing hydrophobicity are gradually eluted as the amount of the organic solvents increases in the mobile phase. Special additives are sometimes mixed to these aqueous eluents for the separation of solutes with ionic groups. Ionic salt solutions or counter ions ("ion-pair" technique) are commonly used.

Reversed-phase chromatography is the LC technique of wider application for trace organic analysis. However, LC includes other separation methods such as direct-phase liquid-liquid chromatography, liquid-solid, size exclusion and ion-exchange. Among these the former is often used as a clean-up technique for the isolation of pollutants in chemical-class fractions prior to analysis by LC reversed-phase or GC. Standard separations are performed with cyano- or, more commonly, aminoalkylsilane columns eluted with n-hexane. Liquid-solid chromatography (generally silica columns eluted with n-hexane) has been also used for the same purpose. However, it has lower retention time reproducibility and more problems of irreversible adsorption than the bonded phases described above. More information on recent developments for LC column separation is given in Ahuja (1986).

3.2 Detectors.

The other major problem for the LC analysis of trace components concerns detection. LC lacks of any completely general detector whose response is proportional to the mass of the eluting solute.

The detection technique used most widely is UV absorption. In principle, the wavelength interval in the 180-210 mn is fairly universal but the lack of transparency of the regular mobile phase solvents at this spectral region makes its use unpractical. 210-260 mn is therefore the UV range of wider application. Sensitivity obviously depends on the molar absorptivity of the solutes at the selected wavelength. At this respect, UV-absorption and UV-fluorescence, are very useful for the determination of some types of trace component such as polycyclic aromatic hydrocarbons (PAHs) (Lee et al. 1981). These compounds are strong absorbers of UV light and they can be selectively analyzed even when present in mixtures containing other components. In addition, repeated analyses of the same mixture at different absorbance wavelengths allows the enhancement of selected aromatic hydrocarbons. Further information on the analysis of PAH is available in Sander and Wise (1986), Bjorseth (1983) and White (1985).

The LC detection problems are not restricted to the lack of a suitable general detector, there is also a shortage of sensitive specific detectors for the case in which the UV absorption and fluorescence spectrophotometers are not applicable. This is especially evident, for instance, in the analysis of organophosphorous pollutants. Many of these components are thermally labile, which recommends LC for their determination, but their UV absorption bands (i.e., malathion, dimethoate, omethoate, parathion, paraoxon,...) occur at wavelengths below 210 nm (Osseton and Smelling 1985) where most elution solvents interfere. This has encouraged the application of selective GC detectors to LC. However, the substitution of the regular GC gases by aqueous and organic solvents at ml/min flow rates has produced numerous technological problems. Many of them have been overcomed by column miniaturization which, as it has been described above, is also useful for the improvement of the separation efficiency. This approach is actually at a level of research. The coupling of microcolumn LC with electron capture, flame photometric and thermionic detectors have been reported in the latter years (McGuffin and Novotny, 1981; Maris et al., 1985; Gluckman et al., 1986, Barceló et al., 1987).

4. GAS CHROMATOGRAPHY-MASS SPECTROMETRY (GC-MS).

GC-MS represents the fruitful combination of an efficient separation technique with a powerful identification tool. Actually it is becoming a routine technique in many environmental laboratories. The use of capillary columns has allowed direct GC-MS coupling without any auxiliary interface representing a great improvement in terms of easiness of operation and avoidance of multicomponent sample discrimination. Lower costs, simplicity in maintenance and repair and the increasing list of trace organic compounds to be monitored have encouraged its extended utilization. The gas chromatographic process has been described in Section 2. Here we will consider the main factors related with mass detection.

Two major aspects determine the mass spectral characteristics: the mode of ionization and the method of mass fragment analysis.

4.1. Modes of ionization.

Ionization may be performed in several ways: electron impact, chemical ionization, atmospheric pressure ionization, field emission or field desorption, fast atom bombardment, californium-252 plasma desorption, etc. According to the mode selected, positive or/and negative ions may be studied. In the analysis of trace pollutants two methods are mostly used: electron impact and chemical ionization.

In **electron impact** the molecules are ionized by inelastic collision with electrons produced from a heated tungsten or rhenium filament. Electron energies of 70 eV are currently used. Extensive fragmentation of the organic molecules is usually produced, their ionization potential being in the order of 7 to 20 eV. The majority of ions generated are positive singly-charged and correspond either to fragments or to the parent molecule. A few multiply-charged ions or negative ions may also be formed. The positive ions are extracted from the ion source by a repeller electrode of positive potential which directs them towards the analyzer. A series of charged plates (called lenses) improve the ion extraction and limit the angular spread of the emerging ion beam. This mode of ionization is the most commonly used because of its stability, ease of operation and the qualitative information that can be deduced from the mass fragmentation process.

In **chemical ionization** the mass spectra are produced by ion-molecule reactions occurring between the sample molecules and a high pressure gas (0.2-2 Torr) of reactive ions which is generated from a gas submitted to electron impact. The chemical ionization sources are usually operated at much higher electron energies (200-500 eV) to ensure penetration of the ionizing electrons into the high pressure gas. In these ion-molecule reactions the electric charge is passed to the sample molecules by chemical interaction, a process involving low energies. Therefore stable molecular ions or molecular ion adducts with little additional fragmentation are usually formed. Most of these reactions are produced by proton transfer which involves that the reagent gas must usually be of lower proton affinity than the analyzed sample. Table 2 shows some of the most commonly used reagent gases, their main ion products and their proton affinities. Because of its relatively low proton affinity and easiness of operation methane is generally selected as first choice gas.

Table 2. Main reagent gases used in chemical ionization mass spectrometry.

Gas	Proton affinity (Kcal/mol)	Predominant ions (at 1 Torr)
Helium	42	He
Methane	127	CH_5^+, $C_2H_5^+$, $C_3H_7^+$
Water	167	H_3O^+
Isobutane	195	$C_4H_9^+$
Ammonia	207	NH_4^+, $(NH_3)_2H^+$, $(NH_3)_3H^+$

Besides ionization by ion-molecule reactions, the high pressure in the ion source induce the formation of thermal electrons which may be captured by molecules with high electron affinities, a process leading to the generation of **negatively-charged molecular or fragment ions**. Due to this high pressure effect the intensity of the negative ion beam is usually three orders of magnitude higher than that obtained by electron impact. The analysis of the negative ions require that the potential of the repeller electrode and the lenses must be reversed. In this sense, the quadrupole instruments may be operated in pulsed positive and negative ion chemical mode for the quasi-simultaneous recording of both ion patterns. In general terms, the positive ion GC-MS profiles parallel FID gas chromatograms and the negative ion profiles parallel the GC traces obtained with ECD.

4.2 Mass analyzers.

Mass analysis in trace organic GC-MS has been usually performed with either magnetic or quadrupolar detectors. Recently, mass analyzers based on the ion trap principles have gone into operation.
In a **magnetic field**, the ions of different mass-to-charge ratio are deflected according to the expression:

$$m/z = r^2 \cdot B^2/2 \cdot V \qquad (3)$$

where m: ion mass
    z: ion charge
    r: radius of the trajectory
    B: magnetic field strength
    V: accelerating voltage

The radius of the circular trajectory is usually fixed so that a particular m/z fragment may only reach the detector at certain B and V. V is usually held constant to give the same kinetic energy to all fragments (otherwise ion discrimination is produced) and therefore mass elucidation is performed by modification of the magnetic field strength (B). The rates at which B may be modified are however limited by magnetic hysteresis. Modern laminated magnets may be scanned at 1-2 s per decade but they usually need a time to reset and stabilize (1-1.5 s). In some cases, these restrictions may result in too low scan speeds for the narrow shape of many capillary gas chromatographic peaks.

Quadrupole mass detectors consist in four parallel rods in a square array connected to a radio frequency (rf) and to direct current voltages (dc). The rf frequency is held constant between 1500-3000 KHz with a peak value ($\sim$6000 V) higher than that of the dc voltage ($\sim$500 V). During operation the adjacent

rods are simultaneously excited with equal but opposite voltages. In
consequence, time variable voltages are generated in the xy plane
perpendicular to the rods and no force is produced along the rod's direction
(the z axis). For a given radio frecuency/direct current voltage ratio only
ions of a specific m/z value may pass through the exit aperture and reach the
detector, the others strike the rods and loss their charge. The quadrupole
acts, in fact, as a mass filter. The mathematic expressions determining which
mass fragment will pass for a given rf and dc are derived from the Mathieu
equations; they are complex and beyond the scope of this chapter. In general
quadrupoles are operated at a fixed radio frequency. Mass selection is usually
obtained by linear scanning of the rf amplitude. At this respect, the
quadrupoles are mainly electrostatic devices which involves very low
inductances and capacitances allowing, consequently, high scan rates (in the
order of milliseconds).

In comparison with the magnets, the quadrupoles are considered to discriminate
at higher masses. This is theoretically supported by the absence of electric
field in the z direction which determines that all mass fragments only depend
on their kinetic energy to reach the detector, as it is indicated by
expression (4).

$$e \cdot V = \tfrac{1}{2}\, m \cdot v^2 \qquad (4)$$

where e: charge of one electron
      V: accelerating voltage
      m: mass of the molecular fragment
      v: the linear velocity

Obviously, in these conditions higher masses will have lower velocities and
their effective transmission may be reduced. However, in our experience,
within the limits of the components able to elute from a gas capillary column,
this effect is not important. It may be easily overcomed by adequate MS
callibration.

Ion trap mass spectrometers have been recently introduced within the field of
GC-MS analysis because of their low cost and easy maintenance. The analyzer
is composed by a quistor, a small "ion bottle" composed of a hyperboloid ring
electrode and two end caps. The end caps are electrically connected and dc and
rf potentials are applied between them and the ring electrode. The values of
the amplitude voltages and radial frequency determine whether the ion motions
will remain stable within the dimensions of the trap or whether they will be
ejected towards the electron multiplier through a hole in one of the end caps.
Mass selection is achieved by adequate ramping of the rf voltage amplitude.
The resolution of the process is increased by the damping effect of helium
which is present within the quistor at a pressure of 10^{-3} Torr.

Several types of ion trap MS are today available. They do not need a separate
ion source and lensing system because the ions are also generated within the
quistor. This obviously results in easier maintenance. After adequate tunning
some of them produce comparable mass spectra to quadrupole or magnetic MS.
Furthermore, under full scan operation, they are even more sensible. Unlike
quadrupoles they do not work as mass filters, they accumulate most of the ions
generated from the sample during the whole scan period, which results in
higher m/z abundances at the detector. Ion trap MS are designed for
chromatographers who are not MS experts and who cannot afford sophisticated
mass spectrometers.

4.3 Modes of operation.

When the GC-MS system is coupled to a computer (the regular case), in adition to the **mass spectra** of the chromatographic peaks, **mass chromatograms** of any desired **mass fragment** within the selected scan range can be generated. However, an alternative way of obtaining such profiles is to focus on one m/z value characteristic of the compound under consideration. This mode of operation is called **single ion monitoring** (SIM) and uses the mass spectrometer as a detector, but a sensitive and selective one. SIM sensitivity is higher than the regular scan mode sensitivity because the mass spectrometer monitors only a few m/z values instead of scanning a wide atomic mass unit range. In another mode of operation the mass spectrometer jumps sequentially over a number of present m/z values. This method, in which more than one ion is monitored, is called **multiple ion detection** (MID). Sensitivity is increased also in this mode because of the longer times spent at each chosen mass.

Further information on recent developments in MS is available from Harrison (1983), Message (1984), Watson (1985), Merritt and McEwen (1980) and Constantin and Schnell (1986).

5. HYPHENATED METHODS.

The hyphenated techniques result from the combination of a separation device with a multiparametric detector. A computer is usually needed for data acquisition and interpretation as well as for instrumental control. The GC-MS represents the most developed example of this type of instrumentation but many other combinations are already commercially available, such as liquid chromatography coupled to mass spectrometry (LC-MS), gas chromatography coupled to infrared spectroscopy (GC-IR), mass spectrometry coupled to mass spectrometry (MS-MS) and others. GC-MS has already been discussed in the previous section. We will briefly consider here these latter three techniques, paying special attention onto their application to trace organic analysis.

5.1 Liquid Chromatography-Mass Spectrometry (LC-MS).

The advantages of coupling LC to MS are evident, especially when considering the lack of suitable detectors in LC outlined in section 3.2. However, important problems of compatibility must be addressed. These concern the vaporization of the LC effluent to overcome the basic mass flow incompatibility between LC and MS and the use of a ionization technique not requiring the thermal vaporization of the sample. Thermosphray and the moving belt are the interface techniques most commonly used for their solution.

Thermospray introduction consists in the controlled heating of a capillary tube for the partial vaporization of the chromatographic eluent as it passes through the capillary. Under adequate heating a supersonic jet of vapor containing a mist of fine droplets or particles is formed. The solute molecules become ionized in the process through a not completely understood mechanism paralleling that of field desorption. In principle no external electric field is needed, the high charge density acquired by the droplets during vaporization and shrinking gives rise to the formation of a high proportion of molecular ions with relatively little fragmentation. The presence of an electrolyte in the mobile phase is critical for this soft ionization technique. Alternatively, the system can be operated under the ionizing electrons emitted by a filament. In this case, it works like a conventional chemical ionization source where the solvent of the mobile phase performs the role of reagent (Voyksner and Haney, 1985; McFadden and Lammert, 1987; Alexander and Kebarle, 1986; Bursey et al, 1985).

Thermospray interfacing has proved to be successful for polar and apolar non volatile components of medium to low molecular weight, such as many herbicides and pesticides containing nitrogen or phosphorous groups (Voyksner et al., 1984a and b; Covey et al., 1985). Both positive and negative ions can be recorded depending on the electron affinity of the solutes (Barceló et al., 1987). The chemical ionization process may be modified by addition of specific components to the mobile phase such as electrolite solutions (ionization enhancement) or electronegative compounds (increase of negative ion yield) (Parker et al., 1986; Geerdink et al., 1987).

In the **moving belt** the sample matrix and the ionization process are independent from the LC solvent and conditions used. This represents an important differential, and often advantageous, feature with respect to the thermospray interface in which the vaporization and ionization environment is strongly affected by the LC effluent. The moving belt was initially used for direct deposition of the mobile phase, thermal vaporization under vacuum and ionization either by electron impact or chemical methods. These type of applications involved that some degree of volatility for both solvents and solutes was required and, in this sense, special difficulties were encountered in the use of aqueous mobile phases. These type of problems, especially those derived from the use of water solutions, are much better resolved by thermospray interfacing. The actually expanding field of application of the moving belts concerns very large molecules which, after spray deposition and solvent evaporation, are ionized by either fast atom bombardment or laser or ^{252}Cf desorption (Hardin et al., 1984a and b). This technique offers a great potential for the analysis of the largely unknown fraction of the unvolatile high molecular weight pollutants which is currently ignored because of lack of adequate analytical instrumentation.

5.2 Mass spectrometry-Mass spectrometry.

Tandem mass spectrometry (MS-MS) is gaining wide acceptance in the analytical community. These instruments are essentially composed by an ion source, two mass analyzers separated by a fragmentation region, and an ion detector. When a mixture of components is introduced into the sample probe characteristic ions representing the individual constituents are formed. These are separated by the first mass analyzer which allow to select an m/z fragment specifically representative of the analyte. This parent ion may be dissociated further in the fragmentation region by collision with neutral gas molecules yielding diverse daughter ions which, in turn, are analyzed by the second mass analyzer.

Several modes of operation are possible, daughter scan, parent scan, neutral loss scan and selected reaction monitoring. In the first two modes one of the mass analyzers is held at constant m/z and the other is scanned over a determined scan range. In neutral loss scan both analyzers are scanned keeping a constant difference in mass, and in selected reaction monitoring a limited number of parent-daughter ion pairs are examined for each analyte. The latter mode is analogous to selected ion monitoring in GC-MS and gives the highest sensitivity in trace organic analysis.

Quadrupole, magnetic sector, electric sector, time-of-flight, ion cyclotron resonance analyzers and their combinations have been used. The most common MS-MS set ups have been the **mass-analyzed ion kinetic energy spectrometer** (MIKES) and the triple-quadrupole mass spectrometer (Yost and Enke, 1979; Beynon et al., 1973). The former consists of a reversed geometry double focussing instrument in which the ions traverse a magnetic sector before the electric sector. The second is formed by three serial quadrupole mass filters where the intermediate, an rf only, acts as a collision chamber and a focussing device. MIKES instruments, however, are of limited application for the analysis of complex mixtures because they produce spectral artifacts and have less-than-unit mass resolution. Conversely, triple quadrupole MS-MS have unit mass resolution for both the parent and daughter ions over the whole mass range not giving rise to spectral artifacts.

The high selectivity that results from the use of MS-MS allows important increases in speed of analysis (sample preparation and chromatographic separation are avoided) and limit of detection (in the order of picograms or femtograms) in complex samples.

5.3 Gas chromatography-Infrared spectrophotometry (GC-IR).

GC-IR has come into operation when a reasonable solution has been given to two main problems: Scan speed and sensitivity. The first has been resolved by application of the already available Fourier Transform IR interferometers instead of the grating monochromators used in current IR spectroscopy. The second has required the development of two new breakthroughs, one devoted to obtain photometers of higher sensitivity and the other to increase the path length of the IR beam within the sample cell. Thus, higher specific photometer sensitivity (about one order of magnitude) has been obtained with the introduction of the liquid nitrogen cooled narrow-range mercury cadmium telluride photodetectors in substitution of the triglycine sulfate pyroelectric bolometers currently used in FT-IR. On the other hand, the IR beam path length has been extended by the use of "light pipes": 1-2 mm i.d. up to 80 cm long borosilicate glass tubes internally coated with gold. Alkali halide windows are fixed to the pipe ends in order to measure the transmitted IR light during the passage of the GC column effluents.

Modern GC-IR instruments can record IR spectra in a range of 4000-750 cm^{-1}(and even in the 4000-500 cm^{-1} range) with a resolution of 8 cm^{-1} in less than 1s with the detection limits below the microgram. A computer controls the data acquisition process and transforms the recorded interferograms into chemically interpretable spectra. Besides the IR spectra of the individual peaks or the GC profile corresponding to a fixed wavelength, it is also possible to obtain a reconstructed chromatogram respresenting the total infrared response of the eluting species. This is called the Gram-Schmidt chromatogram because it is based on the Gram-Schmidt vector orthogonalization method.

GC-IR is a technique giving useful complementary information to GC-MS. The identification of functional groups and positional isomers is often much easier by IR than by MS. Conversely, MS provides information on the molecular weight and skeletal fragments which is also needed for qualitative analysis. In this sense, since IR detection is not destructive, the serial coupling of IR and MS has been developed giving rise to GC-IR-MS. However, the main problem related with the use of GC-IR is still the low sensitivity, an aspect particularly important for trace organic analysis.

A matrix isolation interface, the cryolect, has been recently developed to improve this aspect. With this system the effluent of the GC column is splitted in two parts (He with 1-2% Ar is added). One part goes to a conventional GC detector (i.e. FID) and the other is directed towards a 12° K cooled rotating mirrored cylinder where the Ar atoms and sample molecules condense. This frozen matrix is a cryogenically trapped version of the chromatogram which may be sequentially scanned by FT-IR when the run is over. Each peak may be observed for as long as desired and the computer may be programmed for long time signal average to improve sensitivity. Detection limits in the order or below those of GC-MS are claimed. An additional feature of the cryolect interface is that it provides higher spectral resolution than that corresponding to light pipe systems because at 12°K the band broadening and band shifting effects derived from intermolecular bonding and rotational absorption of the sample molecules are highly diminished.

Further reading on GC-IR is available in Gurka and Betowski (1982), Smith and Adams (1983), Hohne et al. (1981), Gurka et al. (1982), Griffiths et al. (1983) and Gurka et al. (1984).

6. LUMINESCENCE SPECTROSCOPY.

Luminescence encompasses the emission of light by excited chemical species. Depending on the type of energy of excitation and the light emitted several phenomena are produced such as fluorescence, phosphorescence and chemiluminescence. These are besteaded for analytical purposes giving rise to several instrumental techniques in which different types of spectra can be measured (emission, excitation, synchronous, first or second derivative, time-resolved, total luminescence, etc.) using various light sources (mercury-vapor or deuterium lamps, lasers, etc.) for the irradiation of samples dispersed in diverse matrices (dissolved, adsorbed on silica gel, alumina or filter paper) in a wide range of temperatures (12-300°K) (Wehry, 1981). Among all these variants the instrumental applications related with UV-fluorescence are those of more extended use in trace organic analysis.

6.1 UV-fluorescence.

UV-fluorescence is based on the absorption of optical radiation by chemical species and the resultant deactivation of the excited molecules with the release of electromagnetic energy. Accordingly, the emission spectra are produced by recording the emission intensity as a function of the emission wavelength, λ_{em}, when the sample is excited at a fixed wavelength, λ_{ex}. Conversely, the excitation spectra are obtained when λ_{ex} is scanned while the observation is conducted at a fixed λ_{em}. Finally, synchronous excitation spectra consist in the simultaneous scanning of λ_{ex} and λ_{em}, keeping a constant wavelength interval $\Delta\lambda$ throughout the measurement. The utility of UV fluorescence for trace organic analysis is due to its high sensitivity and specificity, allowing quick determinations of hydrocarbon pollutants without the need of extensive sample clean-up. Furthermore, UV-fluorimeters are of relative low cost and easy to operate. However, this technique also presents several drawbacks such as the short range of linearity and the possible interferences in multicomponent samples by signal addition or substraction (quenching) to the fluorescence intensity of the analyte. These and other instrumental or methodological problems are discussed by Miller (1981).

UV-fluorescence is also applied to the analysis of polynuclear aromatic hydrocarbons (PAH). However, as it commonly occurs with most large molecules, the electronic spectra of these components in liquid solution is broad and structureless which limitates the possibilities of individual determination in PAH mixtures. The use of a previous separation step such as liquid chromatography represents a useful solution for avoiding many multicomponent interferences. In fact, UV-fluorimeters are commonly used as LC detectors. Nevertheless, the specific problem of spectral resolution has been overcomed to a large extend by taking advantage of the Shpol'skii effect.

6.2 Shpol'skii UV-fluorescence.

The Shpol'skii effect occurs when a sample is dissolved in certain solvents
(usually n-alkanes) and kept at 77°K or below. A polycrystalline structure of
n-alkane solvent molecules is formed in these conditions and the analyte is
located into substitutional sites. Thus, very similar microenvironments are
produced for each individual sample molecule which is reflected in a large
decrease of the inhomogeneous band broadening. The resulting high spectral
resolution is commonly sufficient to permit unambiguous identification of
individual PAH and therefore Shpol'skii fluorescence is used for the
qualitative analysis of hydrocarbon mixtures. In this sense, the technique is
especially useful for the identification of methyl-substituted PAH isomers
which in many cases display the same room temperature fluorescence and mass
spectra.

7. ATOMIC SPECTROSCOPY.

7.1 Atomic absorption.

Atomic absorption spectrophotometry is by far the most commonly used method
for the analysis of trace metals in environmental samples. The technique
involves the absorption of light of a specific wavelength by atomic species of
the element as it is excited in a flame or other thermal device. The amount of
light observed by the atomic species is proportional to the concentration of
the element present in its ground state. The light source used is a hollow
cathode lamp, with the cathode constructed of the same element as that under
analysis, or an electrodeless discharge lamp which contains a small quantity
of the element to be determined. The latter type is of longer life and
produces a better signal-to-noise ratio, especially for the more volatile
metals, although it needs a special power supply and their conditioning time
is longer.

Various atomization techniques are used, including aspiration of aqueous
solutions into a controlled flame, carbon rods, graphite furnaces and hydride
generation. In general the flameless techniques offer higher sensitivities
than does the flame atomization method (see Table 3). However sensitivity is
limited by the molecular absorption and scattered light background caused by
the portion of the sample matrix that cannot be destroyed during the ashing
step prior to atomization. This effect is compensated by substraction of a
reference background signal coming from either a continuous source (deuterium
lamp) or, better, the magnetically induced distortion of the absorbance
wavelength of the analyte (Zeeman effect).

7.2 Atomic emission.

Besides the measurement of the absorbed light, atomic spectroscopy can also be
operated by monitoring the emission or the fluorescence signals. This
obviously depends on the amount of atomic species in excited or ground state
during analysis. With thermal excitation, the populations of excited (N_j) and
ground (N_o) state atoms may be expressed by the formula

$$\frac{N_j}{N_o} = \frac{P_j}{P_o} \cdot e^{-E_j/KT} \qquad (5)$$

where P_j and P_o are the statistical weights of the excited and ground states.
E_j is the energy of excitation,
K the Boltzmann constant
T the absolute temperature.

Table 3. Detection limits for several elements determined by atomic absorption spectroscopy, by inductively coupled plasma-atomic emission spectrometry (ICP) and by anodic stripping voltammetry (ASV).

Element	Atomic Absorption			ICP	ASV	Element	Atomic Absorption			ICP	ASV
	Flame	Furnace	Hydride				Flame	Furnace	Hydride		
	(a)	(b)	(c)	(a)	(a)		(a)	(b)	(c)	(a)	(a)
Ag	0.9	0.1			0.05	Nb	1500			15	
Al	10	2		15		Nd	600				
As	100	20	0.3	50	0.15	Ni	2	6		10	
Au	5	10				Os	100	2500			
B	1000			4		Pb	10	2		30	0.01
Ba	8	15		0.5		Pd	20	20			
Be	2	0.5				Pr	5000				
Bi	15	10	5		0.005	Pt	40	20			
Ca	1	5				Rb	2	10			
Cd	0.5	0.1		2	0.01	Re	500				
Co	4	2		4		Rh	4	50			
Cr	2	0.3		4		Ru	70				
Cs	2					Sb	20	1	5		0.01
Cu	1	2		3	0.02	Sc	20				
Dy	50					Se	50	50	3	50	0.02
Er	40					Si	20	20			
Eu	20					Sm	60				
Fe	3	2		3		Sn	10	30	4	20	0.2
Ga	50	200				Sr	2	5			
Gd	1200					Ta	1000			15	
Ge	200					Tb	600				
Hf	2000					Te	15	10	5		
Hg	200	100	0.08	100	0.05	Th	10				
Ho	40					Ti	50	50		2	
In	20	20			0.03	Tl	9	10			0.1
Ir	800					Tm	10				
K	2	1				U	8000			50	
La	2000					V	40	2		4	
Li	0.5	5				W	1000			30	
Lu	700					Y	60				
Mg	0.1	0.02				Yb	5				
Mn	1	0.2		0.5		Zn	0.8	0.05		2	0.02
Mo	25	0.2		8		Zr	0.7			4	
Na	0.2	0.2									

a) ng/ml; b) pg; c) ng.

Accordingly, the relative number Nj/No of excited atoms increases exponentially with increasing temperature. Furthermore, since the energy of excitation is inversely proportional to the wavelength, the Nj/No ratio will decrease at decreasing λ. In general, no significant proportion of excited atoms is found below 3000°K for elements with absorption wavelength below 500 nm. On the other hand, the lifetime of the excited atoms is also important for the analytical application of the emitted light. In this sense, the experience has shown that absorption is more sensitive than emission for elements having an excitation potential higher than 3.5 eV.

Spectral interferences represent another aspect to be considered. They are important in emission measurements being practically unknown in absorption spectroscopy. Background emission from the flame and the sample matrix also influences the recorded signal in the emission mode.

In consequence, only few elements are analyzed from their emission lines in flame atomic spectrophotometers. Graphite furnaces have also been used in order to extend the field of application of the emitted light because the background radiation of species such as C_2, CH, CN and OH is smaller than that of the flame. However, the full utilization of the atomic emission spectroscopy has been achieved with another type of instrument, the inductively coupled plasma spectrophotometer.

7.3 Inductively coupled plasma spectrophotometry (ICP).

In this technique the sample is dispersed within a plasma of argon which is formed by ionization in a quartz tube placed inside the induction coil of an rf generator. The center of the plasma reaches temperatures in a range of 6000-8000°K. The combined effect of this high temperature and the relatively long residence time of the sample inside the plasma produces an efficient energy transfer from the ionized gas to the analyte giving rise to its atomization and excitation. Furthermore, at these high temperatures intense emission over the entire spectral range may be produced (see expression (5)). In addition, more excited states are populated so that the number of emission lines increases. Absorption spectra are, in contrast, much simpler because the transition lines originate from the background state.

In consequence ICP instruments are provided with a monochromator for wavelength selection and multielement analysis. This may be placed at a fixed position (simultaneous multielement determinations) or be driven by a stepping motor (sequential analysis). The latter case offers complete flexibility in the choice of lines for element and background measurements being used for quantitative and qualitative trace analysis. In general terms, ICP exhibits better detection limits than flame absorption, especially in the case of a number of refractory elements such as B, Ta, Ti, U, W, etc. (see Table 3). However, atomic absorption spectroscopy presents more than two orders of magnitude lower detection limits than ICP when performed with the graphite furnace. ICP instruments have been recently reviewed by Meyer (1987).

7.4 Atomic fluorescence.

Atomic fluorescence spectrometry parallels at the atomic level the molecular fluorescence technique already described in section 6. The frequencies emitted during radiation at suitable wavelength are characteristic of the atomic species. The intensity of the atomic fluorescence, and therefore the achievable detection limit, depends upon the number of atoms in the ground state and the intensity of the incident radiation source. For this reason, the analytical applications of this phenomenon are closely dependent on the introduction of newer and more intense radiation sources, namely electrodeless discharge lamps and lasers.

Further reading on atomic absorption spectroscopy is available from Price (1979), Pinta (1980) and Welz (1985)

8. ANODIC STRIPPING VOLTAMMETRY (AVS).

ASV involves the electrolytic separation and concentration of metals from aqueous samples to form a deposition or an amalgam on the working electrode, and the ulterior dissolution of the concentrated metal species. The separation (plating) step is performed at controlled potential and time of electrolysis and at reproducible mixing conditions of the test solution. The re-dissolution (stripping) step is usually achieved by application of ramped or pulsed anodic potentials of enough magnitude for the reversal of the electrolysis reactions. The current intensity is monitored and plotted versus voltage during stripping which produces a series of peaks allowing the identification of the metals by their half-wave potentials and quantitation by the heights of the peak current intensities.

This technique requires an instrumental set-up composed by a power source capable of aplying potentials in the range of 0 to 3 V and an electrolysis cell containing a working electrode, a reference half-cell electrode and an auxiliary or counter electrode. The counter electrode serves to perform controlled potential electrolysis and eliminates the problems associated with ohmic drop in the test solution. The plating/stripping cycles are performed in the working electrode. These may be of several types: hanging mercury drop, thin mercury film-graphite, gold, etc. Thin mercury-film graphite electrodes have been applied for trace metal analysis in seawater, fresh waters, wastewaters and drinking waters. Direct and simultaneous determinations of Zn, Cd, Pb, Cu, Sb and Bi in seawater have been performed by differential pulsed ASV with a hanging mercury drop electrode (Guillain and Duyckaerts, 1979). In, Ni, Co, Tl and others can also be analyzed.

The advantage of ASV for the analysis of trace metals in aqueous samples lies on its high sensitivity with respect to other instrumetal methods (see Table 3). The technique is especially useful for the analysis of seawater because the high concentration of chlorine and other ionic species represents an important interference in techniques such as atomic absorption. Conversely, in ASV the presence of dissolved ions is required in order to provide a suitable conductivity to the analyzed samples. In fact, ionic solutions must be added to freshwater samples for their analysis by ASV. Therefore, no preconcentration step or separation of the metal species from the saline solutions are needed with ASV.

In comparison to other electroanalytical techniques ASV has the advantge of acting as a concentration method during the plating step so that the concentration of the metal species into the amalgame is higher than in solution. The limits of detection may be increased by extending the time of electrolysis. In consequence, ASV affords higher sensitivities than direct electroanalytical methods such as polarography or ion-selective potentiometry.

Further reading on ASV and other electrochemical techniques is available in Plambeck (1982), Kissinger and Heineman (1984) and von Rach and Seiler (1985)

9. CONCLUDING REMARKS.

A wide diversity of instrumental techniques is available for trace pollutant analysis but further development is needed since an important number of problems of environmental analysis still remain unresolved or their solution is difficult.

In many cases, however, the problem to be solved consists in the adequate selection of the most suitable technique for the proposed study. It is clear that the technique of choice will not only depend on the component to be analyzed but also on the type of sample and the extraction and clean-up procedures to be used.

Several criteria must be considered such as the required limit of detection, the number, concentration and type of interferences, the needed degree of specificity, the qualitative discrimination power, the quantitative accuracy and precision, the instrumental availability, the time of analysis and the cost. Another aspect to be taken into account is the confirmation of the analytical data obtained that will currently require the use of two different techniques or, at least, the variation of the instrumental method.

Finally, data handling must be considered. The environmental studies easily generate large amounts of analytical results whose correct interpretation relies on the availability of adequate facilities for storage, accessibility and statistical treatment.

10. REFERENCES

Ahuja, S. (1984). Ultrahigh resolution chromatography. ACS Symposium Series. No. 250, Washington, 231 p.

Ahuja, S. (1986). Chromatography and Separation Chemistry. Advances and developments. ACS Symposium Series No. 297. Washington, 304 p.

Alexander, A.J. and Kebarle, P. (1986). Thermospray mass spectrometry. Use of gas-phase ion/molecule reactions to explain features of thermospray mass spectra, Anal. Chem., 58, 471–478.

Barceló, D., Maris, F.A., Frei, R.W., De Jong, G.J. and Brinkman, U.A.Th. (1987). Determination of trialkyl and triaryl phosphates by narrow-bore liquid chromatography with on-line thermionic detection. Intern. J. Environ. Anal. Chem., 30, 95–104.

Barceló, D., Maris, F.A., Geerdink, R.B., Frei, R.W., De Jong, G. J. and Brinkman, U.A. Th. (1987). Comparison between positive, negative and chloride-enhanced negative chemical ionization of organophosphorous pesticides in on-line liquid chromatography-mass spectrometry, J. Chromatogr., 394, 65–76.

Beynon, J.H., Cooks, R.G., Amy, J.W., Baitinger, W.E. and Ridley, T.Y. (1973). Design and performance of a Mass-Azalyzed Ion Kinetic Energy (MIKE) Spectrometer. Anal. Chem., 45 1023–1031A.

Bjorseth, A. (1983). Handbook of Polycyclic Aromatic hydrocarbons, Marcel Dekker, New York, 727 p.

Bursey, M.M., Parker, C.E., Smith, R.W. and Gaskell, S.J. (1985). Gas-phase ionization of selected neutral analytes during thermospray liquid chromatography-mass spectrometry, Anal. Chem., 57, 2597-2599.

Constantin, E. and Schnell, A. (1986). Spectrometrie de Masse. Lavoisier, Paris, 152 p.

Covey, T.R., Crowther, J.B., Dewey, E.A. and Henion, J.D. (1985). Thermospray liquid chromatography/mass spectrometry determination of drugs and their metabolites in biological fluids. Anal. Chem., 57, 474-481.

Dressler, M. (1986). Selective gas chromatographic detectors. Elsevier. Amsterdam, 319 p.

Geerdink, R.B., Maris, F.A., De Jong, G.J., Frei, R.W. and Brinkman, U.A. Th. (1987). Improved detection performance in liquid chromatography-negative chemical ionization mass spectrometry by using halogenated additives in the mobile phase, J. Chromatogr., 394, 51-64.

Gillain, G. and Duyckaerts, G. (1979). Direct and simultaneous determinations of Zn, Cd, Pb, Cu, Sb and Bi dissolved in sea water by DPASV with a hanging mercury drop electrode, Anal. Chim. Acta, 106, 23-37.

Gluckman, J.C., Barceló, D., De Jong, G.J., Frei, R.W., Maris, F.A. and Brinkman, U.A. Th. (1986). Improved design and applications of an on-line thermionic detector for narrow-bore liquid chromatography, J. Chromatogr., 367, 35-44.

Griffiths, P.R., De Haseth, J.A. and Azarraga, L.V. (1983). Capillary GC/FT-IR, Anal. Chem., 55, 1361A.

Grob, K. (1986). Classical split and splitless injection in capillary GC. Huething Verlag, Heidelberg, 324 p.

Grob, K., Grob, G. and Grob, K. (1981). Capillary columns with immobilized stationary phases. I. A new simple preparation procedure, J. Chromatogr., 211, 243-246.

Gurka, D.F. and Betowski, L.D. (1982). Gas chromatographic/Fourier Transform Infrared Spectrometric Identification of Hazardous Waste Extract Components. Anal. Chem., 54, 1819-1824.

Gurka, D.F., Lasaka, P.R. and Titus, R. (1982). The capability of GC/FT-IR to identify toxic substances in environmental sample extracts. J. Chromatogr. Sci., 20, 145.

Gurka, D.F., Hiatt, M. and Titus, R. (1984). Analysis of Hazardous Waste and Environmental extracts by capillary gas chromatography/Fourier Transform Infrared Spectrometry and Capillary Gas Chromatography/Mass Spectrometry. Anal. Chem., 56, 1102-1110.

Hardin, E.D., Fan, T.P. and Vestal, M.L. (1984a). Laser desorption mass spectrometry with thermospray sample deposition for determination of nonvolatile biomolecules. Anal. Chem., 56, 2-7.

Hardin, E.D., Fan, T.P. and Vestal, M.L. (1984b). Direct comparison of secondary ion and laser desorption mass spectrometry on bioorganic molecules in a moving belt liquid chromatography/mass spectrometry system. Anal. Chem., 56, 1870-1876.

Harrison, A.G. (1983). Chemical ionization mass spectrometry, CRC Press, Boca Raton.

Hohne, B.A., Hangac, G., Small, G.W. and Isenhour, T.L. (1981) An On-line class-specific GC/FT-IR reconstruction from interferometric data. J. Chromatogr. Sci., 19, 283.

Jorgenson, J.W. Trends in analytical scale separations (1984). Science, 226, 254-261.

Kissinger, P.T. and Heineman, W.R. (1984). Laboratory techniques in electroanalytical chemistry. Marcel Dekker, New York, 751 p.

Lee, M.L., Novotny, M.V. and Bartle, K.D. (1981). Analytical chemistry of polycyclic aromatic compounds, Academic Press, New York, 462 p.

Lee, M.L., Yong, F.J. and Bartle, K.D. (1984). Open tubular column gas chromatography: Theory and Practice, John Willey, New York, 445 p.

Maris, F.A., Geerdink, R.B. and Brinkman, U.A. Th. (1985). Selection of mobile phases for reversed-phase liquid chromatography with on-line electron-capture detection. J. Chromatogr., 328, 93-100.

McFadden, W.H. and Lammert, S.A. (1987). Techniques for increased use of thermospray liquid chromatography-mass spectrometry, J. Chromatogr., 385, 201-211.

McGuffin, V.L. and Novotny, M.V. (1981), Flame emission detection in microcolumn liquid chromatography. Anal. Chem., 53, 946-951.

McLafferty, F.W. (1984). Trends in Analytical Instrumentation, Science, 226, 251-253.

Merritt, Ch. and McEwen, Ch.N. (1980). Mass spectrometry. Part A and B, Marcel Dekker. New York.

Message, G.M. (1984). Practical aspects of gas chromatography/mass spectrometry, John Wiley, New York, 351 p.

Meyer, G.A. (1987). ICP. Still the panacea for trace metal analysis? Anal. Chem., 59, 1345-1354A.

Miller, J.N. (1981). Standards in Fluorescence Spectrometry, Chapman and Hall, London, 115 p.

Osselton, M.D. and Snelling, R.D. (1985). Chromatographic identification of pesticides. J. Chromatogr. 368, 265-271.

Parker, C.E., Smith, R.W., Gaskell, S.J. and Bursey, M.M. (1986). Dependence of ion formation upon the ionic additive in thermospray liquid chromatography/negative ion mass spectrometry., Anal. Chem., 58, 1661-1664.

Pinta, M. (1980). Spectrométrie d'absorption atomique. Vol. I and II, Masson, Paris 696 p.

Plambeck, J.A. (1982). Electroanalytical chemistry: Basic principles and applications. John Wiley and Sons, New York, 404 p.

Poole, C.F. and Schuette, S.A. (1984). Contemporary practice of chromatography, Elsevier. Amsterdam, 708 p.

Price, W.J. (1979). Spectrochemical analysis by atomic absorption. Heyden and Sons, London, 392 p.

Rach von P. and Seiler, H. (1985). Polarographie und Voltammetrie in der Spurenanalytik. Hüthing Verlag, Heidelberg, 99 p.

Sander, L.C. and Wise, S.A. (1986). Investigations of selectivity in RPLC of polycyclic Aromatic hydrocarbons, in Advances in Chromatography, J.C. Giddings, E. Grushka, J. Cazes and P.R. Brown, eds. Marcel Dekker, New York, 139-218.

Sandra, P. (1985). Sample introduction in Capillary Gas Chromatography, Huething Verlag, Heidelberg, 265 p.

Schoenmakers, P.J. (1986). Optimization of chromatographic selectivity. Elsevier, Amsterdam, 345 p.

Smith, S.L. and Adams, G.E. (1983). Chromatographic performance and capillary gas chromatography-fourier Transform infrared spectroscopy. J. Chromatogr., 279, 623-630.

Tranchant, J. (1982). Manuel practique de chromatographie en phase gazeuse. Masson, Paris, 504 p.

Voyksner, R.D. and Haney, C.A. (1985). Optimization and applications of thermospray high performance liquid chromatography/mass spectrometry, Anal. Chem., 57, 991-996.

Voyksner, R.D., Bursey, J.T. and Pellizari, E.D. (1984a). Postcolumn addition of buffer for thermospray liquid chromatography/mass spectrometry identification of pesticides., Anal. Chem., 56, 1507-1514,

Voyksner, R.D., Bursey, J.T. and Pellizari, E.D. (1984b). Analysis of selected pesticides by HPLC-MS. J. Chromatogr., 312, 221-235 (1984b).

Watson, J.T. (1985). Introduction to Mass Spectrometry, Raven Press, New York, 351 p.

Wehry, E.L. (1981). Modern Fluorescence Spectroscopy 4 volumes. Plenum Press, New York.

Welz, B. (1985). Atomic absorption spectrometry. VCH, Weinheim, 506 p.

White, C.M. (1985). Nitrated polycyclic aromatic hydrocarbons, Huething Verlag, Heidelberg, 376 p.

Yost, R.A., Enke, C.G. (1979). Triple Quadrupole Mass Spectrometry: for Direct Mixture Analysis and Structure Elucidation. Anal. Chem., 51, 1251-1264A.

Sampling, Sample Handling, and Operational Methods for the Analysis of Trace Pollutants in the Marine Environment

Joan O. Grimalt
Department of Environmental Chemistry
CID (CSIC)
Jordi Girona, 18
08034-Barcelona, Spain

1. INTRODUCTION

Marine pollution, as defined in Chapter 1 involves a large variety of environmental aspects including the study of a diversity of physical, chemical and biological parameters which in many cases are not directly indicative of anthropogenic inputs but of mere changes in the natural properties of the aquatic system. Some of these parameters are listed in Table 1 where conservative and non-conservative water properties are included as well as chemical and biological determinations related with productivity, eutrophication, sanitary surveillance, etc. These parameters are usually measured by means of well-known methods described in several handbooks (Head, 1985; Aminot and Chausiepied, 1983).

Another aspect concerns the analysis of the components present at low concentration. In principle, the trace chemicals of interest refer to those included in the Priority Pollutants Lists (Keith and Telliard, 1979) which have already been discussed in Chapter 6. However, those of major concern in monitoring programs consist in the stable compounds or elements which tend to accumulate in the aquatic compartments, namely in sediments and biota, or cause deleterious effects. These are reported in Table 2 where the hydrocarbons, pesticides and metals most currently monitored are presented. Obviously, this is not an exhaustive list. Special site studies or monitoring programs may require the determination of other trace components. Furthermore, as more chemicals are used for newer applications and the level of information on their environmental impact increases more components must be considered in monitoring pollution studies.

From the analytical point of view the determination of trace components in the marine environment is by far more difficult than the study of general parameters such as those reported in Table 1. In this case, besides the use of instrumental techniques of high sensitivity required by the low concentrations to be measured, the complexity of the samples in terms of bulk and molecular composition represents that many problems related with interference effects and analyte recovery must also be addressed. For this reason, environmental analysis is at present essentially concerned with trace components determination. Accordingly, this paper will be focussed onto sampling techniques, sample handling procedures and operational methods devoted to the analysis of trace pollutants. It is not involved, however, that the importance of general parameters for monitoring studies is here underestimated.

TABLE 1

GENERAL PARAMETERS OF POTENTIAL INTEREST
FOR MARINE POLLUTION STUDIES

PHYSICAL: Temperature
Turbidity
Suspended matter

CHEMICAL: Salinity
pH
Dissolved oxygen
Chemical oxygen demand (COD; $KMnO_4$ titration)
Biological oxygen demand (BOD)
Nitrogen content: nitrates
nitrites
ammonia
organic nitrogen
particulate nitrogen
Phosphorous content: reactive phosphate
total phosphorous
polyphosphate
particulate phosphorous
Silicon content: silicates
total silicon
Clorophyll and phaeopigments
Dissolved organic carbon (DOC)
Particulate organic carbon (POC)

BIOLOGICAL: Microorganism determinations: total coliforms
faecal coliforms
faecal streptococci
Staphylococcus aureus
Pseudomonas aeruginosa
salmonella
viruses
Benthic population measurements: recruitment
biochemical stress
Pelagic population measurements

TABLE 2

TRACE COMPONENTS OF POTENTIAL INTEREST FOR MARINE POLLUTION MONITORING STUDIES

Petroleum Hydrocarbons
      Aliphatic petroleum hydrocarbons (ALPHs):
              n-alkanes (including carbon preference index)
              pristane
              phytane
              Unresolved complex mixture (UCM)
      Aromatic petroleum hydrocarbons (ARPHs):
              Total aromatics (UV fluorescence determination)
              Monocyclic
              Resolved
              Unresolved (UCM)
      Polycyclic aromatics
              Resolved
              Unresolved (UCM)

Polycyclic aromatic hydrocarbons (PAHs):

Naphthalene and alkylated homologs	Benzo(ghi)fluoranthene
Biphenyl	Benzo(c)phenanthrene
Acenaphthene	Benzo(j)fluoranthene
Fluorene	Benzo(k)fluoranthene
Phenanthrene and alkylated homologs	Benzo(b)fluoranthene
Anthracene	Benzo(a)pyrene
Dibenzothiophene and alkylated homologs	Benzo(e)pyrene
Pyrene and alkylated homologs	Binaphthyls
Fluoranthene and alkylated homologs	Benzo(ghi)perylene
Benzo(a)fluorene	Indeno(1,2,3,cd)fluoranthene
Benzo(b)fluorene	Indeno(1,2,3,cd)pyrene
Benzo(c)fluorene	Dibenzanthracenes
Chrysene and alkylated homologs	Picene
Benz(a)anthracene	Coronene

Chlorinated Compounds
    Polychlorinated biphenyls (PCBs)
    DDTs:

op'DDE; pp'DDE	Dieldrin
op'DDD; pp'DDD	Aldrin
op'DDT; pp'DDT	Hexachlorobenzene (HCB)
pp'DDMU	Hexachlorocyclohexanes (HCHs)
	α-HCH
	β-HCH
	γ-HCH (lindane)
	Heptachlor
	Heptachlor epoxide

Metals

Mercury	Lead
Cadmium	Nickel
Chromium	Iron
Cobalt	Manganese
Copper	Zinc
Selenium	Arsenic

At this respect some analytical terms should be remembered. **Instrumental techniques** derive from the interaction of fundamental (physics, chemistry, etc.) and applied (electronics, informatics, etc.) sciences. The instrumental techniques of utility for the analysis of trace components in the environment have been reviewed in the preceeding chapter. Their use for the analysis of chemicals (either single compounds or mixtures) give rise to the development of **operational methods.** The application of these methods to the analysis/determination of a compound or a mixture of compounds in environmental samples requires the use of suitable **sample handling procedures.** All these steps as well as the selection of adequate **sampling techniques** must be considered when organizing monitoring programs in marine areas.

As it has been shown in preceeding sections of this book, trace components are usually distributed in various environmental compartments of the aquatic system. Once contaminants are introduced into the sea they undergo complex changes depending on their physico-chemical properties and on their chemical reactivity. Each contaminant will be partitioned between the sediments, the aqueous and particulate phases in the water column, and the biota. Furthermore, they can be incorporated to the atmospheric aerosol by wind action or, in turn, enter into the marine system by aerosol deposition after long-range atmospheric transport. This does not involve that marine pollution monitoring will systematically require the determination of every contaminant of concern in all these compartments. What to determine and where must be defined according to the management and scientific objectives of the monitoring program (Segar et al., 1986). However, from the analytical point of view the determination of trace components cannot be isolated from the type of samples to be examined so that any critical evaluation of the methods available must be extended over as many environmental compartments as possible.

Therefore, in the present paper, sediment, water (aqueous and particulate phases), biota and aerosol samples will be considered. Special attention will be focussed on the compounds described in Table 2, although the determination of other chemicals of recent interest in the marine environment such as organo-phosphorous and organo-metallic components will also be included.

2. SAMPLING TECHNIQUES

Environmental analysis depends upon the successful achievement of a complete series of stages, including sampling, sample handling, analytical measurement and data interpretation. However, sample acquisition is the most crucial step. The errors introduced during sampling are especially significant because they are carried through the entire measurement scheme and constitute the error group of more difficult evaluation. The quality of the data generated in any environmental study is inherently related to the quality of the sample collection process.

Adequate sampling is even more critical when the analysis of trace components is intended. The elements present in small proportion are those whose composition can be easily modified during collection, transport or storage. Furthermore, sample variability is primarily reflected in the non homogeneous distribution of these minor components. In consequence, stringent precautions must be taken during the whole sampling process (from cruise preparation to sample collection and processing) for the attainment of representative samples in the marine environment. All problems that may be encountered during sampling must be anticipated and their solutions foreseen. Also, new unexpected problems must be watched over.

2.1. Sampling requirements.

Unfortunately, a complete list of aspects to be considered for adequate sampling is too long to be included here. However, some general criteria may be provided:

a) All equipment, reagents and solvents must be cleaned in the laboratory prior to field sampling. b) A final rinse of the sampling surfaces should be carried out inmediately before sample collection. c) Blanks must be prepared to control the cleanliness of all sampling operations. d) Sample media should only contact precleaned surfaces or reagents. e) The sampling site must not be influenced by local sources of contamination. f) Collection site and conditions must be appropriated for the attainment of samples truly representative of the system considered. g) The sampling process (and conditions) must be controlled during the whole period of collection, especially in cases where long times are required. h) Appropriate sample containers must be used. Sample frozening if often necessary, and i) The sampling operations must be in charge of trained people, ideally with experience in analytical chemistry. More specific requirements depend on the type of component to be analyzed, the environmental compartment to be studied and the specific purposes of the study.

2.2. Sample variability.

Environmental data will always undergo some degree of uncertainty arising from errors introduced during sampling and analytical measurement, and from the intrinsic variability of the samples. This uncertainty can be expressed in terms of the **accuracy** and the **precision** of the measurements, referring, respectively, to the **systematic** and **random** errors associated to the results. The errors introduced in the analytical process will be considered in Section 5.

The unespecific random errors may be diminished by replicate sampling and analysis. In these cases, the mean ($\bar{X}$), defined by the expression:

$$\bar{X} = \frac{\sum\limits^{n} x_i}{n} \qquad (1)$$

where x_i individual measurement
n number of measurements

and the standard deviation (σ), defined by the expression:

$$\sigma = \sqrt{\frac{\sum\limits^{n}(x_i - \bar{x})^2}{n-1}} \qquad (2)$$

where x_i individual measurement
$\bar{x}$ mean of all measurements
n number of measurements

are used to evaluate the dispersion of the data. In this approach, it is assumed, however, that the population of results approximates a Gaussian distribution

$$F(x) = \frac{1}{\sigma\sqrt{2\pi}}\, e^{-\frac{(x-\bar{x})^2}{2\sigma^2}} \qquad (3)$$

227

where F(x) the frequency with which the value x is obtained
  $\bar{x}$ mean of the measurements
  σ standard deviation of the measurements.

This in turn represents that the underlying causes of variation are truly random and the probability for the occurrence of any given result is inversely proportional to its difference from the mean (large variations will occur less frequently than small). This may not be necessarily the case.

However, in such situation, straightforward integration techniques allow to demonstrate that the common interval $\bar{x} \pm \sigma$ correspond to a probability of 68.26% for the occurrence of any single determination. An interval representing a probability of 95.46% is defined by $\bar{x} \pm 2\sigma$. The correspondence between these probability values and the intervals derived from $\bar{x}$ and σ is statistically significant provided that a good estimate of the standard deviation is available which, in turn, requires the analysis of a large number of replicates. This rarely occurs in the analysis of trace components in the environment so that the problem may be considered in another way, as the odds against obtaining a value outside certain set limits. Thus, it is required the use of the Student t-Factor, a distribution function depending on the desired probability and the size of the population from which the standard deviation is estimated. Now the intervals must be defined as

$$x \pm t \cdot \sigma \qquad (4)$$

where $\bar{x}$ the mean of the measurements
  σ the standard deviation of the measurements
and t the t-Factor corresponding to a given probability and to a number of replicates.

The values of the Student t-Factors are tabulated in most books of statistics (i.e., Himmelblau, 1970). To get some insight into expression (4) we may consider the probability (confidence limit) of 95% referred above. In the case that only two replicates are performed the corresponding t factor amounts 12.7 defining, in consequence, a considerably wider interval of results. Conversely, more than 50 replicates are needed to approach a t-Factor of 2. This is a representative example of a general rule: high confidence limits involve high t-Factors when the number of replicates is small and, in consequence, large intervals of result variability are defined despite that small standard deviations are observed.

Another important aspect in environmental analysis is the confidence that can be placed in the mean of n determinations. The standard deviation of the mean is given by the expression

$$S_{\bar{x}} = \sigma / \sqrt{n} \qquad (5)$$

where $S_{\bar{x}}$ standard deviation of the mean
  σ standard deviation associated with a single determination (calculated according to expression (2)).
  n size of the population.

Since n is always larger than 1, $S_{\bar{x}}$ is always less than σ, indicating lower dispersion for the mean value than for the individual measurements. Another important aspect is that $S_{\bar{x}}$ is inversely proportional to the square root of

228

the number of determinations. That is, to **halve** the dispersion of the mean requires to increase **four** times the number of measurements.

These statements represent that large numbers of replicates are needed for the attainment of statistically significant determinations of environmental trace components within a reasonably small interval of variation and with good confidence limits. This is effectively the situation when the goal of the analytical work is to clarify questions such as what is the concentration of an analyte in a particular place at a specific moment.

However, this type of questions are rather uncommon in environmental studies. Monitoring programs are usually designed for the acquisition of information on spatial or temporal distributions of pollutants. Thus, the main objective is to obtain the trends of variation of diverse components in the area of study, knowing in advance that any single specific value is under the influence of the local variability of the analytes. In this type of studies the sampling strategy may be designed in a probabilistic way (leading to parametric tests and to decisions) or, more often, it may follow more subjective approaches devoted to obtain a definition of the system structures without statistical inferences on the sampled population. In these cases, the samples selected usually includes representative examples of moderate or strong influence by the trace components to be determined and reference samples where no analyte occurrence should be expected. Anyway, these types of monitoring programs are less demanding in terms of replicate determinations than the analysis of single isolated spots.

In short, any relevant sampling scheme is strongly dependent upon the purpose for which the results are acquired. Consequently, a clear and well-defined objective is a prerequisite for the efficient collection of data and to the correct interpretation of results. The following aspects must be considered for the definition of the sampling program: a) purpose, b) population of samples to be acquired, c) components to be analyzed, d) area/period of study and other physical constraints, e) type of sample to be collected (air, water,...), and f) analytical methods of measurement. Ideally, adequate sampling involves an iterative process where analysis and data assessment of the preliminary sets of samples allow to establish the most suitable sampling methodology.

2.3. Atmosphere.

The trace components present in the atmosphere may be grouped for sampling purposes in gases (those in the $C_1 - C_6$ carbon number range or equivalent molecular weight), volatile compounds (approximately those of carbon number between C_4 and C_{15} or equivalent molecular weight) and particulates (those ranging between $C_{10} - C_{40}$ or even larger molecular weight). According to these operational definitions some trace components may be included in both the first and the second group (heavy gases or light volatiles), and, similarly, others belong to the second and to the third (as the heaviest compounds of the volatile fraction or the lightest chemicals in the particulates). This overlapping illustrates that although the air components are unevenly distributed according to their physico-chemical properties, their classification in these three groups is not in correspondence with three well defined environmental compartments and reflects our limitations for atmospheric sampling. Diverse reviews are available where the information on the trace chemical compounds present in the atmosphere is compiled (Graedel, 1978; Simoneit and Mazurek, 1981; Lamb et al., 1980)

Gaseous components

The analysis of gaseous components in the atmosphere encompasses two main types of sampling techniques: those involving specific chemical reactions and those based on concentration methods where chemical reaction is generally undesired.

The techniques involving chemical reactions are currently specific for the component to be analyzed, leading most often to spectroscopic or electrochemical determinations. Usually a known volume of air is bubbled through a reagent solution where some chemical property will be latter measured. Negative and positive interferences coming from other gaseous species are eliminated or deducted in the calculations of the final results. This methodology is generally applied using automatized or semi-automatized sampling devices. Thus, instrumental set-ups for the continuous monitoring of air components such as sulphur dioxide, nitric oxide, nitric dioxide, nitrous and nitric acids, ammonia, ozone, hydrogen peroxide, hydrogen chloride, carbon monoxide, total hydrocarbons, reactive hydrocarbons, acrolein, formaldehyde, acetaldehyde, and others are currently available. These sampling systems have been recently reviewed by Harrison (1984).

The techniques based on concentration stages are usually applied when the atmospheric concentration of the analytes is too low for direct detection or/and when multicomponent analysis is intended. In the latter case gas chromatography is generally used for the separation of the collected components. A wide diversity of adsorbents have been used: silica, alumina, charcoal, Saranex, Teflon, Tenax GC, various Chromosorb packings coated with different stationary phases or treated with diverse chemicals, silver wool or silver gauze, etc. They are essentially selected according to the chemical characteristics of the components of interest. However, solid adsorbents are not useful for the analysis of reactive species, particularly if they are present at low concentration. In these cases the use of cryogenic traps is required. For instance, lead alkyls are collected by cooling at -130°C and volatile sulphur compounds are trapped at -183°C (see Table 3). At this respect, some reduced sulphur components may also be concentrated with solid adsorbents (i.e., dimethyl sulfide with Tenax GC) but this involves the irreversible loss of other more reactive sulphur species in the collection process (i.e., hydrogen sulfide and methyl mercaptane).

The cryogenic loop tubes are usually packed with deactivated surface glass beads. In some particular cases the use of gas-solid chromatographic packings allows to operate at higher temperatures (i.e., the sampling method for ethene developed by De Greef et al (1976) involving the use of Porapak S at -95°C). The temperatures required for cryogenic trapping may be reached with cryocondenser systems but, for practical reasons (especially in field sampling), the use of Dewar flasks containing liquified gases such as nitrogen, oxygen or argon is of widespread application. These are sometimes used in combination with organic solvents, i.e. liquid nitrogen/acetone mixtures. Although liquid nitrogen is the most widely available coolant it has the inconvenient of its too low vaporization temperature (-196°C) giving rise to the condensation of all air gases. The vaporization temperature of liquid oxygen is a bit higher and air condensation is not produced. Therefore, this gas is often selected for sampling at temperatures below -160°C (i.e. in the collection of reduced sulphur species). An alternative solution consists in the immersion of an aluminum block in liquid nitrogen, as described in a method for the determination of C_2-C_{10} hydrocarbons in air by cryotrapping at -120°C (Matuska et al., 1986). When such low temperatures are not needed, the gaseous species may be collected with traps containing mixtures of dry ice and organic solvents (dry ice/acetone: -86°C, dry ice/methanol: -72°C).

TABLE 3

REPRESENTATIVE EXAMPLES OF TRACE GASEOUS COMPONENTS
IN RURAL (a) AND MARINE (b) ATMOSPHERES

COMPONENT	CONCENTRATION	SAMPLING TECHNIQUE	METHOD OF MEASUREMENT
Hydrocarbons	$(ng/l)^a$		
Methane	1000–4000	Tenax GC	Gas chromatography
Ethane	0.05–95	Cryogenic trapping	with flame ioniza-
Propane	12–94	($-120°C$)	tion detection.
Butane	0.01–200		
Ethylene	0.7– 700		
Acetylene	0.2– 230		
Halocarbons	$(ppbv)^b$	Tenax GC or	Gas chromatography
Methyl chloride	0.6	Cryogenic trapping	with electron cap-
Methylene chloride	0.03		ture detection
Chloroform	0.03		or
			Gas chromatography
Carbon Tetrachloride	0.15		coupled to mass
Methyl chloroform	0.10		spectrometry
Trichloroethylene	0.005		
Perchloroethylene	0.005		
Methyl bromide	0.05		
Methyl iodide	0.01		
Fluorotrichloromethane	0.05–0.1		
Dichlorodifluoromethane	0.1–0.2		
Sulphur components	$(ug\ S/m^3)^a$	cryogenic trapping	Gas chromatography
Hydrogen sulfide	0.3–1.8	($-183°C$)	with flame photo-
Carbon diulfide	0.54		metric detection.
Dimethyl sulfide	0.06–0.6		
Dimethyl disulfide	0.02–0.3		
Methyl mercaptane	0.01–0.08		
Carbonyl sulfide	0.73		
Mercury components	$(ng/m^3)^b$	Gold or silver	Atomic absorption
Mercury	0.7	coated glass beads	spectrophotometry
Mercury dichloride	0.35	Silvered Chromosorb P	
Methylmercury chloride	0.30		
Dimethylmercury	0.12		
Lead alkyls	$(ng/m^3)^a$		
Tetramethyllead	1.7	cryogenic trapping	gas chromatography
Trimethylethyllead	0.4	($-130°C$)	coupled to flameless
Dimethylethyllead	<0.1		atomic absorption
Methyltriethyllead	<0.1		spectrophotometry
Tetraethyllead	<0.1		

Another problem related with cryogenic sampling is the amount of water that may accumulate within the traps. In the case of collection of relatively inert species this is overcome by placing an adsorption packing (i.e., sodium sulphate, potassium carbonate) prior to the entrance into the cooled loop. When the analytes are too reactive for this solution the liquid water is removed by transfer of the sample from one cryogenic trap to another, the latter being generally constituted by a capillary tubing. Thus, after connecting the two collection loops in series, the former is immersed in a Dewar flask containing water at 90°C. The sample is rapidly volatilized and recondensed in the second trap whereas most of the water remains in liquid phase in the former.

Table 3 summarizes some of the gaseous species of interest in environmental studies, their typical concentrations in low polluted areas and the methods of sampling and measurement used for their determination.

Volatile components.

Different collection systems have been used for the sampling of the volatile fraction in the marine atmosphere. Polyurethane foam, Florisil and Tenax GC have been the adsorbents of wider application.

Polyurethane foam (PUF) may be used for the collection of specific air components or in general-purpose organic samples. Its porosity allows to keep high flow rates though the packing ($\sim$ 25 m^3/hr), and, therefore, the sampling of large-volumes ($\sim$ 1000 m^3). However, its retention capacity may be limited, especially in the case of collection at high flow rates, which requires the calculation of the breakthrough volumes for the components to be quantitatively determined. These breakthrough volumes may be rather dependent on the ambient temperature at which sampling is produced. On the other hand, PUF also requires extensive solvent cleaning before use, which is particularly important for the analysis of multicomponent mixtures. The usefulness of PUF for the air sampling of aliphatic and aromatic hydrocarbons as well as chlorinated components has been investigated by Atlas and Giam (1981), Burdick and Bidleman (1981), You and Bidleman (1984) and de Raat et al. (1987). Some of their results are displayed in Table 4. Lewis and Jackson (1982) reported improved collection efficiencies for pesticides and semivolatile industrial organic chemicals with dual sorbent cartridges consisting of 25 cm^3 bed of granular sorbents such as Chromosorb 102, Porapak R, XAD-2, Tenax GC and Florisil, sandwiched between two PUF plugs (5 cm thick upstream and 2.5 cm thick downstream).

Florisil is a synthetic coprecipitate of silica and magnesium silicate. It is commonly used in liquid chromatographic separations and for clean up procedures (see H9 in section 3). It has also been applied to air component sampling because it has superior trapping efficiency over PUF for the more volatile organic species. An additional advantage in comparison to PUF consists in the easy precleaning (heating at 320–420°C). However, this adsorbent requires extensive testing when it is used for the analysis of complex mixtures. Chemisorption and irreversible adsorption are commonly produced when collecting polar components. These effects, as well as the general retention efficiency, depend on the degree of water deactivation of the Florisil surface, a parameter which is under the influence of atmospheric moisture. The characteristics of Florisil for the collection of volatile air components have been evaluated by Atlas and Giam (1981) and Adams et al. (1982).

Tenax is a porous polymer resin (poly-(2,6-diphenyl-p-phenylene oxide)) which has been extensively used in air sampling of C_6-C_{18} organic compounds. Because of their volatility, the retained components are thermally desorbed directly into a gas chromatographic column instead than solvent extracted as in PUF or Florisil. This involves some contraints in terms of size and geometry of the adsorbent tube and consequently Tenax samplers operate at low flow rate. The main problems associated to the use of Tenax arise from the formation of resin oxidation products (i.e., benzaldehyde, acetophenone, benzoic acid). Eichmann et al. (1979;1980) have reported the use of this adsorbent for the analysis of n-alkanes in the marine atmosphere. Billings and Bidleman (1980; 1983) have compared the performances of Tenax GC and PUF for high-volume air sampling of chlorinated hydrocarbons. Van Vaeck et al. (1984) have used this adsorbent for an Integrated Gas Phase-Aerosol Sampling System devoted to the quantitative determination of the volatilization losses of organic aerosol constituents in Hi-Vol cascade impactor sampling.

TABLE 4

REPRESENTATIVE EXAMPLES OF ATMOSPHERIC TRACE ORGANIC VOLATILE COMPONENTS

COMPONENT	CONCENTRATION	SAMPLING TECHNIQUE	METHOD OF MEASUREMENT
Aliphatic hydrocarbons	$(ng/m^3)^a$	Adsorption on	Gas chromatography
n-heptadecane	15	polyurethane	coupled to mass
n-octadecane	34	foam	spectrometry
n-nonadecane	45		
n-eicosane	50		
n-heneicosane	30		
n-docosane	20		
n-tricosane	8		
Aromatic hydrocarbons	$(ng/m^3)^a$	Adsorption on	Liquid chromato-
Fluorene	15	polyurethane foam	graphy with UV-
methylfluorene	4		fluorescence
Phenanthrene	85		detection
Anthracene	2		
Fluoranthene	20		
Pyrene	8		
Chlorinated components	$(ng/m^3)^b$	Adsorption on	Gas chromatography
PCB	0.06–1.8	polyurethane foam	with electron
DDE	0.003–0.34		capture detection
Hexachlorobenzene	0.10–0.20		
Dieldrin	0.01–0.07		
Chlordane (γ and α)	0.012–1.3		
Phthalates	$(ng/m^3)^b$	Adsorption on	Gas chromatography
Di-n-butyl phthalate	0.9–18	polyurethane foam	with electron
Di(2-ethylhexyl)phthalate	1.2–17		

a) urban; b) remote, rural and marine.

Particulates.

Sampling air particulates only requires a precleaned glass or quartz fiber filter. If volatile compounds are measured in the same system the filter holder may be connected to an adsorbent cartridge (see for instance Keller and Bidleman, 1984). The sampling assembly is protected by a rain shield and air is sucked by a calibrated high-volume pump.

The particulates may be separated by size using a cascade impactor. The most widely available instrument (Sierra Impactor) separates the particles into the following equivalent cutoff diameters at 50% efficiency (flow rate 68 m^3/h): 1st stage: >7.2 um; 2nd stage: 7.2-3.0 um; 3rd stage: 3.0-1.5 um; 4th stage: 1.5-0.96 um; 5th stage: 0.96-0.5 um; final filter: <0.5 um. Other cascade impactors give 50% cut-off stages down to 0.08 um. However, some air pollutants such as lead have been found to exist primarily as particles of <0.3 um and cascade impactors are of limited efficiency in this size range. The diffusion battery, which separates particles on the basis of their diffusive rather than inertial properties, is then used. Several models have been constructed, the best known consisting of sequences of collimated holes or mesh screens. They may be used in serial connection to cascade impactors.

Several problems are associated with sampling particulate aerosols. Physico-chemical modifications of organic compounds during collection are produced. Thus, the loss of lower molecular weight components and condensation of gases or products from the volatile air fraction onto the particles have been observed. The loss of the lower molecular weight components depends on the air volume collected and the temperature during collection. Van Vaeck et al. (1984), using an integrated gas phase-aerosol sampling system concluded that Hi-Vol sampling technology could only yield accurate particulate phase concentrations for n-alkanes from carbon-numbers C_{25} -C_{27} onward and PAH of molecular weight 252 and higher. Simulation experiments have demonstrated PAH conversion with O_3 or NO_2 on different types of substrates (Coutant et al., 1988). For example, anthracene is more reactive than phenanthrene in the presence of NO_2, and pyrene is more reactive than fluoranthene (Niessner et al., 1985; Pitts et al., 1985). However, the degradation of PAH depends on the nature of the particles (Valerio and Lazzarotto, 1985; Behymer and Hites, 1985). When PAH are associated with soots they appear to be better preserved from alteration. The chemical and photochemical reactions occurring after the components have been collected on the filters have been studied by Peters and Seifert (1980) and Lee et al (1980) among others, who realized that this may represent an additional difficulty when collecting urban aerosols. However, marine aerosols contain much lower amounts of reactive gases and, therefore, the problems associated with filter in-situ degradation reactions may be of minor importance. In any case, alternative collection techniques for airborne particulate have been proposed, such as electrostatic precipitation (Alfheim and Lindskog, 1984).

Recent results using cascade impactors for sampling in the marine atmosphere have been reported by Schneider et al. (1983), Sicre et al. (1987a) and Grimalt et al. (1988). Some of these are reported in Table 5.

2.4. Atmospheric precipitation.

Precipitation samples are usually collected by continuously open, manual collectors. These systems provide composite representations of wet and dry deposition which may be defined as bulk deposition samples. The acquisition of wet-only precipitation samples (i.e., rain, snow, ice, fog, hail or dew)

TABLE 5

REPRESENTATIVE EXAMPLES OF PARTICULATE TRACE COMPONENTS
IN THE MARINE ATMOSPHERE
(PARTICLE SIZE FRACTIONATED WITH A HI-VOL CASCADE IMPACTOR)

COMPONENT	F	5	4	3	2	1	T
Aliphatic hydrocarbons							
n-Pentacosane	55[a]	30	25	35	50	35	230
n-Hexacosane	85	60	55	65	50	30	340
n-Heptacosane	160	130	120	150	170	80	810
n-Octacosane	190	150	110	110	80	35	660
n-Nonacosane	330	210	185	250	360	175	1500
n-Tricosane	235	130	100	95	65	30	650
n-Hentricosane	320	200	160	210	300	180	1400
UCM[b]	8900	4800	4000	4200	4800	2900	30000
Aromatic hydrocarbons							
Phenanthrene	10	7	5	5	7	3	35
Anthracene	3	1.5	0.5	0.5	0.5	0.3	6
Fluoranthene	17	15	12	8	10	2.5	65
Pyrene	19	16	10	5	8	2.5	60
Benzo(e)pyrene	6	5	3	2	2	0.3	20
Benzo(a)pyrene	3	2	1	1	1	0.2	8
Indeno(1,2,3,-cd)pyrene	1.5	1	1	0.5	0.7	0.1	5
Benzo(ghi)perylene	2.5	3	1.5	1	1	0.1	9
Fatty acids							
i-Pentadecanoic acid	90	30	30	15	60	15	240
a-Pentadecanoic acid	200	20	60	40	170	80	570
n-Pentadecanoic acid	600	80	160	130	410	130	1500
n-Hexadecenoic acid	510	190	260	340	520	100	1900
n-Hexadecanoic acid	6900	1900	2000	2300	5200	1700	20000
n-Octadecanoic acid	3700	1500	1000	1400	2700	950	11000
n-Hexacosanoic acid	800	470	270	290	300	85	2200
n-Octacosanoic acid	640	370	250	300	310	83	2000
Fatty acid salts							
i-Pentadecanoate	20	25	60	70	60	40	275
a-Pentadecanoate	55	50	110	170	85	90	560
n-Pentadecanoate	180	140	260	420	260	180	1400
n-Hexadecenoate	340	150	500	470	700	500	2700
n-Hexadecanoate	2900	1800	2700	4300	4800	2000	18500
n-Octadecanoate	2000	980	1200	2000	1900	800	8900
n-Hexacosanoate	180	75	85	90	60	70	560
n-Octacosanoate	90	50	45	110	40	70	400

a) pg/m^3.
b) Unresolved complex mixture

Equivalent diameters (deq) of aerosols corresponding to the impactor stages
are: F, final filter: $0.003 < deq < 0.5$ um; E5: $0.5 < deq < 0.96$ um; E4: $0.96 < deq < 1.5$ um; E3: $1.5 < deq < 3$ um; E2: $3 < deq < 7.2$ um; E1: $deq > 7.2$ um.

235

requires the use of collection devices which are only deployed at the onset of the deposition event and are retired immediately after termination of the episode. In these cases the samples are commonly constituted by composite representations of both in-cloud scavenging processes and below-cloud washout phenomena. Three component fractions may be defined independently of the type of sample (bulk or wet-only): volatile species, "aqueous materials" and particle-associated products. These three fractions will be described in detail in section 2.6.

Although most studies on atmospheric precipitation have been conducted over urban or rural areas (i.e., Van Noort and Wondergem, 1985), the amount of data referring to marine areas is becoming increasingly important. Thus, airborne organic components have found to be a relevant source of organic matter in oceanic oligotrophic sites (Gagosian and Peltzer, 1986; Zafiriou et al., 1985). The precipitation scavenging ratios for various trace metals of aerosol particles over remote marine regions have recently been evaluated (Buat-Menard and Duce, 1986)

The sampling systems designed for collection of atmospheric precipitation events such as rain or snow include dishes, pots, jars, trays, bowls, etc., optionally equipped with reservoirs or storage bottles. Glass, aluminum, stainless steel, polyethylene or Teflon are the most common materials. In general, Teflon is preferred for its chemical inertness and for its low surface activity (Strachan and Huneault, 1984; Pankow et al., 1984). Other type of wet deposition events require special devices. For instance, fog sampling is usually performed by manual removal from rotating arm collectors (Schnell, 1977; Waldman et al., 1982). A throughoutful and extensive review on bulk and wet-only precipitation has recently been made by Mazurek and Simoneit (1986). Particle and gas dry deposition have been reviewed by Sehmel (1980). Table 6 lists the concentration levels, collection techniques and methods of measurement for some trace components reported in atmospheric precipitation samples.

TABLE 6

ORGANIC TRACE COMPONENTS REPORTED IN ATMOSPHERIC PRECIPITATION SAMPLES

COMPONENT	CONCENTRATION	COMPONENT	CONCENTRATION
Aliphatic hydrocarbons		**Chlorinated components**	
n-Tricosane	7 ng/l (rain)	pp'-DDE	1-19 ng/l (rain)
n-Heptacosane	4 ng/l (rain)	pp'-DDT	0-1300 ng/l (rain)
n-Octacosane	4 ng/l (rain)		1-15 ng/l (snow)
n-Nonacosane	4 ng/l (rain)	Dieldrin	0.02-97 ng/l (rain)
			2-6 ng/l (snow)
Aromatic hydrocarbons		δ-HCH	0-9 ng/l (rain)
Anthracene	8-90 ng/l (rain)		0.1-6 ng/l (snow)
Phenanthrene	2-540 ng/l (rain)	Total polychloro-	10-100 ng/l (rain)
Fluoranthene	0-118 ng/l (rain)	biphenyls	5-50 ng/l (snow)
	15-1100 ng/l (snow)	Pentachlorophenol	2-300 ng/l (rain)
Pyrene	4-160 ng/l (rain)		2-14 ng/l (snow)
	1-9 ng/l (snow)		
Benzo(a)pyrene	0-20 ng/l (rain)		
	5-2000 ng/l (snow)		
	330 ng/m^3 (fog)		

236

2.5. Sea surface microlayer.

The chemistry and sampling of the sea surface microlayer has been extensively
reviewed by Liss (1975). The most commonly reported techniques for sea surface
microlayer sampling are based on the immersion of a plate or screen sampler
into the water underlying the sea surface film and withdrawing it slowly
through the microlayer. Glass plates, stainless steel screens, plastic screens,
ceramic rotating drums, germanium prisms, hydrophilic and hydrophobic Teflon
sheets, polyethylene funnels and liquid-nitrogen-cooled freezing probes have
all been used. Hatcher and Parker (1974) have compared the performances of some
of these devices in laboratory tests. These samplers collect about 50-150 um
thick surface layer, although actually the surface microlayer thickness is
estimated in the range of nanometers.

Lipids and other surface-active components are preferentially collected with
hydrophobic materials (i.e. Teflon). However, diverse intercalibration
exercises have shown that each of these techniques has its own advantages and
efficiencies with regard to the chemical and biological species accumulated in
the surface film (Van Vleet and Williams, 1980; Daumas et al., 1976). The
screen technique allows the collection of large volume samples (20 l/h) by calm
up to moderate sea conditions. Marty et al. (1979) and Ho et al. (1982) have
used a stainless steel screen made of 360 um diameter wire with 1.25 mm square
openings in the mesh, obtaining a water film of about 440 um thick. The
collected film is then drained into a glass container.

Three component fractions may be defined in these surface water samples: the
volatile species, the aqueous and the particulate materials. They will be
considered in detail in section 2.6. Table 7 reports the ranges of
concentration of aliphatic hydrocarbons and fatty acids found in the aqueous
and the particulate phases of surface microlayer waters corresponding to
diverse studies in the Mediterranean Sea.

2.6. Water.

In oceanography, the water components have been traditionally classified as
those present in the "dissolved" or in the "particulate" phases. The problem of
distinction between these two water fractions is purely operational: what
passes or is retained by filters of less than one micron. In principle, it is
assumed that particles smaller than one micron are under the influence of the
brownian movements so that their probabilities of sedimentation are small.
However, in terms of water composition, this one micron boundary is arbitrary,
not reflecting any specific particle size distribution. In fact, the
"dissolved" water phase contains a wide array of "particulate" material in the
form of colloids and molecular aggregates. These organic materials as well as
humic and fulvic acids determine that the "dissolved" phase is far more complex
that it would correspond to a solution of purely dissolved species. Seawater
must be considered as a complex dynamic system containing truly dissolved
organics and a continuous range of particle sizes, from the smallest colloids
up to larger organic aggregates, bacteria, and plankton. The crossed
interactions between some of these materials are very relevant in terms of
solubilities (Whitehouse, 1985; McCarthy and Jimenez, 1985; Chiou et al., 1986;
Morehead et al., 1986) or sediment sorption processes (Gschwend and Wu, 1985;
Voice et al., 1983; Caron et al., 1985 and Baker et al., 1986). In order to
avoid misleading terms the "dissolved" phase will be referred here as the
"aqueous" phase.

TABLE 7

CONCENTRATIONS OF TRACE COMPONENTS IN THE SEA SURFACE MICROLAYER

COMPONENT	AQUEOUS PHASE	PARTICULATE PHASE
Aliphatic hydrocarbons		
n–Heptadecane	2–6 [a]	20–25
n–Octadecane	4–8	6–11
n–Nonadecane	4–7	3–22
n–Eicosane	3–12	6–30
n–Heneicosane	2–20	8–50
n–Docosane	7–85	20–65
n–Tricosane	8–110	20–90
n–Tetracosane	10–160	30–100
n–Pentacosane	10–180	30–140
n–Hexacosane	10–210	30–120
n–Heptacosane	10–200	50–150
n–Octacosane	8–190	40–110
n–Nonacosane	8–160	50–250
n–Triacontane	4–120	30–85
n–Hentriacontane	20–90	30–200
n–Dotriacontane	5–60	20–70
n–Tritriacontane	3–40	20–75
Fatty Acids		
n–Dodecanoic acid	10–170	30–420
n–Tridecanoic acid	10–30	20–70
n–Tetradecanoic acid	60–950	900–6000
n–Pentadecanoic acid	10–350	350–650
n–Hexadecanoic acid	350–4200	4100–27000
n–Heptadecanoic acid	15–90	170–390
n–Octadecanoic acid	200–1600	1600–4400
n–Nonadecanoic acid	2–40	40–160
n–Eicosanoic acid	8–250	190–2500
n–Heneicosanoic acid	2–30	20–60
n–Docosanoic acid	8–140	140–600
n–Tricosanoic acid	5–40	40–100
n–Tetracosanoic acid	20–180	180–670
n–Pentacosanoic acid	6–45	45–160
n–Hexacosanoic acid	30–100	100–320
n–Heptacosanoic acid	6–30	10–50
n–Octacosanoic acid	15–60	60–130
n–Nonacosanoic acid	2–20	10–30
n–Triacontanoic acid	20–80	20–100

a) ng/l

On the other hand, a portion of the organic components present in the aqueous phase have too high vapor pressure to be concentrated by stages involving solvent evaporation (either at atmospheric pressure or under vacuum). The study of these components, labelled as the volatile organic fraction, requires the use of specially designed sampling and analytical procedures. Head space, purge and trap or, better, closed loop stripping techniques are commonly used. Again, the concept of "volatile organic species" is introduced by operational requirements and not by observation of a well defined class of seawater components. In a general sense, "volatile" components are those of high vapor pressure, low molecular weight and low water solubility that may be easily stripped from the sea surface by wind action, turbulence, etc. At this respect, water solubility is an important aspect determining whether or not a component may be found in the "volatile" fraction. High vapor pressure products will not be stripped by gases from water solutions if they are highly hydrophilic.

The volatile organic fraction

The volatile components are separated from water solutions by taking advantage of their high vapor pressure. Therefore, the sampling procedures are devoted to the extraction of the gaseous phase in equilibrium with the water sample. Diverse methods have been developed, their usefulness for organic pollutant analysis has been recently reviewed by Wegman and Melis (1985). For trace organics, closed loop stripping analysis is the technique of choice due to its low detection limits. Even species of relatively high water solubility can be analyzed at the ng/L level.

Closed loop stripping analysis was developed by Grob (1973). It essentially consists in a closed circuit in which a gas (nitrogen) is continuously bubbled into a water sample ($\sim$ 2 liters), passed through an adsorption tube (1.5 mg of activated carbon) and recycled towards the aqueous solution. The adsorption column is extracted with 10 ul of CS_2 when the stripping process has ended ($\sim$ 2-12 hours). The effects of extraction solvent, stripping time, stripping temperature, pH and salt content on the efficiency of the system have been studied by diverse authors (Grob, 1983; Gomez-Belinchon and Albaigés, 1987; Grob et al., 1984). To this end, Gomez-Belinchón and Albaigés (1987) have indicated that stripping at 45°C during 0.5 hours gives better performances than the more conventional 35°C/2 hrs conditions. The CLSA technique allows the routine screening of large numbers of compounds. The analysis of the extracts by gas chromatography using diverse detectors (both general-purpose and specific as well as mass spectrometry) allows to obtain quantitative and qualitative information on a wide diversity of volatile natural and anthropogenic trace chemicals. Some representative components of the CLSA extracts corresponding to coastal open marine waters are reported in Table 8.

A problem which have been addressed by several authors in relation with the routine use of CLSA is the generation of many contaminant blank peaks after a period of operation (3-6 months). Grob (1983) and Grob et al. (1984) have indicated that the major problems for blank contamination arise from particles generated from damaged adsorbent filters which may be blown into the system and be deposited on the internal surfaces of the pump and the lines. Gomez-Belinchon and Albaigés (1987) agree with this hypothesis, and stress the need for careful wipping of the internal lines of the pump for blank recovery.

Another drawback of the closed-loop system is that the most volatile compounds are masked by the solvent peak in the GC run. At this respect, Graydon et al. (1984) have introduced thermal desorption CLSA. However, the quantitative aspects of this modification become very critical because important

239

breakthroughs are observed for the most volatile compounds. The heavier products displace the lighter components from the filters which obligates to select a set of operating conditions for a narrow range of compounds (adjustment of the gas volume during adsorption and desorption and optimization of the temperature of the carbon filter). Purge and trap methods specific for the analysis of some of these highly volatile components have been recently proposed (i.e., Holdway and Nriagu, 1988).

TABLE 8

REPRESENTATIVE COMPONENTS FOUND IN THE VOLATILE FRACTION
OF COASTAL WATERS (SAMPLED BY CLSA)

COMPONENT	CONCENTRATION	COMPONENT	CONCENTRATION
Aliphatic hydrocarbons		**Aromatic hydrocarbons**	
n-Octane	1–3[a]	Toluene	2–35
n-Nonane	0.3–1	Styrene	0.5–5
n-Decane	0.5–10	Xylene	1–10
n-Undecane	0.3–5	Trimethylbenzenes	0.5–5
n-Dodecane	1–20	Ethyltoluene	0.5–10
n-Tridecane	1–10	Naphthalene	0.1–10
n-Tetradecane	0.5–70	Methylnaphthalenes	0.1–15
n-Pentadecane	3–20	Dimethylnaphthalenes	0.5–30
n-Hexadecane	1–6	Trimethylnaphthalenes	0.1–15
n-Heptadecane	2–15	Decylbenzenes	0.1–5
n-Octadecane	1–7	Undecylbenzenes	0.5–10
n-Nonadecane	1–2	Dodecylbenzenes	0.5–5
n-Eicosane	1–6	Tridecylbenzenes	0.5–10
Pristane	0.6–5		
Phytane	0.1–1.4		
		Volatile phosphates	
Chlorinated hydrocarbons		Tributylphosphate	0.1–20
Chlorobenzene	0.1–1		
Dichlorobenzenes	2–15	**Phthalates**	
Trichlorobenzenes	0.1–20		
Chloronaphthalenes	0.1–5	Diethyl phthalate	1–3
Dichloronaphthalenes	0.1–10		

a) ng/l

240

The aqueous organic fraction

The analysis of organic pollutants in water is a subject which has deserved an intense analytical effort over the last twenty years. Excellent extended summaries of this field are published every 2 years in Analytical Chemistry. However, despite all this analytical work many uncertainties are still limiting the reliability of trace component analysis in water, especially in the case of marine water. The problem arises from the complexity of the aqueous matrix. The performance of the analytical methods has generally been tested with spiked waters by using low ionic strength and particle colloid free aqueous samples, i.e. distilled, deionized, tap or finished waters. These obviously represent rather different matrices from natural waters and particularly from seawater. The comparison of the available methodology by parallel sampling of real seawater with different techniques is required. However, this type of evaluations have rarely been carried out. De Lappe et al. (1983) reported the comparison of two systems (liquid-liquid extraction and polyurethane foam adsorption) in open sea waters collected from a research vessel. Gómez-Belinchón et al (1988) have recently summarized the results from an intercomparison study of liquid-liquid extraction and adsorption on polyurethane and Amberlite XAD-2 for the analysis of hydrocarbons, polychlorobiphenyls and fatty acids in the aqueous seawater fraction.

Although almost all available methodology has been used for the analysis of organic pollutants in water (i.e., liquid-liquid extraction, adsorption on various types of resins, steam distillation, freeze-drying techniques, reverse osmosis, ultrafiltration, etc.), the analysis of trace organic components essentially encompasses liquid-liquid extraction and adsorption.

Liquid-liquid extraction is in principle a simple and conventional technique. Extraction at pH 7 can be inefficient for some classes of compounds such as phenols, thus pH must be adjusted in some cases. Low solvent/water ratios may lead to emulsions and low kinetics to reach equilibrium, especially for high molecular weight organic compounds. If high solvent/ratios are needed very pure solvents must be used. Liquid-liquid extractions may be performed in batch mode, shaking or stirring 1 or 2 liters of seawater with a small volume of solvent as described in the sample handling procedure H4 (section 3). This type of extraction is suitable for bulk determinations of classes of trace components such as the PAH by UV-fluorescence (analytical method M2, section 4).

A continuous liquid-liquid extractor has been constructed by Ahnoff and Josefsson (1974) for field sampling of large volumes of water. Two extraction units based on the mixer-settler principle are used in parallel. These may be loaded with organic solvents either heavier or lighter than water (200-250 ml). Good extraction recovery depends on the constant control of the stirring process ($\sim$ 900 rpm) and the use of a low water flow rate (50 ml/min). These requirements as well as the relative large size of the extractors difficult the practical utilization of this design. This situation has promoted the development of adsorption techniques for the isolation of trace components from large volumes of water.

The use of adsorption resins for the analysis of organic components in water has been extensively reviewed by several authors (Dressler, 1979; Moody and Thomas, 1982; Wegman and Melis, 1985), the reader is referred to these sources for a general description of the available polymers, their properties and their diverse applications. Table 9 shows the main adsorbents used in the analysis of trace organics in seawater as well as the most likely bonding mechanisms by which the resins interact with the analytes.

241

TABLE 9

ADSORBENTS USED FOR THE ISOLATION OF TRACE ORGANIC COMPONENTS FROM SEAWATER

SORBENT	COMPOSITION	LIKELY BONDING MECHANISM*	REPRESENTATIVE COMPOUNDS ANALYZED
Amberlite XAD-1	Styrene divinylbenzene polymer (100 m^2/g, 200 Å)	A,B	Lindane, DDT, Endrin, Malathion, Triton X-100, Rhodamine B
Amberlite XAD-2	Styrene divinylbenzene polymer (330 m^2/g, 90 Å)	A,B	PCB, Alkylbenzenes Phthalates Carbamates, Azaarenes, Pesticides
Amberlite XAD-4	Styrene divinylbenzene polymer (750 m^2/g, 50 Å)	A,B	Chloroethers, Haloalkanes, Acids, Phenols
Amberlite XAD-7	Methylmethacrylate polymer (450 m^2/g, 80 Å)	A,C	PCB, PAH, Aminoacids, Chlorinated compounds, Fenitrothion
Amberlite XAD-8	Methylmethacrylate polymer (140 m^2/g, 250 Å)	A,C	Humic acids, Phenols, Pyridine Quinoline, Acridine
Polyurethane Foam	Polyurethane (0.0285 g/cm^3)	A,C,B,	Hydrocarbons, PAH, PCB Chlorinated components
Graphitized Carbon black	Graphite	A,B	Hydrocarbons, Malathion, PAH Humic compounds, Phenols
Sep-pack C$_{18}$	Octadecylsilane bonded to silica	A	PAH, Aromatic amines, Phenols Carbofuran, Thiram, PCB
Tenax GC	2,6-diphenyl-p-phenylene oxide polymer (19-30 m^2/g 720 Å)	B,A	Hexachlorobenzene, PAH, Organohalides, Naphthalene

*) A, Hydrophobic effect; B, π-electron interactions; C, hydrogen bonding (listed in order of estimated importance)

242

These interaction mechanisms are effectively observed in studies where spiked distilled water solutions are analyzed. However, the situation changes considerably for real seawater. First, as it has been stated above, only a portion of the components from the aqueous fraction are present in true solution. Others, especially the hydrophobic components such as aliphatic, chlorinated or polycyclic aromatic hydrocarbons, are associated to large macromolecules, like humic and fulvic acids. In this situation, the physico-chemical behaviour of these species in aqueous solution is strongly modified (i.e. their solubilization in water increases and their adsorption onto solid surfaces decreases; Gschwend and Wu, 1985; Voice et al., 1983; Caron et al., 1985; Baker et al., 1986). Likewise, the interaction mechanisms with the adsorption resins are also modified. In fact, the possibility of retention of many hydrophobic species may depend on the capacity of the absorbents to retain the humic and fulvic acids to which they are associated. In this sense, Gomez-Belinchon et al (1988) observed a low capacity of XAD-2 for the retention of n-alkane homologs of higher molecular weight ($> n\text{-}C_{20}$)which was interpreted to result from the poor adsorption capacity of this resin to retain humic and fulvic acids.

Another effect concerns the competitive interaction between resin adsorbed species (i.e., analytes) and other organic components present in the aqueous fraction (i.e. humic and fulvic acids). As it is described in Table 9, the adsorption mechanisms of the resins involves primarily hydrophobic bonding. Thus, in the case of adsorption of humic or fulvic compounds, polar groups of the molecule are left oriented in the opposite direction, towards the stream of water. After passing large volumes of water through the column, important amounts of these macromolecules may be retained, leaving an important proportion of the hydrophobic sites of the adsorbent covered by polar groups. This, in turn, may result in a major change of its surface characteristics, and as soon as the degree of polarity is increased, a lower proportion of hydrophobic analytes will be retained. Furthermore, in non polar polymers such as XAD-2, these adsorption processes are reversible (Dressler, 1979), and since retention of the humic and fulvic acids generally involve more than one hydrophobic site, smaller apolar species already adsorbed in the resins may be displaced. Such interchanges involve a gain in entropy resulting from the extension of the hydrophobic interaction to the maximum available apolar sites of the macromolecule binded onto the adsorbent. That is, an interaction paralleling the "octopus" effect described for the sorption of water-soluble oligomers on sediments (Podoll et al., 1987).

In this sense, Gomez-Belinchón et al. (1988), in their intercomparison study with real seawater samples, have observed that no constant level of hydrocarbons was established in the adsorption columns despite the fact that water with constant concentration of hydrocarbons was collected, being apparent that after a period of accumulation further sampling resulted in desorption of n-alkanes. The phenomenon have been observed well before that the adsorption capacity of the resins for hydrocarbons (in pure solutions of distilled water) was reached. In fact, to a large extent it was independent of the concentration of hydrocarbons in the water analyzed. Such desorption effect of hydrophobic trace components has been attributed by these authors to competitive interaction with larger macromolecules.

It is important to stress all these effects related with the occurence of complex mixtures of organic polymeric materials in water because they are simply ignored in most application of adsorption resins to water analysis. It is usually taken for granted that the results obtained in the laboratory with spiked solutions can be extrapolated to real waters. To this end, the study of Gomez-Belinchon et al (1988) showed that polyurethane foam adsorption is giving

243

TABLE 10

CONCENTRATIONS OF DIVERSE TRACE COMPONENTS IN OPEN SEAWATER (AQUEOUS PHASE [a])

ELEMENT	CONCENTRATION	ELEMENT	CONCENTRATION
Metals [b]			
Arsenic	1.4–1.7	Lead	0.04–0.05
Cadmium	0.029–0.020	Manganese	0.02–0.24
Chromium	0.08–0.2	Molibdenum	10
Cobalt	0.002–0.004	Nickel	0.2–0.26
Copper	0.1–0.12	Zinc	0.16–0.40
Iron	0.20–0.40	Mercury	0.03
Aliphatic hydrocarbons [c]			
n-Tetradecane	1–120	n-Tetracosane	10–2000
n-Pentadecane	5–500	n-Pentacosane	10–2000
n-Hexadecane	10–700	n-Hexacosane	10–2000
n-Heptadecane	5–2000	n-Heptacosane	10–2000
Pristane	10–400	n-Octacosane	10–2000
n-Octadecane	10–2000	n-Nonacosane	10–1500
Phytane	20–300	n-Triacontane	10–1500
n-Nonadecane	20–1500	n-Hentriacontane	10–1000
n-Eicosane	10–1500	n-Dotriacontane	10–500
n-Heneicosane	10–1500	n-Tritriacontane	10–500
n-Docosane	10–2000	n-Tetratriacontane	10–500
n-Tricosane	10–2000	UCM [e]	10–40000
Chlorinated components [c]			
op'DDE	0–300	Polychlorobiphenyls	0–3000
pp'DDE	0–500	Aldrin	0–150
op'DDT	0–400	Hexachlorobenzene	0–200
pp'DDT	0–200	Lindane	0–20
Fatty Acids [d]			
n-Dodecanoic acid	0–1.5	n-Nonadecanoic acid	0–5
n-Tridecanoic acid	0–1.5	n-Eicosanoic acid	0–5
i-Tetradecanoic acid	0–2.9	n-Heneicosanoic acid	0–2
n-Tetradecanoic acid	0–90	n-Docosanoic acid	0–15
i-Pentadecanoic acid	0.1–7	n-Tricosanoic acid	0–5
a-Pentadecanoic acid	0.1–15	n-Tetracosanoic acid	0–10
n-Pentadecanoic acid	0.1–25	n-Pentacosanoic acid	0–15
i-Hexadecanoic acid	0–60	n-Hexacosanoic acid	0–10
n-Hexadecenoic acid	1.5–270	n-Heptacosanoic acid	0–2
n-Hexadecanoic acid	5–500	n-Octacosanoic acid	0–10
i-Heptadecanoic acid	0–10	n-Nonacosanoic acid	0–5
a-Heptadecanoic acid	0–10	n-Triacontanoic acid	0–10
n-Heptadecanoic acid	0–7	n-Hentriacontanoic acid	0–5
n-Octadecenoic acids	0–600	n-Dotriacontanoic acid	0–5
n-Octadecanoic acid	1.5–50		

a) Collected with polyurethane foam
b) ug/l; c) pg/l; d) ng/l
c) Unresolved complex mixture of aliphatic hydrocarbons

closer qualitative results to liquid-liquid extraction than adsorption on Amberlite XAD-2. Table 10 reports the concentration ranges observed for aliphatic hydrocarbons, chlorinated components and fatty acids in the aqueous fraction of seawater collected in moderately polluted open areas. Sampling was performed with polyurethane foam.

Metals in the aqueous fraction

The analysis of trace metals in seawater has long been a challenge for marine chemists. The comparison of review trace metal concentrations from the middle seventies with respect to those obtained from intercomparison exercises and reference seawater materials in the early eighties (Table 10) show important decreases for the average levels of many metals in open seawater (Berman and Yeats, 1985). Thus, the average concentrations of As and Cr is open seawater are now considered to be approximately 2.4 times lower than in the seventies. These correction factors are even higher for other metals. For instance, 4-4.5 for Cd and Cu; ~ 7 for Fe and Ni; ~ 12.5 for Cu and in the order of 18 for Zn. The improved reliability of the results arises from the recent general improvement of trace metal separation chemistry and analytical instrumentation as well as from the general reduction of contamination during sampling, storage and analysis. The conditions for adequate seawater sampling for trace metal determinations have been carefully compiled by Berman and Yeats (1985).

As a general criteria, the methods allowing direct measurements of metal concentrations in water are preferred for its simplicity. These usually involve an acidic digestion of the water sample prior to instrumental determination. To this end, sample handling procedures (H13-14, H17-18) and operational methods (M6-30) for trace metal analysis are respectively described in sections 3 and 4. Procedures for their chemical speciation are also included (H23).

However, in cases that particular metal chemical species must be selectively analyzed or a metallic component is present in very low concentration, diverse extraction or concentration procedures may be needed. These usually encompass a wide diversity of techniques such as liquid-liquid extraction, dialysis, electrophoresis, ultrafiltration, centrifugation, and ion exchange chromatography. It must be stressed, however, that they are actually used for very particular applications because, as described in the preceeding chapter, the general trend in trace metal analysis is to solve this type of problems by improvement of the instrumental sensitivity and selectivity taking advantage of the good performances of atomic absorption spectrophotometry, inductively coupled plasma-atomic emission spectroscopy and anodic stripping voltammetry. The procedures for metal concentration are in each case closely dependent on the ionic species to be analyzed and the substrate where they have to be determined. Therefore, it is out of the scope of the present review to proceed to their detailed description. The reader is remitted to two comprehensive reviews where concentration methods for most metals and metallic ionic species are described (Florence and Batley, 1980; Zolotov et al., 1982).

The particulate fraction

The elements for sampling the particulate fraction are in principle rather simple: A glass fiber filter mounted on a filter holder and a pump pushing the water through the filter. Non contaminating pumps should be used, as for instance the Teflon impeller, magnetically-driven pumps utilized by De Lappe et al. (1983). The glass fiber filters must be precleaned by kiln-firing at 350°C. It should be observed that glass filters are not constituted by truly

TABLE 11

CONCENTRATIONS OF DIVERSE TRACE COMPONENTS IN OPEN SEAWATER
(PARTICULATE PHASE)

ELEMENT	CONCENTRATION	ELEMENT	CONCENTRATION
Metals [a]			
Cadmium	1–50	Cobalt	50–250
Zinc	300–5000	Nickel	100–1000
Lead	50–1000	Manganese	400–8000
Chromium	200–1000	Iron	1000–25000
Copper	10–800	Titanium	100–2000
Aliphatic hydrocarbons [b]			
n-Tetradecane	0–100	n-Tetracosane	5–2000
n-Pentadecane	10–3000	n-Pentacosane	10–5000
n-Hexadecane	20–600	n-Hexacosane	5–2000
n-Heptadecane	30–20000	n-Heptacosane	10–15000
Pristane	10–2500	n-Octacosane	5–2500
n-Octadecane	10–1000	n-Nonacosane	5–25000
Phytane	10–1000	n-Triacontane	5–4000
n-Nonadecane	10–10000	n-Hentriacontane	10–15000
n-Eicosane	5–5000	n-Dotriacontane	5–4000
n-Heneicosane	5–3500	n-Tritriacontane	5–3000
n-Docosane	5–2000	n-Tetratriacontane	5–1500
n-Tricosane	5–3000	UCM [d]	5–80000
Chlorinated components [b]			
op'DDE	1–1500	Polychlorobiphenyls	2–10000
pp'DDE	1–1500	Aldrin	1–400
op'DDT	1–1000	Hexachlorobenzene	1–2000
pp'DDT	1–1000	Lindane	1–200
Fatty acids [c]			
n-Dodecanoic acid	1–200	n-Nonadecanoic acid	0.5–20
n-Tridecanoic acid	0.1–1	n-Eicosanoic acid	0.5–200
i-Tetradecanoic acid	0.5–15	n-Heneicosanoic acid	0.5–2
n-Tetradecanoic acid	1–1000	n-Docosanoic acid	0.5–100
i-Pentadecanoic acid	1–200	n-Tricosanoic acid	0.5–50
a-Pentadecanoic acid	1–100	n-Tetracosanoic acid	0.5–150
n-Pentadecanoic acid	1–150	n-Pentacosanoic acid	0.5–50
i-Hexadecanoic acid	1–80	n-Hexacosanoic acid	0.5–150
n-Hexadecenoic acid	1–900	n-Heptacosanoic acid	0.5–50
n-Hexadecanoic acid	1–4000	n-Octacosanoic acid	0.1–100
i-Heptadecanoic acid	0.1–2	n-Nonacosanoic acid	0.1–5
a-Heptadecanoic acid	0.1–2	n-Tricontanoic acid	0.1–50
n-Heptadecanoic acid	0.5–50	n-Hentriacontanoic acid	0.1–5
n-Octadecenoic acids	1–1000	n-Dotriacontanoic acid	0.1–5
n-Octadecanoic acid	1–1000		

a) ug/l;   b) pg/l;   c) ng/l
d) Unresolved complex mixture of aliphatic hydrocarbons

reticulated materials but by braided fibers resulting in an average pore size distribution. This involves that the cut-off limits concerning what may pass or be retained by the filter are a bit diffuse. In general, filters with rated pore size smaller than the micron (i.e., 0.5 um) are selected in order to ensure that the particles having a diameter equal or higher than one micron will be effectively retained in the particulate fraction. Table 11 reports the concentration ranges observed for metals, aliphatic hydrocarbons, chlorinated components and fatty acids in the particulate phase of seawater collected in moderately polluted coastal areas.

Another type of particles that may be studied concerns the sedimenting particulate material. This is collected with particle-interceptor traps. The reader is referrer to specialized references for the technical aspects involved in the use of these devices (i.e., Suess, 1980; Landing and Freely, 1981; Soutar et al., 1977; Wefer et al., 1982). Particle-interceptor traps have also been used for the study of the ascending particles in oceanic waters (Simoneit et al., 1986).

2.7. Sediment.

Sediments accumulate natural and anthropogenic products from the overlying water and integrate different inputs introduced into the marine environment. Thus, the distribution of trace components in surficial sediments, especially pollutants, may be indicative of geographical specific sources of contamination, and the analysis of the horizontal profiles of concentration in the sedimentary column may provide a hystorical record of pollutant inputs in a particular area. Furthermore, the study of the molecular sedimentary composition provides information on the transport mechanisms of these components as well as on their fate in the marine environment. Sediments may be collected with grabs or cores.

In general grabs are constituted by two or more articulated elements in the form of a jaw that penetrate by gravity within the sediments in open position. Closing is induced by adequate springs or by the tension of the cable when lifting. Grabs must fulfill at the best the following conditions: a) minimum disturbance of the sediment when sampling, b) avoid leaching during lifting, and c) facilitate subsampling at the top of the collected sediment. The grabs in which closing operates through the tension of the cable usually must be handled with a crane. The most common type is the Van Veen. Others such as the Petersen or the Smith-McIntyre are designed in order to facilitate a direct access to the sediment for subsampling. On the other hand, the Shipek grab constitute a representative example of the models with automatic closing when reaching the bottom whereas the Ekman grab is mechanically closed by the action of messengers.

Cores are devoted to obtain representative samples of the sedimentary column. They encompass two main types of devices: cyliders and boxes. Diverse cylinder cores are available, including different modes of penetration into the sediment (gravity, piston, etc.) or bottom obturation (diaphragm, sphincter, pneumatic valves, etc.). The most common type of box core is the Reineck type, a bottomless box (side 20-30 cm, height 60-80 cm) allowing to obtain undisturbed core samples suitable for close examination of compound profiles.

Whatever sampling system is selected, particular attention must be given to contamination from the sampler material and from the procedures used to transfer the sediment from the sampler to a storage container. As soon as the subsamples are retrieved, they should be placed in glass jars and immediately

TABLE 12

CONCENTRATIONS OF DIVERSE TRACE COMPONENTS IN COASTAL SEDIMENTS

COMPONENT	CONCENTRATION	COMPONENT	CONCENTRATION
Metals[a]			
Mercury	0.05–3	Manganese	200–500
Cadmium	0.1–0.3	Zinc	50–120
Lead	10–50	Nickel	20–55
Selenium	<0.01–0.04	Chromium	8–20
Copper	5–25		

COMPONENT	CONCENTRATION	COMPONENT	CONCENTRATION
Aliphatic hydrocarbons[b]			
n-Tetradecane	0.1–20	n-Tetracosane	3–100
n-Pentadecane	0.1–30	n-Pentacosane	3–250
n-Hexadecane	0.1–40	n-Hexacosane	2–250
n-Heptadecane	1–250	n-Heptacosane	5–450
Pristane	0.5–90	n-Octacosane	5–300
n-Octadecane	2–50	n-Nonacosane	10–700
Phytane	0.5–25	n-Triacontane	4–350
n-Nonadecane	2–50	n-Hentriacontane	10–650
n-Eicosane	2–70	n-Dotriacontane	3–200
n-Heneicosane	2–150	n-Tritriacontane	5–400
n-Docosane	2–150	n-Tetratriacontane	2–50
n-Tricosane	3–250	UCM[c]	5–30000

COMPONENT	CONCENTRATION	COMPONENT	CONCENTRATION
Chlorinated components[b]			
op'DDE	0–1500	Polychlorobiphenyls	0.5–1000
pp'DDE	0–2000	Aldrin	0–100
op'DDT	0–1500	Hexachlorobenzene	0–150
pp'DDT	0–1000	Lindane	0–20

COMPONENT	CONCENTRATION	COMPONENT	CONCENTRATION
Fatty Acids[b]			
n-Dodecanoic acid	1–700	n-Nonadecanoic acid	1–400
n-Tridecanoic acid	1–200	n-Eicosanoic acid	10–800
i-Tetradecanoic acid	1–700	n-Heneicosanoic acid	1–800
n-Tetradecanoic acid	5–4000	n-Docosanoic acid	10–2000
i-Pentadecanoic acid	1–2000	n-Tricosanoic acid	10–1000
a-Pentadecanoic acid	1–3000	n-Tetracosanoic acid	20–5000
n-Pentadecanoic acid	20–1500	n-Pentacosanoic acid	10–300
i-Hexadecanoic acid	5–500	n-Hexacosanoic acid	10–3000
n-Hexadecenoic acid	100–5000	n-Heptacosanoic acid	0.1–200
n-Hexadecanoic acid	600–30000	n-Octacosanoic acid	0.1–2000
i-Heptadecanoic acid	1–400	n-Nonacosanoic acid	0.1–90
a-Heptadecanoic acid	1–500	n-Triacontanoic acid	0.1–500
n-Heptadecanoic acid	1–600	n-Hentriacontanoic acid	0.1–30
n-Octadecenoic acids	100–8000	n-Dotriacontanoic acid	0.1–200
n-Octadecanoic acid	40–2000		

a) ug/g; b) ng/g
c) Unresolved complex mixture of aliphatic hydrocarbons

frozen at -20°C for later analysis. Sediments may be stored at room temperature after freeze-drying. Table 12 reports the concentration ranges observed for metals, aliphatic hydrocarbons, chlorinated components and fatty acids in sediments collected off the coast of the Ebre Delta (Northwestern Mediterranean) which is considered a moderately polluted environment.

Finally, the problems derived from sediment heterogeneousness must also be considered. Sediments, especially those of coastal areas, are composed of sands, silts and clays and, as shown by several authors (Thompson and Eglinton, 1978; Brassell and Eglinton, 1980; Grimalt et al., 1984), these fractions accumulate hydrophobic components differently. At this respect, the coarser fractions contain mainly detritus from autochthonous organisms and sands, which exhibit a low adsorption capacity for such chemical species. On the other hand, silts and clays tend to concentrate most of the allochthonous, mainly pollutant, components. A study encompassing the analysis of different grain-size fractions of coastal sediments (Grimalt et al., 1984) has given further evidence of these trends. A striking difference in hydrocarbon concentrations between the coarser (> 66 um) and the finer (< 66 um) fractions was observed. This difference was paralleling the range which separates sands from silts and clays and was indicative of distinct mineral-organic interactions corresponding to each material. The aromatic hydrocarbons were strongly concentrated in the finer fraction indicating a selective association of allochthonous anthropogenic inputs with the silt/clay fractions. This was in turn consistent with the higher relative concentration of aliphatic unresolved hydrocarbons (a mixture indicative of petrogenic hydrocarbons) in the same fraction. Therefore, grain-size distributions have to be considered in the interpretation of analytical results from sedimentary samples.

2.8. Organisms.

The study of trace pollutants in organisms involves two main scopes: a) the measurement of biological effects; that is, the hormonal, biochemical and morphological changes encompassing stress (or adaptation) syndromes (Bayne et al., 1975; 1978), and b) the use of some species as "sentinels" or "bioindicators" of marine pollution in order to determine and intercompare pollutant trends from wide geographical areas (Goldberg, 1980; Farrington et al., 1983).

In any case, the sampling procedures for marine organisms are rather straightforward. Perhaps, more important than the sampling technique is the selection of adequate marine species. Burns and Smith (1981) and Farrington et al (1983) have summarized the rationale making the selection of mussels and oysters useful for the monitoring of pollution in coastal and estuarine areas. The general concepts are also presented in chapter 6 of this book. In this sense, Table 13 reports the concentration ranges observed for metals, aliphatic and aromatic hydrocarbons and chlorinated components in mussels collected in different parts of the world representing levels of moderate pollution. Fishes, marine birds and mammals have also been considered for the monitoring of specific pollutants in open sea waters.

The amount of lipid present in a tissue or an organism exerts a powerful influence on the effective accumulation of many lipophilic components, such as the organochlorines. Specific types of lipids and their relative proportions also affect the ability for the retention of hydrophobic species. In contrast, these relationships are not so clearly observed for other components like metals. Several extraneous variables (e.g., sex, age, seasonality) are also important for the accumulation capacity of many organisms although these aspects are sometimes wholly or partially explained by the covariance of the amount of lipids in the body of the organism or its tissues. The factors affecting the processes of accumulation of lipophilic components in organisms have been compiled and discussed by Phillips (1980).

TABLE 13

REPRESENTATIVE LEVELS OF TRACE POLLUTANTS IN MUSSELS (Mytilus sp.)
COLLECTED IN MODERATELY POLLUTED COASTAL AREAS

ELEMENT	COMPONENT	ELEMENT	COMPONENT
Metals[a]			
Cadmium	0.1–5	Mercury	0.1–2
Chromium	0.1–15	Manganese	0.5–400
Copper	0.1–20	Nickel	0.5–20
Iron	50–3000	Lead	0.1–30
Zinc	50–600		
Aliphatic hydrocarbons[a]			
Unresolved	0.1–2000		
Resolved	0.5–200		
Total	1–2000		
Aromatic monocyclic hydrocarbons[a]		Aromatic polycyclic hydrocarbons[a]	
Unresolved	0.1–800	Unresolved	0.1–800
Resolved	0.5–200	Resolved	0.1–200
Total	0.5–1000	Total	0.5–1000
Chlorinated components[b]			
pp'DDE	0.1–300		
pp'DDD	0.1–500		
pp'DDT	0.1–1000		
op'DDT	0.1–400		
Polychlorobiphenyls	1–1500		

a) ug/g; b) ng/g

250

3. SAMPLE HANDLING PROCEDURES

The sample handling procedures are intermediate stages of the general analytical process devoted to the transfer of the analytes present in the samples into a suitable form for instrumental measurement. They obviously depend on the type of samples to be analyzed, the chemical properties of the analytes of interest, and the available instrumentation.

In the case of the analysis of metals they are relatively simple (see procedures H12-18, H20), generally involving acidic digestions. When chemical speciation is intented the procedures become more elaborated encompassing successive treatments of increasing chemical strength (see procedures H22-24 and also Obiols et al.,1986). The chemical speciation procedures for trace metals have been reviewed by Zolotov et al. (1982) and Florence and Batley (1980).

When organic species are to be analyzed the situation becomes more complicated because they must be detected within complex mixtures constituted by a high diversity of components of similar chemical properties. Unlikely to metals high specific and sensitive methods for the determination of particular organic components in such mixtures are not currently available. As described in the preceeding chapter such a problem may only be mastered by tandem mass spectrometry, but this type of instrumentation is not of common use for monitoring purposes.

The sample handling procedures for trace organic components are more or less elaborated depending on the final objective of the whole analytical process. To some extent, they are relatively simple when only bulk determinations of groups of components are intended. In these cases the procedures are aimed to the extraction of the group of analytes from the sample and to the preparation of adequate solutions for operational methods based on the measurement of a physico-chemical property characterizing the species of interest. Thus, H4 describes a procedure for the analysis of hydrocarbons present in marine water ($\sim$ 2 1) consisting in the extraction of the whole hydrocarbon mixture from the aqueous matrix and its concentration into a solution of n-hexane suitable for UV-fluorescence measurement (Method M2 in Section 4). However, if hydrocarbons must be individually determined other sampling techniques and handling procedures must be used (H2 and H3).

The procedures consisting in the mere extraction of determined classes of analytes sometimes include the formation of colored complexes with the species of interest for ulterior spectrophotometric determination. This is the case, for instance of procedures H25 and H26 devoted, respectively, to the analysis of anionic and cationic detergents in water. Methylene blue is added in the former case and bromphenol blue in the latter. Many other dyes have also been applied to the analysis of surfactants (i.e., methyl green, azure, ferroin, rhodamine, etc.). Diverse procedures for the analysis of anionic, cationic and non-ionic surfactants in water have been compiled by Leithe (1973) and Swisher (1987).

The individual analysis of organic trace components requires more elaborated procedures generally encompassing two main stages: a) the extraction of the analytes from the sample, and b) the isolation of the components of interest in fractions of chemical products with the same functionality. The complexity of these fractions will in turn require their analysis by means of instrumental techniques involving chromatographic separation (as described in the operational methods M1, M3, M5).

Different systems of extraction have been reported in the literature and their

251

performances have been compared. These include soxhlet extraction (Wise et al., 1977). Soxhlet extraction with internal thimble stirring (Durand et al., 1970), direct reflux, with or without alkaline hydrolysis (Hilpert et al., 1978) and steam distillation (Bellar et al., 1980). Lower temperature and shorter-time extraction procedures have also been applied in order to diminish the risk of losses for low-boiling or labile components. Among these are ball-mill tumblers (Hilpert et al., 1987), mechanical shaking and sonication (Wise et al., 1977; Bellar et al., 1980). Sonication and Soxhlet extraction have recently been evaluated for the analysis of aliphatic, aromatic and chlorinated hydrocarbons in sediments (Grimalt et al., 1984). Diverse solvents, sediment/solvent ratios and sample pretreatment systems have been included. Freeze-drying and sieving of the sediment through a 250 um filter before extraction have been strongly recommended by these authors.

The isolation of the analytes in chemical fractions is devoted to remove components of different polarity or size that could interfere in the subsequent instrumental chromatographic analysis by either masking or changing the perfomance of the capillary column. Partition or adsorption processes are used. In the former case, a destructive stage (i.e., alkaline or acidic digestions) for the elimination of undesired chemical products is often included. The adsorption processes may be carried out by thin layer or column chromatography although the latter is usually selected for its easy sample recovery which facilitates the achievement of quantitative data.

Many different adsorbents have been used but silica gel, alumina and Florisil are the packings of wider application. The emphasis of these column separation procedures is placed on fractionation rather than on individual component resolution which is left for subsebquent instrumental analysis. The activity of the adsorbent is an important property for the column processes being usually controlled by addition of known amounts of water after fully activation by adequate heating.

Silica is the most widely used chromatographic adsorbent. Among other advantages it has an unsurpassed capacity for both linear and nonlinear isotherm separations and almost complete inertness toward labile samples. Chromatographic silicas (empirical formula SiO_2 x H_2O) are amorphous, porous solids which can be prepared in a wide range of surface areas (200-800 m^2/g) and average pore diameters. Silica may be regarded as a typical polar adsorbent. The mechanisms by which retention occurs are not understood with absolute certainty but there is a general agreement in relating its selectivity with the presence of reactive surface hydroxyl groups. These interact with the adsorbed molecule by hydrogen bonding, the absorbate molecule usually acting as an electron donor. In consequence, an important aspect for sample fractionation is the degree of hydroxylation of the adsorbent. This is adjusted by activation at temperatures of 150-200°C (sufficient to drive off most of the adsorbed water without significant loss of surface hydroxyls) and latter addition of a known proportion of water (i.e., 5% as shown in procedure H1).

Alumina is the second most popular adsorbent used in column chromatography (empirical formula Al_2O_3 x H_2O). It is prepared by low temperature (< 700°C) dehydration of alumina trihydrate and is a mixture of the crystalline form γ -alumina with perhaps a small amount of the less active α-alumina and sodium carbonate. Adsorption interactions on alumina are primarily electrostatic, being recognized three types of active sites: a) acidic or positive field sites, b) basic or proton acceptor sites, and c) electron acceptor (charge transfer) sites. Most sample molecules adsorb on alumina sites of acidic character (a) but acidic molecules (pK < 13) adsorb with proton transfer to basic sites (b). Polycyclic aromatic hydrocarbons adsorb on alumina as charge

252

transfer complexes (c sites) permitting their separation by ring number. Like silica, alumina may be regarded as a typical polar adsorbent having a similar elution order for the components according to their functionality. Adsorbent activity is also controlled by addition of known amounts of water (i.e. 5% in procedure H1) after activation at ≈ 350°C. This involves the formation of layers of chemisorbed water, not hydroxyl groups. Alumina is commercially available in three forms: neutral (pH 6.9-7.1), that of most widely use, basic (pH 10-10.5), used to separate acid-labile substances, and acid (pH 3.5-4.5), prepared by acid-washing of neutral alumina, used for the separation of base-sensitive samples or relative strong acids that tend to chemisorb on neutral alumina.

Florisil (magnesium silicate) is prepared by coprecipitation of silica and magnesia. It has intermediate separation properties between silica and alumina which are also dependent on the presence of variable amounts of sodium sulfate. Florisil is permanently acidic which leads to many chemisorption effects on basic solutes (basic nitrogen compounds may not be eluted) and on relatively neutral components such as many polycyclic aromatic hydrocarbons. The activity of Florisil is controlled by heating at 300°C overnight and deactivation with a known amount of water (i.e. 6% in procedure H10).

More information on these and other adsorbents is given in Snyder (1968). Their use for the separation of petroleum and chlorinated hydrocarbon fractions has been recently reviewed (Aceves et al., 1988).

The sample handling procedures presented here allow the analysis of aliphatic and aromatic hydrocarbons, chlorinated components and metals of environmental concern in aerosols, water, sediments and organisms. They are based on the experience accumulated in our laboratory on environmental analytical chemistry. Obviously, other alternative procedures or variations from those reported here can also be sucessfully used. Their consideration is, however, out of the scope of the present paper.

H1. Hydrocarbons in sediments (aliphatic and aromatics)

20-50 g of sediment are extracted with (2:1) methylene chloride-methanol in a Soxhlet apparatus for 24 hours. The extracts are vacuum evaporated near to dryness and saponified overnight with 10 ml of 6% methanolic KOH. Neutrals are recovered by extracion with hexane (3 x 15 ml) and optionally desulphurized with activated copper. After new vacuum evaporation down to a volume of about 0.5 ml the mixture is introduced into a column (25 x 0.9 cm i.d.) filled successively with 8 g of silica (bottom) and 8 g of alumina (top), both deactivated with 5% of water. Then, three fractions are obtained by elution with hexane (20 ml), 10% of methylene chloride in hexane (20 ml) and 20% methylene chloride in hexane (40 ml), corresponding, respectively, to the saturated, monocyclic and polycyclic hydrocarbons. These fractions are again concentrated by vacuum evaporation to a suitable volume and submitted to instrumental analyses.

H2. "Particulate" hydrocarbons in water (aliphatics and aromatics)

A Teflon impeller pump pushes the water through a pre-combusted glass fiber filter (500°C overnight) with a rated pore size of 0.3 um (Gelman corp., type A/E). Once enough water (100-1000 l) is passed through the filter this is torn into 1 cm^2 pieces with stainless steel tweezers, placed into a round bottom flask and reflux-extracted for 18 h with two successive 750 ml volumes of methanol—methylene chloride (1:2). These extracts are combined, vacuum evaporated down to about 0.5 ml and saponified overnight with 10 ml of 6% methanolic KOH. Hydrocarbons are recovered by extraction with hexane (3 x 15 ml) and again vacuum evaporated to about 0.5 ml. They are then submitted to the column chromatography fractionation process described in procedure H1.

H3. "Aqueous" hydrocarbons in water (aliphatics and aromatics)

A Teflon impeller pump pushes the water through a pre-combusted glass fiber filter (500°C overnight) with a rated pore size of 0.3 um (Gelman corp., type A/E). The filtered water is passed through a teflon column (30.5 x 6.35 cm o.d., 5.08 cm i.d.) containing five polyurethane foam plugs which had initially been cleaned by reflux extraction (minimum 72 h) with acetone in a large Soxhlet apparatus. After the plugs were packed into the Teflon columns, cold hexane and acetone were successively passed through, reduced to a small volume and analyzed by gas chromatography until satisfactory blanks were obtained. Once enough water (100-1000 l) is passed through the polyurethane foam column, this is eluted successively with 500 ml of acetone and 500 ml of hexane.

The combined extracts are reduced to about 700 ml by vacuum rotary evaporation and partitioned in a 2 l separatory funnel with 700 ml of purified water, the hydrocarbons are collected in the hexane layer. The acetone-water mixture is back-extracted once with an additional 300 ml of hexane, previously passed through the column. The combined hexane extracts are reduced to a volume of approximately 0.5 ml by vacuum rotary evaporation and saponified overnight with 10 ml of 6% methanolic KOH. The hydrocarbons are recovered by extraction with hexane (3 x 15 ml) and again vacuum evaporated to about 0.5 ml. Then, they are submitted to the column chromatography fractionation process described in procedure H1.

H4. Total hydrocarbons in water.

50 ml of carbon tetrachloride are added to the water sample ($\sim$ 2 liters) and the mixture is vigorously shaken for 3 min. After separation of the two liquids the organic layer is collected with a pipette and introduced into a flask containing anhydrous sodium sulphate. The water is re-extracted with additional 50 ml of carbon tetrachloride which are combined with the former extract. The solvent is then decanted, the sodium sulphate is rinsed with more carbon tetrachloride and the combined solution of extracts and rinses concentrated by vacuum rotary evaporation. The sample is finally reconstructed to adequate volume with n-hexane.

H5. Hydrocarbons in organisms (aliphatic and aromatics)
The tissue to be analyzed is separated from the rest of the organism under appropriately clean conditions and homogenated. Known weights of these homogenates (8-10 g) are placed in centrifuge tubes together with 10 ml of 6N aqueous solution of NaOH. The tubes are closed with Teflon-lined caps and shaken for several minutes, then maintained at 30°C for 18 h. The mixture is extracted three times with ethyl ether (15-10-10 ml) and the combined extracts are vacuum evaporated near to dryness. The residue is dissolved in hexane (0.5 ml) and submitted to the column chromatography fractionation process described in procedure H1.

H6. Hydrocarbons in aerosol (aliphatics and aromatics)

The collection filters are torn into 1 cm^2 pieces with stainless steel tweezers, placed into a round bottom flask and Soxhlet extracted 30 hours with methylene chloride. The extract is vacuum evaporated near to dryness and saponified under inert atmosphere with 2N methanolic KOH for two hours. The neutrals are recovered by extraction with hexane (3 x 15 ml) and again concentrated to a small volume ($\sim$ 0.5 ml) by vacuum rotary evaporation. Then, the mixture is introduced into a column (4 x 0.5 cm i.d.) filled with 2 g of fully activated silica. Two fractions are collected, aliphatic hydrocarbons (6 ml of hexane) and aromatic hydrocarbons (3 ml of 0.2% ethyl acetate in hexane + 3 ml of 0.05% ethyl acetate in hexane + 2 ml of hexane). These two fractions are submitted to instrumental analysis.

H7. Chlorinated compounds in sediments

20-50 g of freeze-dried sediment are extracted with (2:1) methylene chloride-methanol in a Soxhlet apparatus for 24 h. The extracts are vacuum evaporated near to dryness and submitted to the column chromatography fractionation process described in procedure H10.

H8. "Particulate" chlorinated compounds in water

They are filtered and concentrated following the same procedure described for the aliphatic and aromatic hydrocarbons (H2) but no saponification step is performed. Then, they are submitted to the column chromatography fractionation process described in procedure H10.

H9. "Aqueous" chlorinated compounds in water

They are extracted from water and concentrated following the same procedure described for the aliphatic and aromatic hydrocarbons (H3) but no saponification step is performed. Then, they are submitted to the column chromatography fractionation process described in procedure H10.

H10. Chlorinated compounds in organisms

Aliquots of approximately 70 g of homogenate are freeze-dried, mixed with anhydrous sodium sulphate and Soxhlet-extracted with methylene chloride for 24 h. After removal of the solvent by rotary evaporation, lipids are separated by florisil column chromatography (Supelco 60/100 mesh florisil; activated at 300°C overnight and deactivated with 6% water). A column of 25 x 0.5 cm i.d. is filled with 5 g of absorbent suspended in hexane. Two fractions are collected, the former (28 ml of hexane) containing PCBs, DDEs, DDDs, aldrin heptachlor and hexachlorobenzene, and the later (20 ml of 15% ethyl ether in hexane) containing DDTs, HCHs, dieldrin and, endrin. These fractions are concentrated by vacuum evaporation to a suitable volume and submitted to instrumental analysis.

H11. Chlorinated compounds in aerosols

The collection filter is torn into 1 cm^2 pieces with stainless steel tweezers, placed into a round bottom flask and Soxhlet extracted 30 hours with methylene chloride. The extract is vacuum evaporated near to dryness and, optionally, submitted to the column chromatography fractionation process described procedure H10 before instrumental analysis.

H12. Metals in sediments (total concentration)

1 gr of freeze dried 250 um sieved sediment is introduced in a 40 ml centrifuge tube provided with a screw Teflon stopper. 5 ml of conc. HNO_3 are added drop by drop and the tube is closed and heated to 95°C for 1 h. Then, the mixture is diluted to 25 ml with deionized water, centrifuged (15 min at 4000 rpm) and the liquid is decanted. 25 ml of deionized water are added again for washing and after centrifugation the liquid is decanted, combined with the initial solution and submitted to instrumental analysis.

H13. Metals in water (total concentration)

0.6 ml of conc. HNO_3 and 0.25 ml of 70% $HClO_4$ are added to 100 ml of unfiltered or filtered water. The samples are then evaporated in covered beakers in a $HClO_4$ fume hood until evolution of white $HClO_4$ fumes, diluted to a final volume of 25 ml and submitted to instrumental analysis.

H14. Metals in water suspended particles

About 1L of water is filtered through a 0.4 um Nucleopore filter. This is then placed into a Pyrex tube and 15 ml of 0.1 N HNO_3 are added. The filter is digested at 50-55°C during 2 h. After cooling the liquid solution is decanted, the filters washed twice with the same reagent and all fractions added together. Instrumental analysis is performed on this final solution.

H15. Metals in organisms

0.3 gr of homogenized freeze dried tissue are mixed with 2 ml of conc. HNO_3. The mixture is kept overnight at room temperature and then heated to 130°C for 4 h in a pressure container. After cooling deionized water is added (to a volume of 15 ml) before instrumental analysis.

H16. Metals in aerosols (total concentration)

An adequate portion of the collection filter is broken into fragments and placed into a centrifuge tube of 80 ml capacity provided with a screw thread stopper. 5 ml of conc. HNO_3 are added drop by drop, following 2 ml of conc. HCl and 20 ml of H_2O. The resulting solution is heated at 95°C for 90 min. Then, the mixture is centrifuged (15 min at 4000 rpm) and the liquid is decanted, ashing twice with the same reagent and adding together all fractions. Instrumental analysis is performed on this final solution.

H17. "Aqueous" mercury in water

The water sample is filtered as soon as possible through a 0.4 um filter (Nucleopore or similar). After filtration the water samples are acidified with 2 ml of H_2SO_4 per 100 ml of sample. For the determination of total "aqueous" mercury (organic + inorganic mercury compounds) the water samples sould be photo-oxidized with a 500 watt U.V. lamp for 8 hours.

H18. "Particulate" mercury in water

The loaded filter (0.4 um, Nucleopore) is rinsed with about 50 ml of bi-distilled water to remove salt. Then, it is transferred to a Pyrex flask and 10 ml of conc. H_2SO_4, 5 ml of conc. HNO_3 and 10 ml of 5% $KMnO_4$ are added. The flask is placed in a water bath at ambient temperature and digested for 12 h. After digestion 10 ml of 3% hydroxylamine hydrochloride solution are added. Mixing is performed until the solution is clean and the precipitated MnO_2 is dissolved. Then, this solution is centrifuged at 3000 rpm for 10 min to eliminate the remaining particulate matter. The supernate is transferred to a 100 ml volumetric flask and brought to volume with bi-distilled water.

H19. Methylmercury in organisms

5 g of homogenized tissue are introduced in a 200 ml centrifuge tube with 60 ml of distilled water, 14 ml of conc. HCl and 10 g of NaBr. After mixing, 70 ml of toluene are added and the tube is closed and hand shaken for 15 min. The toluene extract is transferred into a separating funnel after centrifugation. 8 ml of cysteine solution (0.5% of cysteine hydrochloride and 0.34% of sodium acetate in (3:1) ethanol-water) are added and the mixture is vigorously shaken for 2 min. After further centrifugation the aqueous layer is decanted and 5 ml of 6N HCl and 2 gr of NaBr are added for the de-complexation of methylmercury from the thiol-compound with a strong acid (HBr). The mixture is extracted with 3 x 20 ml of toluene and the organic extract is then dried over anhydrous sodium sulphate and brought to volume for gas chromatographic analysis.

H20. Beryllium in sediments

0.1-0.25 g of sediment are introduced into a PTFE-lined decomposition vessel along with 2 ml of conc. HNO_3 and 4 ml of conc. HF. The container is closed and heated at 225°C for 16 h. After cooling the vessel content is transferred to a platinum dish and evaporated to dryness. Then, 4 ml of 20N H_2SO_4 are poured and the mixture is evaporated to fumes of sulphuric acid. After cooling once more, 4 ml of 3M HCl and 7 ml of 0.3 N EDTA solution are added together with ammonia to bring the pH to 8. The mixture is transferred to a separating funnel with a solution of 0.0005% of bromphenol blue and 0.0004% of NaOH in water. More ammonia (12M) is poured dropwise until the indicator colour remains a permanent blue. Two drops of ammonia are added in excess to set pH > 8. 10 ml of (1:9) acetylacetone-xylene are used for the extraction of beryllium by 6 min shaking. The aqueous layer is discarded and the extracted beryllium is stripped from the organic layer by shaking with 4 ml of 3M HCl for 10 min. The acidic solution is brought to volume with 3M HCl for instrumental analysis.

H21. Arsenic in organisms

About 0.5 g of dry weight sample are introduced into a homogenizer along with an ashing slurry (50 g of magnesium oxide powder suspended in 1 liter of a solution of 3% magnesium nitrate) making a combined weight of 15 g. The mixture is throughly homogenized and dried at 80°C for one hour. Then an aliquot (10 g) is introduced into a 50 ml nickel crucible which is placed in a well ventilated muffle furnace. The mixture is heated at 200°C (1 h), 300°C (1 h) and 500°C (8 h). After cooling to room temperature the residue is dissolved in 15 ml of 6 M HCl and the resulting solution transferred to a quartz vessel. Two more rinsings of the crucible with 15 ml of distilled water are also added to the quartz vessel. Then 2 ml of conc. $HClO_4$ and 0.5 ml of conc. H_2SO_4 are introduced into the quartz vessel and this is heated at 31°C until fumes of SO_3 appear. After cooling, the solution is transfered into a 25 ml volumetric flask and brought up to volume with distilled water.

H22. Metals in sediments (chemically speciated in six fractions)

2 gr of freeze dried 250 um sieved sediment are introduced in a 80 ml centrifuge tube provided with a screw teflon stopper and then extracted with the following set of succesive reagents: Fraction I (soluble metal-present in interstitial water). 10 ml of deionized water. Fraction II (exchangeable metal). 8 ml of 1 M NH_4 OAc. Mechanical agitation during 1 h at room temperature. Fraction III (metal associated to carbonates). 8 ml of 1 M NaOAc and then HOAc until pH=5 (measured with pH meter). Mechanical agitation for 6 h at room temperature. Fraction IV (metal associated to oxides). 15 ml of 0.04 M NH_2OH:HCl and slow heating until total evaporation. New addition of 10 ml of 0.04 M NH_2OH:HCl and return to dryness with gentle heating. Redissolution with 10 ml of 1 N HNO_3. Fraction V (metal associated to organic matter and sulfides). 5 ml of 0.02 M HNO_3 and 5 ml of 30% H_2O_2 or Na_2CO_3. Heating to 85°C for 30 min. Add 3 ml of 30% H_2O_2 and heat again at 85°C for 30 min. Finally, addition of 5 ml of 3 M NH_4OAc and stir for 30 min. Fraction VI (residual metal). 10 ml of (4:1) conc. $HNO_3/HClO_4$. Heating to 90°C for 1 h. In each fraction, after the extraction process, the mixture is centrifuged (~ 15 min at 4000 rpm) and the liquid is decanted, washing twice with the same reagent and adding together the rinses and the fraction. Then the solid residue is extracted for the next fraction in the same tube. The solutions containing the extracts are submitted to instrumental analysis.

H23. Metals in filtered water (chemically speciated in three fractions)

Fraction A: About 20 ml of 0.4 um Nucleopore filtered water are acidulated to pH 4.8 with 0.1 N HCl. Dissolved oxygen is removed by bubbling nitrogen for 30 min and then the mixture is submitted to electrochemical analysis. Fraction B: Another aliquot of about 100 ml is irradiated with ultraviolet light (medium pressure Hg lamp) for 6 hours, then acidulated to pH 4.8 and submitted to electrochemical analysis following the procedure previously described. Fraction C: About 100 ml of the same water are again irradiated with UV light for 6 hours, then 20 ml of this solution are separated and 0.4 ml of 12 N HCl are added. Nitrogen is bubbled for 30 min and the mixture is analyzed by an electrochemical method. Fraction A is considered to be the labile metal (non complexed). Fraction B minus Fraction A is considered the metal associated to organic matter. Fraction C minus Fraction A is considered to be the metal strongly associated to inorganic components (forming stable complexes).

H24. Metals in aerosols (chemically speciated in four fractions)

An adequate portion of the collection filter is broken into fragments and placed into a centrifuge tube of 80 ml capacity provided with a screw thread stopper, and then extracted with the following set of successive reagents: Fraction I (soluble and exchangeable metal). 30 ml of 1% NaCl. Mechanical agitation during 30 min at room temperature. Fraction II (carbonates and oxides). 20 ml of 0.04 M NH_2OH:HCl in 25% AcOH. Heating to 95°C for 1 h, agitating occasionally. Fraction III (metal bound to organic matter). 20 ml of 0.02 M HNO_3 + 10 ml 30% H_2O_2. Heating to 85°C for 1 h agitating occasionally, further 3 ml 30% H_2O_2 and then 5 ml of 3.2 M NH_4OAc in 20% HNO_3, continuous agitation at room temperature. Fraction IV (residue metal). 5 ml of conc. HNO_3,

2 ml of conc. HCl and 20 ml H_2O. Heating to 95°C for 90 min. In each fraction, after the extraction process, the mixture is centrifuged ($\sim$15 min at 4000 rpm) and the liquid is decanted, washing twice with the same reagent and adding together all the fractions. Then the solid residue is extracted for the next fraction in the same tube. The solutions containing the extracts are submitted to instrumental analysis.

H25. Anionic detergents in water

About 100 ml of water are introduced into a separately funnel with 10 ml of 1.25% $Na_2HPO_4 \cdot 2H_2O$ in water (pH adjusted to 10 with a dilute solution of NaOH), 5 ml of 3.5% of methylene blue in water, and 15 ml of chloroform. The mixture is shaken for 1 min. The decanted chloroform layer is shaken in a second separatory funnel with a mixture of 100 ml distilled water and 5 ml of acidified (a few drops of conc H_2SO_4) methylene blue in water (3.5%). The mixture is allowed to settle and the chloroform layer is separated. Extraction in both separatory funnels is repeated two more times, each time with 10 ml chloroform; the chloroform extracts are combined and brought to 50 ml by vacuum rotary evaporation. Measurement is performed in a spectrophotometer at 650 um using tetrapropylenebenzenesulfonate as standard.

H26. Cationic detergents in water.

A water sample of about 100 ml is shaken for 3 min in a separatory funnel with 10 ml citrate buffer solution, 5 ml 0.1 N HCl, 2 ml bromphenol blue solution and 50 ml chloroform. After setting the chloroform layer is separated and measured spectrophotometrically at 416 nm using adequate solutions of cetyltrimethylammonium bromide as standard. The citrate buffer solution is prepared by dilution of 21 g citric acid in 200 ml 1N Na OH and dilution to 1 l with water. 310 ml of this solution are diluted to 1 l with 0.1 N HCl. The bromphenol blue solution is prepared by dissolution of 0.15 g bromphenol blue in 200 ml 0.01 N NaOH and treated with 42 ml 0.1 N HCl.

4. OPERATIONAL METHODS.

As it has been stated above, the analysis of complex mixtures of trace organic pollutants requires the use of chromatographic techniques. This is effectively the case for the aliphatic and aromatic hydrocarbons as well as for the chlorinated components (see M1 and M5 in the list of methods reported below). Whenever possible these techniques are used in combination with specific detectors affording higher sensitivity for a defined group of elements such as in the analysis of chlorinated components by capillary gas chromatography with electron capture detection (M5). Likewise, the analysis of sulphur or nitrogen containing PAH requires, respectively, the use of flame photometric or thermionic detectors (White, 1985; Lee et al., 1981). The use of multiparametric detectors in hyphenated couplings represents another way for the achievement of high specificity and sensitivity. An application of this approach is given in M3 for the analysis of PAH by GC-MS.

The selective detectors are also used in liquid chromatography. In this sense, the use of specific UV bands in absorption or fluorescence spectrophotometry for the analysis of PAH is well-known. Another interesting application concerns the determination of trace organo-phosphorous components. This is currently performed by GC with thermionic, flame photometric or mass spectrometric detection (Kjolholt, 1985; Zenon-Roland et al., 1984; Singh et al., 1986) and now the field of application of these methods is being extended to the thermally unstable products by combination of these detection systems as well as electron capture with liquid chromatography (Maris et al., 1985; Gluckman et al., 1986; Barceló et al., 1987a and b).

The chromatographic techniques used in combination with specific detectors are also adequate for the analysis of many organometallic components. Thus, methods based on gas chromatography with electron capture or flame photometric detection have been reported for the analysis of tributyltin chloride (Junk and Richard, 1987) and butyltin and butylmethyltin species (Matthias et al., 1986; Unger et al., 1988), respectively. Gas chromatography with electron capture detection is also the technique of choice for the determination of methylmercury (M28). In some cases more sophisticated detectors are required, such as in the analysis of volatile alkyllead species, where gas chromatography coupled to an atomic absorption spectrophotometer is used (De Jonghe et al., 1981; Chakraborti et al., 1987).

However, bulk determinations of a group of pollutants based on specific physico-chemical properties are also possible. This is illustrated in the analysis of total aromatic petroleum hydrocarbons by UV fluorescence emission (M2). The same approach is also useful for the determination of individual components such as benzo(a)pyrene (M4).

Metals are also determined in bulk mixtures taking advantage of specific properties such as their absorption or emission spectra. As indicated in the preceeding chapter, inductively coupled plasma-atomic emission spectroscopy and atomic absorption spectrophotometry are the techniques of current choice, the latter encompassing three types of atomization, flame, graphite furnace and hydride formation. These are selected according to the physico-chemical properties of the studied metals and their concentration in the environmental compartment analyzed. In cases where interferences produced by important matrix effects limitate the application of these techniques (i.e. chlorine content in marine water), methods based on other instrumental applications such as anodic stripping voltammetry (M29) and potentiometric stripping analysis (M30) are used.

The operational methods presented here correspond to those currently used in our laboratory for monitoring purposes. They are based on the analytical experience that we have accumulated during our monitoring and research work on marine pollution. They obviously admit many variations which are not included here for practical reasons. We have discussed some of them elsewhere (Grimalt and Albaigés, 1982; Albaigés and Grimalt, 1982; Bayona et al., 1983; Sicre et al., 1987b). The following basic list allows the determination of aliphatic and aromatic hydrocarbons, chlorinated components and metals of interest for monitoring purposes in sediments, water, organisms and aerosols.

M1. Aliphatic or aromatic hydrocarbons

Capillary gas chromatography. Detection by flame ionization. Injection: splitless or on-column. Column: 25 m x 0.25 mm i.d. coated with SE-54 (surface film thickness 0.15 um). Carrier gas: hydrogen or helium. Temperatures: oven, from 60°C to 310°C at 6°C/min; injector, 300°C; detector, 330°C.

M2. Total aromatic hydrocarbons

UV-fluorescence. Emission mode. Excitation wavelength 300 nm. Slit 20 nm. Emission spectra recorded from 320 nm up to 450 nm. Slit 2.5 nm. Quantitation at 360 nm.

M3. Polycyclic aromatic hydrocarbons

Gas chromatography coupled to mass spectrometry. Injection: splitless or on-column. Column: 25 m x 0.25 mm i.d. coated with SE-54 (surface film thickness 0.15 um). Carrier gas: helium. Temperatures: oven, from 60°C to 310°C at 6°C/min; injector, 300°C; transfer line, 300°C; ion source, 200°C. Detector used in the electron impact mode. Emission current 0.2 ma. Electron energy 70 eV. Scanning: from 50 to 500 daltons at 1 s per decade. The concentrations of individual PAH are obtained by mass fragmentography from the molecular ions of the compounds present in the samples. For example: m/z 178 anthracene and phenanthrene, m/z 202 fluoranthene and pyrene, m/z 228 chrysene plus tryphenylene and benzo(a)anthracene, m/z 252 benzo(a)pyrene, benzo(e)pyrene and benzofluoranthenes, m/z 276 benzo(ghi)perylene and indeno(1,2,3-cd)pyrene. Perdeuterated anthracene-d_{18} (m/z 188) is used as internal standard.

M4. Benzo(a)pyrene

UV-fluorescence. Synchronous excitation mode. Slits: 5 nm. Constant difference between emission and excitation wavelenghts: 20 nm. Quantitation at 402 nm (emission units).

M5. Chlorinated components

Capillary gas chromatography. Detection by electron capture. Injection: Splitless or on-column. Column: 25 m x 0.25 mm i.d. coated with SE-54 (surface film thickness 0.15 um). Carrier gas: helium. Temperatures: oven, from 60°C to 290°C at 6°1C/min; injector, 300°C; detector, 300°C.

M6. Mercury

Cold vapor atomic absorption spectrophotometry. Reducing agent: $NaBH_4$ (solution of 3% $NaBH_4$ in 1.5% NaOH). Cell temperature: 200°C. Carrier gas: nitrogen. Determination: 3 ml of the mineralized solution of the sample are introduced in the reaction vessel. Air is removed with nitrogen for 5 s. Then 1-5 ml of the reactive are added and the absorbance is measured at 253.7 nm. The circuit is cleaned with nitrogen during 30 s after the signal goes to ground level.

M7: Mercury in water

Flameless atomic absorption spectroscopy after concentration on a gold trap. 100 ml of an adequate aliquot coming from either "aqueous" or "particulate" mercury in water are placed into a flask along with 5 ml of 15% $SnCl_2$:$2H_2O$, mercury-free nitrogen (cleaned by bubbling in a solution of $KMnO_4$) is bubbled into the solution at a flowrate of 120 ml/min for 10 min. The gas passes through a gold collector (gold thread cut up into small pieces) for amalgamation of the liberated gaseous mercury. When the aeration is stopped the gold collection is heated with a heating coil (Ni-Cr heating wire with variable transformer) to a red colour to liberate gaseous elemental mercury. After a short time the gas flow is resumed entraining the gaseous elemental mercury into the measuring cell where absorbance at 253.7 nm is registered.

M8. Cadmium

Atomic absorption spectrophotometry. Atomization in a graphite furnace with the following program of temperatures: drying, to 120°C in 1 s holding 15 s; charring, to 250°C in 1 s holding 20 s; atomization, to 1100°C (ballistic heating) holding 10 s (nitrogen flow stopped and measurement of absorbance at 228.8 nm); cleaning, to 2500° C in 1 s holding 5 s (nitrogen flow re-started).

M9. Chromium

Atomic absorption spectrophotometry. Atomization with a reducing acetylene-air flame. Measurement at 357.9 nm.

M10. Chromium

Atomic absorption spectrophotometry. Atomization in a graphite furnace with the following program of temperatures: drying, to 120°C in 1 s holding 15 s; charring, to 1000°C in 10 s and to 1200°C in 1 s holding 10 s; atomization, to 2500°C (ballistic heating) holding 5 s (nitrogen flow stopped and measurement of absorbance at 375.9 nm); cleaning, down to 30°C in 1 s holding 5 s (nitrogen flow re-started).

M11. Cobalt

Atomic absorption spectrophotometry. Atomization with an (1:1) acetylene-air flame. Measurement at 240.8 nm.

M12. Cobalt

Atomic absorption spectrophotometry. Atomization in a graphite furnace with the following program of temperatures: drying, to 100°C in 1 s holding 10 s and to 120°C in 1 s holding 5 s; charring, to 1000°C in 10 s holding 10 s; atomization, to 2200°C (ballistic heating) holding 10 s (nitrogen flow stopped and measurement of absorbance at 241.1 nm); cleaning, to 2500°C in 1 s holding 3 s (nitrogen flow re-started).

M13. Copper

Atomic absorption spectrophotometry. Atomization with an (1:1) acetylene-air flame. Measurement at 324.8 nm.

M14. Copper

Atomic absorption spectrophotometry. Atomization in a graphite furnace with the following program of temperatures: drying, to 120°C in 1 s holding 15 s; charring, to 800°C in 10 s and to 900°C in 1 s holding 10 s; atomization, to 2300°C (ballistic heating) holding 8 s (nitrogen flow stopped and measurement of absorbance at 324.7 nm); cleaning, to 2500°C in 1 s holding 1 s (nitrogen flow re-started).

M15. Lead

Atomic absorption spectrophotometry. Atomization with an (1:1) acetylene-air flame. Measurement at 283.3 nm.

M16. Lead

Atomic absorption spectrophotometry. Atomization in a graphite furnace with the following program of temperatures: drying, to 120°C in 1 s holding 15 s; charring, to 400°C in 15 s and to 500°C in 1 s holding 10 s; atomization, to 1400°C (ballistic heating) holding 10 s (nitrogen flow stopped and measurement of absorbance at 283.3 nm); cleaning, to 2500°C in 1 s holding 3 s (nitrogen flow re-started).

M17. Nickel

Atomic absorption spectrophotometry. Atomization with an (1:1) acetylene-air flame. Measurement at 232.0 nm.

M18. Nickel

Atomic absorption spectrophotometry. Atomization in a graphite furnace with the following program of temperatures: drying, to 120°C in 1 s holding 15 s; charring, to 700°C in 15 s and to 1200°C in 1 s holding 10 s; atomization, to 2500°C (ballistic heating) holding 5 s (nitrogen flow stopped and measurement of absorbance at 232.0 nm); cleaning, to 20° C in 1 s holding 3 s (nitrogen flow re-started).

M19. Iron

Atomic absorption spectrophotometry. Atomization with an (1:1) acetylene-air flame. Measurement at 248.3 nm.

M20. Manganese

Atomic absorption spectrophotometry. Atomization with an (1:1) acetylene-air flame. Measurement at 280.0 nm.

M21. Manganese

Atomic absorption spectrophotometry. Atomization in a graphite furnace with the following program of temperatures: drying, to 120°C in 1 s holding 15 s; charring, to 800°C in 15 s and to 1000°C in 1 s holding 10 s; atomization, 2100°C (ballistic heating) holding 7 s (nitrogen flow stopped and measurement of absorbance at 279.5 nm); cleaning, to 2500°C in 1 s holding 3 s (nitrogen flow re-started).

M22. Zinc

Atomic absorption spectrophotometry. Atomization with an (1:1) acetylene-air flame. Measurement at 213.9 nm.

M23. Zinc

Atomic absorption spectrophotometry. Atomization in a graphite furnace with the following program of temperatures: drying, to 120°C in 1 s holding 15 s; charring, to 400°C in 10 s holding 10 s; atomization, to 1200°C (ballistic heating) holding 7 s (nitrogen flow stopped and measurement of absorbance at 213.9 nm); cleaning, to 2500°C in 1 s holding 5 s (nitrogen flow re-started).

M24. Selenium

Cold vapor atomic absorption spectrophotometry. Reducing agent: $NaBH_4$ (5% of $NaBH_4$ in 0.2 N NaOH). Cell temperature: 900°C. Carrier gas: nitrogen. Determination: 10 ml of the mineralized solution of the sample are introduced in the reaction vessel. Air is removed with nitrogen for 5 s. Then 5 ml of the reactive are added and the absorbance is measured at 196.0 nm. The circuit is cleaned with nitrogen during 30 s after the signal goes to ground level.

M25. Arsenic

Cold vapor atomic absorption spectrophotometry. Reducing agent $NaBH_4$ (solution of 5% $NaBH_4$ in 0.2N NaOH). Cell temperature: 900°C. Carrier gas: nitrogen. Determination: 10 ml of the mineralized solution of the sample are introduced in the reaction vessel together with 1 ml of (1:3) conc. H_2SO_4/H_2O . Air is removed with nitrogen for 5 s. Then 5 ml of the reactive are added and the absorbance is measured at 193.7 nm. The circuit is cleaned with nitrogen during 30 s after the signal goes to ground level.

M26. Beryllium

Atomic absorption spectrophotometry. Atomization with a nitrous oxide-acetylene flame. Measurement at 234.9.

M27. Beryllium

Atomic absorption spectrophotometry. Atomization in a graphite furnace with the following program of temperatures: drying, to 110°C in 1 s holding 20 s; charring, to 800°C in 15 s holding 20 s; atomization, to 2700°C (ballistic heating) holding 6 s (nitrogen flow stopped and measurement of absorbance at 234.9 nm).

M28. Methylmercury

Gas chromatography. Detection by electron capture. Column: 5 m x 0.32 cm i.d. packed with 5% phenyldiethanolamine succinate on Chromosorb W or similar, 60/80 mesh, and lithium chloride. Carrier gas: nitrogen (60-75 ml/min). Temperatures: oven, 175°C; injector 200°C; detector 205°C. Column preparation: 0.5 g of lithium chloride and 1.5 g of phenyldiethanolamine succinate are dissolved in

50 ml of a mixture of 5% ethanol in acetone. 10 g of Chromosorb W, 60/80 mesh, are added. Air bubbles are eliminated from the suspension under vacuum. After gentle stirring for 10 min the mixture is transferred to a filter funnel and the liquid removed by suction. The stationary phase is air-dried on filter paper before packing. Column conditioning: The packed column is heated at 210°C for 18 h (25-30 ml/min N_2) without attaching the end of the column to the detector. Then, the gas chromatograph is brought to operational conditions of temperature and carrier flow rate and the column packing is saturated with methylmercury by injection of repeated aliquots (∿ 10 ng each) of methylmercury chloride in toluene. If the sensitivity is not sufficient and/or peak shape is poor the column can be improved by injection of aliquots of methylmercury iodide in toluene. This solution is prepared before use by adding sodium iodide to aqueous solutions of methylmercury chloride and extracting with toluene.

M29. Zinc, Cadmium, Lead and Copper by anodic stripping voltammetry

A three-electrode potentiostat equipped with a platinum foil electrode (counter), a saturated calomel electrode (reference) and a glassy carbon electrode (working) is used. These are screwed into the lid of a tritation vessel where the sample is agitated by a constant rate mechanical stirrer. Before each determination the working electrode is first polished with 3-um diamond paste for 30-60 s and then cleaned carefully with acetone. The sample is acidified to a suitable pH, 10-50 mg/l of mercury (II) is added and stirring is started. The plating potential is adjusted to -0.95 V .vs. standard counter electrode and the plating time to 1 min. A minimum of four plating/stripping cycles, each cycle comprising 60 s of plating and 15 s of stripping are then performed. Following this conditioning the electrode is ready for use in an analytical run. Background and sample determinations are carried out successively by applying the plating potential for a selected time determined by the sought-for concentration levels. Then stirring is stopped and after a 15 s rest period the metals are stripped from the mercury film by applying an anodic potential scan. The scan is stopped at + 0.05 V, and this potential is maintained for 30 s before the next determination is performed. The mercury film is removed at the end of a series of experiments by holding the electrode at + 0.5 V for 20 min.

M30. Zinc, Cadmium, Lead and Copper by potentiometric stripping analysis

A three-electrode potentiostat equipped with a platinum foil electrode (counter), a saturated calomel electrode (reference) and a glassy carbon electrode (working) is used. These are screwed into the lid of a titration vessel where the sample is agitated by a constant rate mechanical stirrer. Before each determination the working electrode is first polished with 3-um diamond paste for 30-60 s and then cleaned carefullly with acetone. The sample is acidified to a suitable pH, 10-50 mg/l of mercury (II) is added and stirring is started. The plating potential is adjusted to -0.95 V .vs. standard counter electrode and the plating time to 1 min. A minimum of four plating/stripping cycles, each cycle comprising 60 s of plating and 15 s of stripping are then performed. Immediately after completion of the last plating/stripping cycle, the plating potential and the plating time are set to the values to be used during subsequent analysis and the potentiometric stripping controller is initiated. For example, the simultaneous determination of zinc, cadmium, lead and copper can be achieved by plating for 10-60 min at -1.30 V. Stripping may occur at voltage .vs. standard counter electrode of about -1.0 (zinc), -0.6 (cadmium), -0.4 (lead) and -0.2 (copper).

5. CONCLUDING REMARKS

As it is well-known no unique analytical methodology nor reference materials are today available in environmental trace component analysis. Nevertheless, in table 14 diverse combinations of the sample handling procedures listed in section 3 and the operational methods listed in section 4 are indicated, allowing the determination of the trace elements of table 2 in sediments, water, organisms and aerosols. Again, it must be stated that these resulting specific methods are not presented here as the "best" possible methodology but as a group of methods based on our experience in environmental analysis allowing to obtain reasonably accurate, precise and reproducible data. For instance, they may be compared with the available reference methods developed by UNEP for the participants in the Regional Seas Programmes (referenced in table 15). Some of them are essentially coincident and others contain significant differences. The "correctness" of the analytical values obtained can only be established by means of quality control and quality assurance tests (Taylor, 1981; Dux, 1986; Albaigés and Grimalt, 1987) and on a comparative basis through intercalibration exercises performed on field sample replicates and trying to identify the sources of disagreement by further studies. These evaluations should not only be restricted to mere analytical methodology but also embrace sampling, running of blanks, isolation, separation and identification.

Diverse intercalibration exercises have been reported (Gearing et al., 1978; Hilpert et al., 1978; Wise et al., 1980; MacLeod et al., 1982; Ducreux and Bodennec, 1985; Albaigés and Grimalt, 1987). Obviously, the dispersion of the results is greater in the exercises performed, without a previous definition of an analytical protocol. Thus, the comparison of the analysis of alkanes in sediments among diverse laboratories, each using its particular methodology, show a scattering of values between three orders of magnitude in the order of 1 ug/g (Hilpert et al., 1987; MacLeod et al., 1982). The differences are smaller when similar methods are used and the concentrations are higher. Accordingly, Ducreux and Bodennec (1985) in an inter-laboratory comparison in which Soxhlet extraction and IR or UV-fluorescence were respectively used for quantitation of hydrocarbons in the range of 300 ug/g, obtained standard deviation values (2) in the order of 44-54% of the mean (1)

Intra-laboratory precisions are generally better. Hilpert et al. (1978) and MacLeod et al. (1982) reported standard deviations for the GC quantitation of different families of hydrocarbons in the order of 10-30% of the mean. Gearing et al. (1978) reported relative standard deviations of 16-24% for the analysis of sedimentary hydrocarbons by GC (average concentrations ~140 ug/g).

Albaigés and Grimalt (1987) have reported the results corresponding to two exercises for the evaluation of one method of Table 14: the IOC Manual and Guides No. 11 for the analysis of hydrocarbons in sediments. These were performed in Barcelona (November 1984 and October 1986) with the participation of 10 and 11 scientists, respectively, from 7 and 9 Mediterranean countries. Relative standard deviations of 17% (n-alkanes), 28% (unresolved complex mixture of aliphatic hydrocarbons) and 26% (aromatics) were obtained after correction for recovery ratios calculated with internal standards. Intra-laboratory precision for five replicate analyses performed in parallel to these determinations was situated between 14-18% for saturated and aromatic hydrocarbons. All participants analyzed subsamples of a single large sediment sample collected at the coast off Barcelona for these exercises. Sampling errors and sample variability were not, therefore, included in these dispersion values.

TABLE 14

COMBINATION OF THE ANALYTICAL METHODS AND PROCEDURES DESCRIBED IN THE TEXT
FOR THE ANALYSIS OF TRACE ORGANIC COMPOUNDS AND METALS IN MARINE SAMPLES.

	Sediment	Water "particulate"	Water "aqueous"	Organisms	Aerosols
ALPHs*	H1/M1	H2/M1	H3/M1	H5/M1	H6/M1
ARPHs	H1/M1	H2/M1	H3/M1	H5/M1	H6/M1
PAHs	H1/M3	H2/M3	H3/M3	H4/M3	H6/M3
PCBs	H7/M5	H8/M5	H9/M5	H10/M5	H11/M5
DDTs	H7/M5	H8/M5	H9/M5	H10/M5	H11/M5
Dieldrin	H7/M5	H8/M5	H9/M5	H10/M5	H11/M5
Aldrin	H7/M5	H8/M5	H9/M5	H10/M5	H11/M5
HCB	H7/M5	H8/M5	H9/M5	H10/M5	H11/M5
HCHs	H7/M5	H8/M5	H9/M5	H10/M5	H11/M5
Heptachlors	H7/M5	H8/M5	H9/M5	H10/M5	H11/M5
Hg	H12/M6	H18/M7	H17/M7	H16/M6	H15/M6
Cd	H12/M8	H14/M29-30	H13/M29-30	H16/M8	H15/M8
Cr	H12/M10	H14/M10	H13/M10	H16/M10	H15/M10
Co	H12/M12	H14/M12	H13/M12	H16/M12	H15/M12
Cu	H12/M14	H14/M29-30	H13/M29-30	H16/M14	H15/M13
Pb	H12/M15	H14/M29-30	H13/M29-30	H16/M16	H15/M15
Ni	H12/M17	H14/M18	H13/M18	H16/M18	H15/M18
Fe	H12/M19	H14/M19	H13/M19	H16/M19	H15/M19
Mn	H12/M20	H14/M21	H13/M21	H16/M21	H15/M20
Zn	H12/M22	H14/M29-30	H13/M29-30	H16/M22	H15/M23
Se	H12/M24	H14/M24	H13/M24	H16/M24	H15/M24
As	H12/M25	H14/M25	H13/M25	H16/M25	H21/M25
Be	H20/M26				

* Full names are given in table 2.

TABLE 15

UNEP REFERENCE METHODS
FOR TRACE POLLUTANT ANALYSIS IN COASTAL AND ESTUARINE WATERS

	Sediment	Water	Organisms	Aerosol
ALPHs*	No**11	No**13		
ARPHs	No**11	No**13		
PCBs	No 17	No 16	No 14	
DDTs	No 17	No 16	No 14	
Dieldrin	No 17	No 16		
Aldrin	No 17	No 16		
HCB	No 17	No 16		
HCH	No 17	No 16		
Heptachlors	No 17	No 16		
Hg	No 26		No 8	
Cd	No 27	No 18	No 11	
Cr	No 31			
Co	No 32			
Cu	No 33		No 11	
Pb	No 34		No 11	
Ni	No 35			
Fe	No 37			
Mn	No 38			
Zn	No 39		No 11	
Se			No 10	
As			No 9	

 * Full names in table 2.
** IOC method.

268

An important aspect derived from all these observations is the assessment of the significance of the analytical results obtained with these methodologies. This can be evaluated using the t test for comparison of mean values. In the case that diverse replicate analyses per sample have been performed the problem consists in the ascertainment of the minimal difference that must be observed between the average analyte concentrations of two distinct samples for their recognition as different within the level of dispersion of the analytical methods used for quantitation. Albaigés and Grimalt (1987) have addressed this problem assuming the same number of replicate analyses per sample and selecting a confidence limit of 10% (α = 5). The results have obviously been dependent on the number of replicate determinations and the standard deviation of the method. Thus, assuming relative standard deviations of 14% (best intra-laboratory) and 30% (best inter-laboratory) it has been obtained that for four replicate analyses (a rather unrealistic situation) the two average results could be differentiated if they would differ in more than 20% in the former case and more than 50% in the second. For duplicate analyses, the two average concentrations could be significantly differentiated provided that they differed in more than 35% (intra-laboratory) or 100% (inter-laboratory). In the case of one single determination per sample (the most regular situation) the minimal percentages amount to 87% in the former case and 880% in the latter (almost one order of magnitude).

Similarly to that described in section 2.2, perhaps the most important feature of this study is the strong difference in terms of statistical significance that is observed between single and duplicate analyses. On the other hand, relative standard deviations in the range of 10–16% and 20–30% may probably be considered as the best that can be obtained, respectively, in intra- and inter-laboratory comparisons. This is an important aspect to be examined for the assessment of spatial or temporal trends. In any case, the implementation of an integrated data base for marine pollutants requires an uninterruped effort for the continued revision of all these methods and procedures by means of intercalibration exercises and quality assurance studies.

REFERENCES

Aceves, M., Grimalt, J., Albaigés, J., Broto, F., Comellas, L. and Gassiot, M. (1988). Analysis of hydrocarbons in aquatic sediments. II. Evaluation of common preparative procedures for petroleum and chlorinated hydrocarbons. J. Chromatogr., 436, 503–509.

Adams, J, Atlas, E. and Giam, C.S. (1982). Ultra-trace determination of vapor-phase nitrogen heterocyclic bases in ambient air, Anal. Chem., 54, 1515–1518.

Ahnoff, M. and Josefsson, B. (1974). Simple apparatus for on-site continuous liquid extraction of organic compounds from natural waters, Anal. Chem., 46, 658–663.

Albaigés, J. and Grimalt, J. (1982). Fingerprinting of environmental PAH by high speed HPLC. Chromatogr. Newslett., 10, 8–11.

Albaigés, J. and Grimalt, J. (1987). A quality assurance study for the analysis of hydrocarbons in sediments, Intern. J. Environ. Anal. Chem., 31, 281–293.

Alfheim, I. and Lindskog, A. (1984). A comparison between different high volume sampling systems for collecting ambient airborne particles for mutagenicity testing and for analysis of organic compounds, The Sci. Tot. Environ., 34, 203-222.

Aminot, A. and Chausiepied, M. (1983). Manual des analyses chimiques en milieu marin, CNEXO, Brest, 395 p.

Atlas, E. and Giam, C.S. (1981). Global transport of organic pollutants: Ambient concentrations in the remote marine atmosphere, Science, 211, 163-165.

Baker, J.E., Capel, P.D. and Eisenreich, S.J. (1986). Influence of colloids on sediment-water partition coefficients of polychlorobiphenyl congeners in natural waters. Environ. Sci. Technol., 20, 1136-1143.

Barceló, D., Maris, F.A., Frei, R.W., De Jong, G.J. and Brinkman, U.A.Th. (1987a). Determination of trialkyl and triaryl phosphates by narrow-bore liquid chromatography with on-line thermionic detection., Intern. J. Environ. Anal. Chem., 30, 95-104.

Barceló, D., Maris, F.A., Geerdink, R.B., Frei, R.W., De Jong, G.J. and Brinkman, U.A. Th. (1987b). Comparison between positive, negative and chloride-enhanced negative chemical ionization of organophosphorous pesticides in on-line liquid chromatography-mass spectrometry, J. Chromatogr., 394, 65-76.

Bayne, B.L., Gabbott, P.A. and Widdows, J. (1975). Some effects of stress in the adult on the eggs and larvae of Mytilus edulis L., J. Mar. Biol. Assoc. U.K.), 55, 675-689.

Bayne, B.L., Holland, D.L., Moore, M.N., Lowe, D.M. and Widdows, J. (1978). Further studies on the effects of stress in the adult on the eggs of Mytilus edulis, J. Mar. Biol. Assoc. (U.K.), 58, 825-841.

Bayona, J.M., Grimalt, J., Albaigés, J., Walker II W. de Lappe, B.W. and Risebrough, R.W. (1983). Recent contributions of high resolution gas chromatography to the analysis of environmental hydrocarbons, J. High Res. Chromatogr., 6, 605-611.

Behymer, T.D. and Hites, R.A. (1985). Photolysis of polycyclic aromatic hydrocarbons adsorbed on simulated atmospheric particulates, Environ. Sci. Technol., 19, 1004-1006.

Bellar, T.A., Lichtenberg, J.J. and Lonneman, S.C. (1980). Recovery of organic compounds from environmentally contaminated bottom materials. In Contaminants and Sediments, Vol. 2,. R.A. Baker, ed. Ann Arbor Sci, Michigan, pp. 57-70.

Berman, S.S. and Yeats, P.A. (1985). Sampling of seawater for trace metals. CRC Crit. Rev. Anal. Chem., 16, 1-14.

Billings, W.N. and Bidleman, T.F. (1980). Field comparison of polyurethane foam and Tenax-GC resin for High-Volume Air sampling of chlorinated hydrocarbons, Environ. Sci. Technol., 14, 679-683.

Billings, W.N. and Bidleman, T.F. (1983). High volume collection of chlorinated hydrocarbons in urban air using three solid adsorbents. Atmos. Environ., 17, 383-391.

Brassell, S.C. and Eglinton, G. (1980). Environmental Chemistry, -An Interdisciplinary Subject. Natural and Pollutant organic compounds in contemporary aquatic environments, In Analytical Techniques in Environmental

Chemistry, J. Albaigés, ed. Pergamon Press, Oxford, pp. 1-22.

Buat-Menard, P. and Duce, R.A. (1986). Precipitation scavenging of aerosol particles over remote marine regions, Nature, 321, 508-510.

Burdick, N.F. and Bidleman, T.F. (1981). Frontal movement of hexachlorobenzene and polychlorinated biphenyl vapors through polyurethane foam. Anal. Chem., 53, 1926-1929.

Burns, K.A. and Smith, J.L. (1981). Biological monitoring of ambient water quality: the case of using bivalves as sentinel organisms for monitoring petroleum pollution in coastal waters. Estuarine Coast. Shelf. Sci., 13, 433-443.

Caron, G., Suffet, I.H. and Belton, T. (1985). Effect of dissolved organic carbon on the environmental distribution of non polar organic compounds. Chemosphere, 14, 993-1000.

Chakraborti, D., Van Cleuvenbergen, R. and Adams, F. (1987). Speciation of ionic alkyllead in aerosols by gas chromatography-atomic absorption spectrometry, Intern. J. Environ. Anal. Chem., 30, 233-242.

Chiou, C.T., Malcolm, R.L., Brinton, T.I. and Kile, D.E. (1986). Water solubility enhancement of some organic pollutants and pesticides by dissolved humic and fulvic acids. Environ. Sci. Technol., 20, 502-508.

Coutant, R.W., Brown, L., Chuang, J.C., Riggin, R.M. and Lewis, R.G. (1988). Phase distribution and artifact formation in ambient air sampling for polynuclear aromatic hydrocarbons. Atmos. Environ., 22, 403-409.

Daumas, R.A., Laborde, P.L., Marty, J.C. and Saliot, A. (1976). Influence of sampling method on the chemical composition of water surface film. Limnol. Oceanogr., 21, 319-326.

De Greef, J., De Proft, M. and De Winter, F. (1976). Gas chromatographic determination of ethylene in large air volumes at the fractional parts-per-billion level. Anal. Chem., 48, 38-44.

De Jonghe, W.R.A., Chakraborti, D. and Adams, F.C. (1981). Identification and determination of individual tetraalkyllead species in air, Environ. Sci. Technol., 15, 1217-1222.

De Lappe, B.W., Risebrough, R.W. and Walker, II, W. (1983). A large-volume sampling assembly for the determination of synthetic organic and petroleum compounds in the dissolved and particulate phases of seawater. Can. J. Fish Aquat. Sci., 40, 322-336.

De Raat, W.K., Shulting, F.L., Burghardt, E. and de Meijere, F.A. (1987). Application of polyurethane foam for sampling volatile mutagens from ambient air. The Sci. Total Environ., 63, 175-189.

Dressler, M. (1979). Extraction of trace amounts of organic compounds from water with porous organic polymers. J. Chromatogr., 165, 167-206.

Ducreux, J. and Bodennec, G. (1985). Exercice interlaboratoire sur les dosages des hydrocarbures dans l'eau et les sédiments. Rev. Inst. Fr. Petrol., 40, 355-374.

Durand, B., Espitalie, J. and Oudin, J.L. (1970). Analyse Géochimique de la matière organique extraite des roches sédimentaires. III. Accroissement de la rapidité du protocole opératoire per l'amélioration de l'appareillage. Rev. Inst. Franç. du Pétrole, 25, 1268-1279.

Dux, J.P. (1986). Handbook of quality assurance for the analytical chemistry laboratory, Van Nostrand, New York, 123 p.

Eichman, R., Ketseridis, G., Schebeske, G., Jaenick, R., Hahn, J., Warnek, P. and Junge C. (1980). n-Alkane studies in the troposphere. 2: Gas and particulate concentrations in Indian Ocean Air. Atmos. Environ., 14, 695-703.

Eichmann, R., Neuling, P., Ketseridis, G., Hahn, J., Jaenicke, R. and Junge, C. (1979). n-Alkane studies in the trophosphere. 1: Gas and particulate concentrations in North Atlantic Air. Atmos. Environ., 13, 587-599.

Farrington, J.W., Goldberg, E.D., Risebrough, R.W., Martin, J.H. and Bowen, V.T. (1983). U.S. "Mussel Watch" 1976-1978: An overview of the trace metal, DDE, PCB, hydrocarbon, and artificial radionuclide data. Environ. Sci. Technol., 17, 490-496.

Florence, T.M. and Batley, G.E. (1980). Chemical speciation in natural waters. CRC Crit. Rev. Anal. Chem., 10, 219-296.

Gagosian, R.B. and Peltzer, E.T. (1986). The importance of atmospheric input of terrestrial organic material to deep sea sediments. Org. Geochem., 10, 661-669.

Gearing, J.N., Gearing, P.J., Lytle, T.F. and Lytle, J.S. (1978). Comparison of thin-layer and column chromatography for separation of sedimentary hydrocarbons. Anal. Chem., 50, 1833-1836.

Gluckman, J.C., Barceló, D., De Jong, G.J., Frei, R.W., Maris, F.A. and Brinkman, U.A.Th. (1986). Improved design and applications of an on-line thermionic detector for narrow-bore liquid chromatography, J. Chromatography, 367, 35-44.

Goldberg, E.D. (1980). The International Mussel Watch., U.S. National Academy of Sciences, Washington, 248 p.

Gomez Belinchón, J.I. and Albaigés, J. (1987). Organic pollutants in Water. I. Optimization of operational parameters of the CLSA Technique. Intern. J. Environ. Anal. Chem., 30, 183-195.

Gómez-Belinchón, J.I., Grimalt, J.O. and Albaigés, J. (1988). Intercomparison study of liquid-liquid extraction and adsorption on polyurethane and Amberlite XAD-2 for the analysis of hydrocarbons, polychlorobiphenyls, and fatty acids dissolved in seawater, Environ. Sci. Technol., in press (No. 5).

Graedel, T.E. (1978). Chemical compounds in the atmosphere. Academic Press, New York, 440 p.

Graydon, J.W., Grob, K., Zuercher, F. and Giger, W. (1984). Determination of highly volatile organic contaminants in water by the closed-loop gaseous stripping technique followed by thermal desorption of the activated carbon filters, J. Chromatogr., 285, 307-318.

Grimalt, J. and Albaigés, J. (1982). Oil spill identification by High speed HPLC, J. High Res. Chromatogr., 5, 255-260.

272

Grimalt, J., Albaigés, J. Sicre, M.A., Marty, J.C. and Saliot, A. (1983). Aerosol transport of polynuclear aromatic hydrocarbons over the Mediterranean Sea, Naturwissenschaften, 75, 39–42.

Grimalt, J., Marfil, C. and Albaigés, J. (1984). Analysis of hydrocarbons in aquatic sediments. I. Sample handling and extraction, Intern. J. Environ. Anal. Chem., 18, 183–194.

Grob, K. (1973). Organic substances in potable water and its precursor. I. Methods for determination by gas-liquid chromatography, J. Chromatogr., 84, 255–273.

Grob, K. (1983). Modified stripping technique for the analysis of trace organics in water, J. Chromatogr., 260, 527.

Grob, K., Grob, G. and Habrich, G. (1984). Overcoming background contamination in closed loop stripping analysis (CLSA), J. High Resol. Chromatogr. Chromatogr. Commun., 7, 340.

Gschwend, P.M. and Wu, S.C. (1985). On the constancy of sediment-water partition coefficients of hydrophobic organic pollutants. Environ. Sci. Technol., 19, 90–96.

Harrison, R.M. (1984). Recent advances in air pollution analysis. CRC Crit. Rev. Anal. Chem., 15, 1–61.

Hatcher, R.F. and Parker, B.C. (1974). Laboratory comparisons of four surface microlayer samplers. Limnol. Oceanogr., 19, 162–165.

Head, P.C. (1985). Practical estuarine chemistry, Cambridge University Press, Cambridge, 337 p.

Hilpert, L.R., May, W.E., Wise, S.A., Chesler, S.N. and Hertz, H.S. (1978). Interlaboratory comparison of determinations of trace level petroleum hydrocarbons in marine sediments. Anal. Chem., 50, 458–463.

Himmelblau, D.M. (1970). Process Analysis by Statistical Methods, John Wiley, New York, 463 p.

Ho, R., Marty, J.C. and Saliot, A. (1982). Hydrocarbons in the Western Mediterranean Sea, 1981. Intern. J. Environ. Anal. Chem., 12, 81–98.

Holdway, D.A. and Nriagu, J.O. (1988). A purge and trap gas chromatographic method for dimethyl sulfide in freshwater, Intern. J. Environ. Anal. Chem., 32, 177–186.

IOC Manual and Guides, No. 13. Monitoring oil and dissolved/dispersed petroleum hydrocarbons in marine waters and on beaches (IOC/UNESCO, Paris, 1984) 35 pp.

Junk, G.A. and Richard, J.J. (1987). Solid phase extraction, GC separation and EC detection of tributyltin chloride. Chemosphere, 16, 61–68.

Keith, L.H. and Telliard, W.A. (1979). Priority pollutants. I- A perspective view. Environ. Sci. Technol., 13, 416–423.

Keller, C.D. and Bidleman, T.F. (1984). Collection of airborne polycyclic aromatic hydrocarbons and other organics with a glass fiber-polyurethane foam system. Atmos. Environ., 18, 837–845.

Kjolholt, J. (1985). Determination of trace amounts of organophosphorous pesticides and related compounds in soils and sediments using capillary gas chromatography and a nitrogen-phosphorous detector. J. Chromatogr., 325, 231-238

Lamb, S.I., Petrowski, C., Kaplan, I.R. and Simoneit, B.R.T. (1980). Organic compounds in urban atmospheres: A review of distribution, collection and analysis, APCA Journal, 30, 1098-1115.

Landing, W.M. and Freely, R.A. (1981). The chemistry and vertical flux of particles in the northeastern Gulf of Alaska, Deep-Sea Res., 28A, 19-37.

Lee, M.L., Novotny, M.V. and Bartle, K.D. (1981). Analytical Chemistry of polycyclic aromatic compounds, Academic Press, New York, 462 p.

Lee, F.S.C., Pierson, W.R. and Ezike, J. (1980). The problem of PAH degradation during filter collection of airborne particulates, in an evaluation of several commonly used filter media. In Polynuclear Aromatic Hydrocarbons: Chemical and Biological Effect, Bjorseth, A. and Dennis, A.J., eds, Batelle Press, Columbus OH, p. 543.

Leithe, W. (1973). The analysis of organic pollutants in water and waste water. Ann Arbor Sci., Michigan. 213 p.

Lewis, R.G. and Jackson, M.D. (1982). Modification and evaluation of a high volume air sampler for pesticides and semivolatile industrial organic chemicals. Anal. Chem., 54, 592-594.

Liss, P.S. (1975). Chemistry of the sea surface microlayer. In Chemical Oceanography. Riley, J.P. and Skirrow, G., 2nd. ed. vol. 2. Academic Press, London, pp. 193-244.

MacLeod, W.D., Prohaska, P.G., Gennero, D.D. and Brown, D.W. (1982). Interlaboratory comparisons of selected trace hydrocarbons from marine sediments. Anal. Chem., 54, 386-392.

Maris, F.A., Geerdink, R.B. and Brinkman, U.A.Th. (1985). Selection of mobile phases for reversed-phase liquid chromatography with on-line electron-capture detection, J. Chromatogr, 328, 93-100.

Marty, J.C., Saliot, A., Buat-Menard, P., Chesselet, R. and Hunter, K.A. (1979). Relationship between the lipid compositions of marine aerosols, the sea surface microlayer, and subsurface water. J. Geophys. Res., 84, 5707-5716.

Matthias, C.L., Bellama, J.M., Olson, G.J. and Brinkman, F.E. (1986). Comprehensive Method for Determinations of Aquatic butyltin and butylmethyltin species at ultratrace levels using simultaneous hydridization/Extraction with gas chromatography-flame photometric detection. Environ. Sci. Technol., 20, 609-615.

Matuska, P., Koval, M. and Seiler, W. (1986). A high resolution GC-analysis method for determination of C_2-C_{10} hydrocarbons in air samples. J. High Resol. Chromatogr., 9, 577-584.

Mazurek, M.A. and Simoneit, B.R.T. (1986). Organic components in bulk and wet-only precipitation. CRC Crit. Rev. in Environ. Control, 16, 1-140.

McCarthy, J.F. and Jimenez, B.D. (1985). Interactions between polycyclic aromatic hydrocarbons and dissolved humic material: binding and dissociation. Environ. Sci. Technol, 19, 1072-1076.

Moody, G.J. and Thomas, J.D.R. (1982). Chromatographic separation and extraction with foamed plastics and rubbers. Marcel Dekker, New York, 139 p.

Morehead, N.R., Eadie, B.J., Lake, B., Landrum, P.F. and Berner, D. (1986). The sorption of PAH onto dissolved organic matter in Lake Michigan Waters. Chemosphere, 15, 403-412.

Niessner, R., Klockow, D., Bruynseels, F. and Van Grieken, R. (1985). Investigation of heterogeneous reactions of PAHs on particle surfaces using laser microprobe mass analysis, Intern. J. Environ. Anal. Chem., 22, 281-295.

Obiols, J., Devesa, R. and Sol, A. (1986). Speciation of heavy metals in suspended particulates in urban air, Toxicol. Environ. Chem., 13, 121-128.

Pankow, J.F., Isabelle, L.M. and Asher, W.E. (1984). Trace organic compounds in rain. I. Sampler design and analysis by adsorption/thermal desorption (ATD), Environ. Sci. Technol., 18, 310-318.

Peters, J. and Seifert, B. (1980). Losses of benzo(a)pyrene under the conditions of high-volume sampling. Atmos. Environ., 14, 117-119.

Phillips, D.J.H. (1980). Quantitative aquatic biological indicators. Applied Science Publishers, London, 488 p.

Pitts, J.N. Jr., Sweetman, J.A., Zielinska, B., Atkinson, R., Winer, A.M. and Harger, W.P. (1985). Formation of nitroarenes from the reaction of polycyclic aromatic hydrocarbons with dinitrogen pentaoxide. Environ. Sci. Technol., 19, 1115-1121.

Podoll, R.T., Irwin, K.C. and Brendlinger, S. (1987). Sorption of water-soluble oligomers on sediments. Environ. Sci. Technol., 21, 562-568.

Schneider, J.K., Gagosian, R.B., Cochran, J.K. and Trull, J.W. (1983). Particle size distributions of n-alkanes and ^{210}Pb in aerosols off the coast of Peru, Nature, 304, 429-432.

Schnell, R.C. (1977). Ice nuclei in seawater, fog water and marine air off the coast of Nova Scotia. Summer 1975., J. Atmos. Sci., 34, 1299.

Segar, D.A. and Stamman, E. (1986). Strategy for design of marine pollution monitoring studies, Wat. Sci. Tech., 18, 15-26.

Sehmel, G.A. (1980). Particle and gas dry deposition a review. Atmos. Environ., 14, 983-1011.

Sicre, M.A., Marty, J.C., Saliot, A., Aparicio, X., Grimalt, J. and Albaigés, J. (1987a). Aliphatic and aromatic hydrocarbons in different sized aerosols over the Mediterranean Sea: Occurrence and Origin. Atmos. Environ., 21, 2247-2259.

Sicre, M.A., Marty, J.C., Saliot, A., Aparicio, X., Grimalt, J. and Albaigés, J. (1987b). Aliphatic and aromatic hydrocarbons in the mediterranean aerosol, Intern. J. Environ. Anal. Chem., 29, 73-94.

Simoneit, B.R.T. and Mazurek. M.A. (1981). Air pollution: The organic components. CRC Crit. Rev. Environ. Contr., 11, 219-276.

Simoneit, B.R.T., Grimalt, J.O., Fisher, K. and Dymond, J. (1986). Upward and downward flux of particulate organic material in abyssal waters of the Pacific Ocean, Naturwissenschaften, 73, 322-325.

Singh, A.K., Kewetson, D.W., Jordon, K.C. and Ashraf, M. (1986). Analysis of organophosphorous insecticides in biological samples by selective ion monitoring gas chromatography-mass spectrometry. J. Chromatogr., 369, 83-96.

Snyder, Ll.R. (1968). Principles of adsorption chromatography. Marcel Dekker, New York, 413 p.

Soutar, A., Kling, S.A., Crill, P.A., Duffrin, E. and Bruland, K.W. (1977). Monitoring the marine environment through sedimentation, Nature, 266, 136-139.

Strachan, W.M.J. and Huneault, H. (1984). Automated rain sampler for trace organic substances. Environ. Sci. Technol., 18, 127-130.

Suess, E. (1980). Particulate organic carbon flux in the oceans -surface productivity and oxygen utilization. Nature, 288, 260-263.

Swisher, R.D. (1987). Surfactant biodegradation, 2nd. ed. Marcel Dekker, New York, 1085 p.

Taylor, J.K. (1981). Quality Assurance of chemical measurements. Anal. Chem., 53, 1569-1588A.

Thompson, S. and Eglinton, G. (1978). The fractionation of a recent sediment for organic geochemical analysis. Geochim. Cosmochim. Acta., 42, 199-207.

Unger, M.A., MacIntyre, W.G., Greaves, J. and Huggett, R.J. (1988). GC determination of butyltins in natural waters by flame photometric detection of hexyl derivatives with mass spectrometric confirmation, Chemosphere, 15, 461-470.

Valerio, F. and Lazzarotto, A. (1985). Photochemical degradation of polycyclic aromatic hydrocarbons (PAH) in real and laboratory conditions, Intern. J. Environ. Anal. Chem., 23, 135-151.

Van Noort, P.C.M. and Wondergem, E. (1985). Scavenging of airborne polycyclic aromatic hydrocarbons by rain, Environ. Sci. Technol., 19, 1044-1048.

Van Vaeck, L., Van Cauwenberghe, A. and Janssens, J. (1984). The gas-particle distribution of organic aerosol constituents: measurement of volatilization artifact in high volume cascade impactor sampling, Atmos. Environ., 18, 417-430.

Van Vleet, E.S. and Williams, P.M. (1980). Sampling sea surface films: a laboratory evaluation of techniques and collecting materials. Limnol. Oceanogr., 25, 764-770.

Voice, T.C., Rice, P.C. and Weber, W.J. (1983). Effects of solids concentration on the sorptive partitioning of hydrophobic pollutants in aquatic systems. Environ. Sci. Technol., 17, 513-518.

276

Waldman, J.M., Munger, J.W., Jacob, D.J., Flagan, R.C., Morgan, J.J. and Hoffman, M.R. (1982). Chemical composition of acid fog. Science, 218, 667.

Wefer, G., Suess, E., Balzer, W., Liebezeit, G., Muller, P.J., André, C., Ungerer, A. and Zenk, W. (1982). Fluxes of biogenic components from sediment trap deployment in circumpolar waters of the Drake Passage, Nature, 299, 145-147.

Wegman, R.C.C. and Melis, P.H.A.M. (1985). Organic pollutants in water. C.R.C. Crit. Rev. Anal. Chem., 16, 281-321.

White, C.M. (1985). Nitrated polycyclic aromatic hydrocarbons. Hüthing Verlag, Heidelberg, 376 p.

Whitehouse, B. (1985). The effects of dissolved organic matter on the aqueous partitioning of polynuclear aromatic hydrocarbons. Estuarine Coastal Shelf. Sci., 20, 393-402.

Wise, S.A., Chesler, S.N., Guenther, F.R., Hertz, H.S., Hilpert, L.R., May, W.E. and Paris, R.M. (1980). Interlaboratory comparison of determinations of trace level hydrocarbons in mussels. Anal. Chem., 52, 1828-1833.

Wise, S.A., Chesler, S.N., Gump, B.H., Hertz, H.S. and May, W.E. (1977). Interlaboratory calibration for the analysis of petroleum levels in sediment. In Fate and Effects of Petroleum Hydrocarbons in Marine Ecosystems and Organisms, D.A. Wolfe, ed. Pergamon Press, Oxford, pp. 345-350.

You, F. and Bidleman, T.F. (1984). Influence of volatility on the collection of polycyclic aromatic hydrocarbon vapors with polyurethane foam. Environ. Sci. Technol., 18, 330-333.

Zafiriou, O.C., Gagosian, R.B., Peltzer, E.T. and Alford, J.B. (1985). Air to sea fluxes of lipids at Enewetak Atoll., J. Geophys. Res., 90, 2409-2423.

Zenon-Roland, L., Agnessens, R., Nanginot, P. and Jacobs, H. (1984). Analysis of pesticide residues by high resolution gas chromatography. Part 1: Comparison between packed and capillary columns in pesticide residue determinations. Practical considerations for routine use of capillary gas chromatography. J. High Resol. Chromatogr. Chromatogr. Commun., 7, 480-484.

Zolotov, Y.A., Bodnya, V.A. and Zagruzina, A.N. (1982). Applications of extraction methods for the determination of small amounts of metals, CRC Crit. Rev. Anal. Chem., 14, 93-174.

Principles and Methods in Environmental Management of Coastal Marine Waters

ANDERS RANDLØV, STEEN Ø. DAHL, and EARLING POULSEN
COWIconsult
Consulting Engineers and Planners As
45, Teknikerbyen, DK-2830 Virum, Denmark

INTRODUCTION

Human activities stress the environment of coastal waters and this calls for operational principles and effective methods in the environmental management.

Different approaches to environmental management have developed in different countries. In this paper some of these management strategies will be discussed and the interaction between the methods and tools of environmental investigations and coastal management will be presented for two cases:

Case 1. Eutrophication of marine waters and measures adopted for rehabilitation in Denmark

Case 2. Avoidance reaction in fish - a method in marine environmental management.

MANAGEMENT STRATEGIES

Strategies for environmental management have recently been a major task for regulating authorities both at national and international levels. Various approaches have been adopted, but they remain subject to intensive discussion. The basic management principles can be divided roughly into two categories:

i) Source related strategies, i.e. management and control is directed primarily at emissions from pollution sources, with only minor attention to the receiving waters.

ii) Environment related strategies, i.e. management relies on quality objectives, quality criteria, impact assessment, and receiving waters monitoring. This forms the basis for stipulating the emission requirements.

Both principles have their merits and drawbacks. There seems to be a tendency to shift towards or include part of the second approach in the EEC countries (1). This may be because of sufficient improvements in pollution reduction have not been forthcoming.

Supporters of the two strategies favour either one or the other based on how the principles have been and can be put into practice, rather than disagreeing over basic theory. In the end, political rethoric must be followed by a willingness to implement effective measures and pay the bill whatever the strategy. This is perhaps the real controversy at the heart of the debates in international forums.

In Fig. 1 the most common terms for the principles have been classified in the group of "source related" and "environment related" principles. However, there is a great deal of overlap, and a sharp distinction between the strategies is difficult to define even in between the principles.

Source Related Strategies

The source related principles comprise:

- Uniform Emission Standards (UES)
- Percentage Reduction Programmes
- Prevention Principles, based on
  o Zero discharge
  o Best technical means
  o Best technical means, economically available.

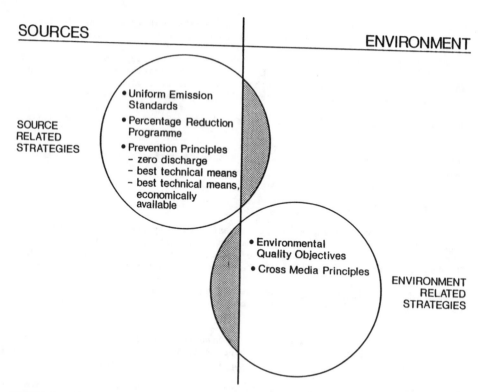

FIGURE 1.   Management strategies and their common principles

<u>Uniform Emission Standards (UES)</u>. These are a very common principle and set limits to pollutants, e.g. in waste water outlets. The advantage of this principle is equal license for all polluters, simple administration and easy control to meet the set standards. The difficulty is to find the level for these standards without knowing the actual condition of the receiving water.

The sensitivity to pollutants varies in coastal waters. As depicted in Fig. 2, standards defined for the majority of coastal waters (named UES - Level 1) will perhaps be insufficient to protect highly sensitive areas, e.g. spawning and nursery areas, marshes etc. More stringent emission standards (named UES-Levels 2) will protect the entire coastline but would of course have economic implications.

The UES approach cannot be applied to non-point sources of pollution. In addition concentration of many point sources can occur in one area, each of which individually complies with the UES level, but in combination they overload the environment.

<u>Percentage Reduction Programme</u>. This approach is often introduced when signs of heavy pollution have been picked up in the environment. The approach consists of a general reduction in pollutants from existing sources by a specified percentage. This is relevant in regions where concentrations of point sources each individually fulfilling the UES levels create environmental problems. The approach can also include non-point sources when the pathway and course-effect relationships have been identified. This could typically be enrichment of nutrients in coastal waters resulting in limitations on the use of fertilizer and manure on adjacent farmland (see case 1).

<u>Prevention Principle</u>. This is based on the philosophy that prevention is better than cure. The precise effects of individual pollutants on various aspects of the environment is not known, but certain effects are anticipated, and so can be prevented. The introduction of 'black lists' of harmful substances, the use of which is prohibited or strongly restricted, are examples of a preventive policy.

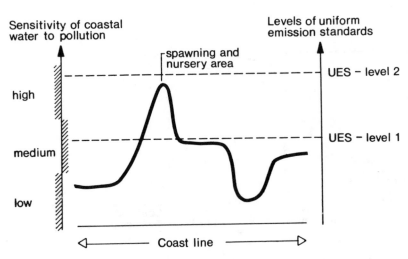

FIGURE 2. Two levels of Uniform Emission Standards with different protection of the coastal environment.

Despite considerable research to clarify the complex relationships between pollutants and their impact on the environment, there are still many uncertainties. Because of this, the prevention principle puts emphasis on the anticipated risks to the environment, and seeks to prevent these by limiting the potential polluter.

The prevention principle includes the terms 'best technical means' and 'best technical means economically available'.

The first term is a theoretical one only as pollutant levels are lowered progressively until, in theory, the 'zero discharge' target has been reached (see Fig. 3). This depends solely on resources allocation, so economical considerations determine how far along the curve shown in Fig. 3 pollution prevention can progress.

The prevention principle can be introduced as retrospective prevention aimed at curing proven symptoms, or to prevent anticipated pollution from certain sources.

Genuine prevention is an effort to prevent environmental problems from occurring which have a high degree of certainty. This could be taken to exclude chronic adverse influences on the environment, or situations where human activity and the use of technology are a potential risk to the environment only through accidents. An example of genuine prevention against potential risks is the development of contingency plans against spills, blow-outs etc during the offshore exploration of oil.

In general the prevention principle is very much a political question, namely: 'what is acceptable cost to society in avoiding environmental risks?'

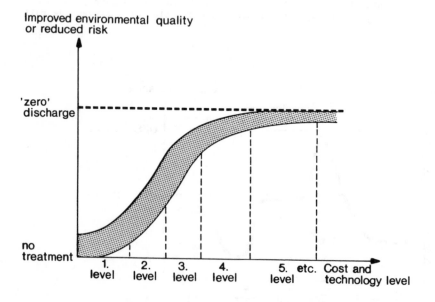

FIGURE 3.   Relationship between cost/technology levels and improved environment.

The Environment Related Strategy

The environment related principles comprise

- Environmental Quality Objectives (EQO)
- Cross-media principles.

Environmental Quality Objectives (EQO). These are based on the principle that
the quality of the ambient coastal water be related to the proposed use. The use
of coastal waters is often formulated in qualitative terms as 'objectives', e.g.
areas of natural scientific interest and research; fishery areas; recreational
areas; receiving waters (waste water discharge zones); exploitation of natural
resources (e.g. sand and gravel) etc. To these objectives, a number of more
specific criteria can be defined either on biological grounds or physico-
chemical parameters. The concentration of certain specified water constituents
then define the emission standards for the sources of pollution. Obviously the
emission standards depend on the assimilative capacity of coastal waters. This,
as stated earlier, varies from place to place, so the emission standards will be
'Non-Uniform Emission Standards' (NUES).

The EQO strategy requires an extensive knowledge of the coastal waters, and
especially of cause-effect relationships. This can be difficult to obtain. The
merits are that non-point sources and concentrations of point sources are taken
into account before the emission standards are settled. However, the EQO
principle has been critisized because the difficulty of establishing 'scientific
proofs' of the relationships between deterioration of the environment and the
pollution load has been used as an excuse to postpone the implementation of
improved treatments.

The procedures used to define acceptable pollution levels in the receiving
waters have been criticized. One approach is to define over-concentrations
relative to a 'background level' measured at reference stations out of the
influence of the pollution. Selection of such background locations is not easy
in coastal areas, and Fig. 4 shows how trends in background level can lead to
erroneous conclusions. Since the acceptable levels are set relative to
background levels there are no signs that pollution is actually increasing at
the control station, as the concentration above the reference level remains the
same due to rising background concentrations.

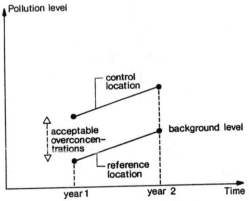

FIGURE 4. Effect of increasing background pollution levels on the perceived
pollution of the environment.

283

Conflicts between objectives increase the demand for overall planning to define the interests and the various uses of coastal zones. Therefore the 'Environmental Quality Objectives' principle is linked to environmental quality planning for the coastal zones. This is a fundamental part of many management systems such as that used in Denmark.

The environmental quality planning process may be described in four steps (2), the four M's:

- Managing (regulations, licenses, action planning)
- Monitoring (control on sources, discharges and environment)
- Mapping (registration of interests, conflicts and sensitive areas)
- Modelling (impact assessment, cause-effect relations)

The four steps are interlinked and form a feedback loop underlining the continuous process (Fig. 5). The managing activity is the initiator and directs the frames for the other activities in the loop. Feedback promotes new actions and adjustments to the policy and regulations based on the latest knowledge. Therefore, environmental quality planning must be a dynamic and continuous process.

The cross-media principle. This is based upon the view that pollutants can be transported through different media such as air, water and soil and that actual pollutant levels in waste water discharges or run-off through rivers do not arise exclusively from sources directly linked to them.

Seepage from solid waste dump sites, nutrient losses from farmland etc. are examples of important pollution sources transported through soil.

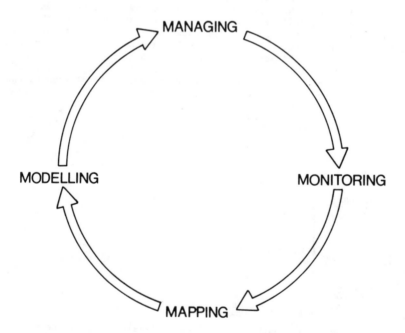

FIGURE 5. The process of environmental quality planning (2).

Air pollution and the resultant deposits from air to marine areas can contribute significant proportions of the total load of, for example, nitrogen, heavy metals, polynuclear aromatic hydrocarbons. Therefore it is important that air pollution management should form part of the sea protection policy. Airborne pollutants can be transported over long distances, providing an international dimension to the problem. For example, the acidification of Scandinavian lakes is contributed to by SO_2 emissions from the combustion of fossil fuels in the United Kingdom and central Europe. These sources may also constribute to the eutrophication of Danish coastal waters through NO_x emissions.

The management strategy should not therefore be limited to one separate part of the environment. All sources and their pathways through water, soil and air have to be considered together, hence management of coastal waters will, indeed, consist of more than regulations on waste water discharges alone.

The cross-media principle will probably become increasingly important in future strategies adopted to reduce environmental pollution. The difficulty with this concept is the massive effort needed to gather complex information on substances, pathways and sources before administrative arrangements can be devised and political decisions taken on a national and international basis.

SUMMARY

Disadvantages are associated with adopting either the source related approach or the environment related approach to marine pollution problems. Experience over the last decade has shown that unforeseen adverse environmental effects can still occur despite the strategy used. Typically the strategies adopted are incapable of handling all environmental problems, due to the complexity of the ecosystem, insufficient knowledge of cause-effect relationships and of the multimedia pathways involved.

Future strategies should consist of an integration of the two approaches. For example the management strategy could be based on uniform standard emissions for some substances with proper consideration to prevention principles combined with downwards revision for discharges to specified sensitive areas i.e. the environmental quality objectives principle. The cross-media approach must also be taken into account as to provide a holistic view.

However, the success of any future strategy depends fundamentally on the political willingness to introduce effective anti-pollution measures combined with the proper allocation of sufficient resources.

REFERENCES

1. International Conference on Environmental Protection of the North Sea 24-27 March 1987, section six - Environmental management of the North Sea, England, 1987.

2. Somer, E., Marine pollution control legislation and coastal zone management, WHO Western Pacific Regional Centre for the Promotion of Environmental Planning and Applied Studies (PEPAS), Kuala Lumpur, Malaysia, 1985.

Case 1:
Eutrophication of marine waters and measures adopted for rehabilitation in Denmark

INTRODUCTION

In August 1981 several areas of marine open water were hit by severe oxygen deficiency, causing massive fish death. Until then oxygen deficiency in Danish marine areas had been observed only in lagoons and other areas with limited water exchange. Monitoring by the National Agency of Environmental Protection (NAEP) had provided no warning as to the serious state of eutrophication. Since 1981 similar situations have occurred in 1983 and 1986, usually following periods of calm wind in the autumn months of August and September.

In the autumn of 1986, the area affected by oxygen deficiency in the open Kattegat was sufficiently large that the Danish government passed a resolution in demanding a 50% reduction in nitrogen discharges and an 80% reduction in phosphorus discharges.

An account of the observed effects of eutrophication in the open marine areas and the measures which will be taken to rehabilitate the areas follows. Finally, the implications for monitoring and management strategies are discussed.

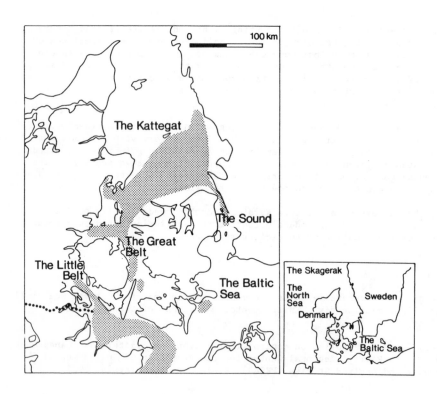

FIGURE 1. Areas in the Kattegat and the Belt Sea affected by oxygen deficiency in autumn 1986.

Observations

The earliest reports indicating oxygen deficiency are usually given by fishermen observing dead or dying fish and bottom fauna in their trawl. In fact, fishermen have observed that this situation is often preceded by short period of exceptionally good catches of Norway lobster, as these animals abandon their oxygen depleted borrows in the sediment, at the very beginnings of the oxygen deficit.

The extent of areas with oxygen deficiency in the autumn of 1986 is shown in Fig. 1.

Decrease in Fisheries

The observed incidents of oxygen deficiency in the Kattegat and the Belt Sea have been preceded by a steady decline in the catch of commercially important bottom living fish species in the area (Fig. 2). The yearly catch of plaice has been reduced from more than 15,000 t in early 1970s to about 5,000 t in recent years. It is now believed that this decrease may be explained partly by the pro-gressively deteriorating oxygen conditions in the Kattegat resulting in oxygen depletion in 1981.

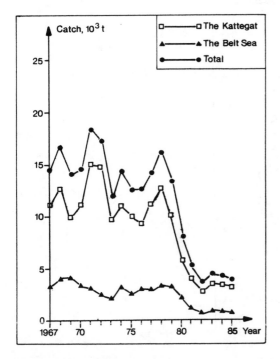

FIGURE 2. Danish catches of the bottom living plaice from the Kattegat and Belt Sea, 1967-1985

Monitoring Programmes

In the period 1975-79 the NAEP carried out an intensive monitoring programme of the water quality and the biological conditions of the Kattegat and the Belt Sea with the objective of assessing the need for waste water treatment (1). The investigation comprised determination of primary production and nutrient concentration on more than 50 stations at about once a month (Fig. 3).

Neither this programme nor the following two years of routine monitoring were able to predict the severe eutrophication effects occurring in 1981. The 1975-79 investigation concluded that 'problems related to eutrophication do not occur in open Danish waters'(1).

This conclusion was based primarily on measurement of primary production. An increase in primary production had been observed in the Great Belt, the Sound and the southern and western Kattegat (Fig. 4).

CAUSES OF EUTROPHICATION

The Hydrography of the Kattegat

The Kattegat (and the Belt Sea south of the Kattegat) is a transition zone between the brackish Baltic Sea and the more saline Skagerak. The result is a

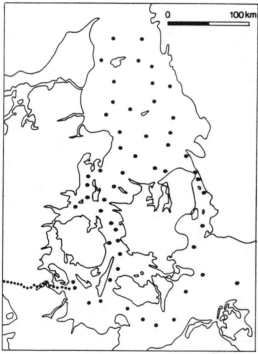

FIGURE 3. Stations at which nutrient concentration and primary production were monitored monthly from 1975-79 (1).

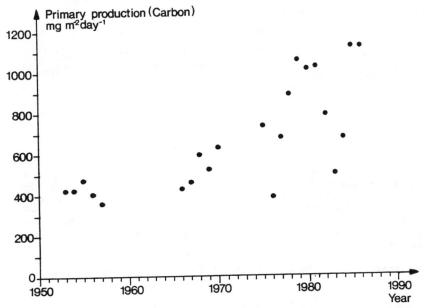

FIGURE 4. Mean daily primary production (July - September) in the Great Belt,
1953-1986 (Data from ref. 2 and Gunni Ærtebjerg, pers. comm.).

stratification of the water column with north-flowing low-saline Baltic water
above and south-flowing high-saline Skagerak water below. Typically, this causes
a halo line (Fig. 5). The Kattegat can therefore be considered a vast estuary.
However, it is affected by the meteorological conditions over the entire
Skagerak-Baltic area, resulting in frequently changing hydrography.

The oxygen necessary for bottom mineralisation processes is mainly transported
with the bottom current of Skagerak water which has not been in contact with the
atmosphere during its transport from the Skagerak. Vertical transportation of
oxygen only takes place during storms. The result is a net loss of oxygen from
bottom water, making the area vulnerable to oxygen deficiency induced by
eutrophication. Oxygen deficiency is most likely to occur in the autumn after
prolonged periods with calm weather; in such situations vertical transportation
of oxygen is insignificant but oxygen consumption is high due to the high
sedimentation rate at that time of the year.

There has been some speculation that hydrographic variations could have tipped a
delicate oxygen balance in the area. These variations, however, do not show any
trend and is therefore not believed to be the cause of the observed effects (3).

Nutrients

It is now generally accepted that an increased input of nitrogen (N) and
phosphorous (P) compounds to the open water environment is the cause of the
eutrophication effects observed. Most Danish scientists also agree that N is the
limiting factor for primary production in Danish open waters (4).

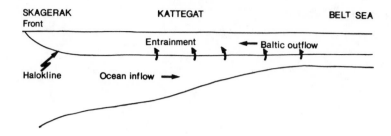

SKAGERAK Front   KATTEGAT   BELT SEA

Entrainment   Baltic outflow

Halokline   Ocean inflow →

Since the 1981 incident, the NAEP has carried out research with the purpose of assessing the magnitude of the different nutrient inputs in all Danish waters. The figures given in Table 1 are the preliminary results of this work, and form the basis of the discussion which has led to parliament action against specific sources. The general conclusion is that about 90% of the N-compounds originate from agriculture and about 70% of the P-compounds originate from urban and industrial waste waters.

The yearly input of N to the Kattegat and the Belt Sea is shown on Fig. 6.

TABLE 1.
Input of N and P to the Danish Fresh and Marine Waters (ref. 4).

Source	N t/year	P t/year
Agriculture	260,000	4,400
Waste water	25,000	7,200
Industry	5,000	3,400
From atsmosphere	?	

NATIONAL ACTION PLAN

The action plan drawn up by the Government requires a reduction in the discharge of N by 50% and P by 80%.

The plan stipulates that this reduction is to be achieved by enforcing specific requirements on agriculture, municipalities and industry:

- Agriculture: Leaching of N and P to be reduced 49% and 91%, respectively, by the improved use of manure and systematic planning of green areas.

- Municipal Treatment Plants: Disharge of N and P to be reduced by 60% and 72%, respectively, by issuing uniform emission standards (8 mg N/1 and 1.5 mg P/1).

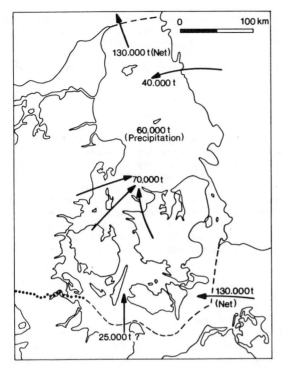

FIGURE 6. Estimate of annual nitrogen input to the Kattegat and Belt Sea (4).

- Industries: Special emission must reduce N and P by 60% and 82% respectively.

The total investment costs of the plan have been estimated at $1.7 billion over a 6 year period.

IMPLICATION FOR MONITORING AND MANAGEMENT STRATEGIES

Monitoring on a Scientific Basis

Many scientists and environmentalists view the lack of success of the Danish monitoring programme in predicting the severe eutrophication as a major weakness of the Environmental Quality Objectives approach; they have demanded a shift toward a more source related strategy.

Whichever strategy is adopted, there is an independant requirement for a monitoring programme to enable an understanding of the major processes in the marine environment, and to predict possible changes.

New research into the cycling of nutrients and organic matter in stressed marine ecosystems shows that there is no simple relationship between nutrient load, primary production and oxygen deficiency at the sea bottom. Primary production is not, therefore, a suitable monitoring parameter for eutrophicated marine ecosystems.

291

A relevant monitoring strategy for eutrophicated marine ecosystems is one which monitors general aspects of the cycling of nutrients and organic matter as a whole, including an assessment of the fate of primary production.

Monitoring of the cycling of carbon by isotope tracer technique seems promising in this respect.

Future Management Strategy

The setting of uniform emission standards for municipal waste water treatment plants implies a shift in the Danish management strategy. However, the new strategy is a combination of the Uniform Emission Standards approach and the Environmental Quality Objectives approach. Receiving waters with limited assimilative capacity receive discharge based on the Environmental Quality Objectives. This requires lower emissions standards than those set by the Uniform Emission Standards approach. In other words, Uniform Emission Standards have been introduced in order to protect open marine areas from unpredicted and hard-to-monitor eutrophication effects, whereas the Environmental Quality Objectives approach is maintained as the management principle for local receiving waters.

SUMMARY

Management of Danish marine areas using the Environmental Quality Objectives approach has proved unsuccessful due to inadequate monitoring programmes designed on the basis of an insufficient understanding of the functioning of marine ecosystems. The Danish authorities have now made a shift in the strategy combining Uniform Emission Standards with Environmental Quality Objectives.

REFERENCES

1. Nielsen, G.Æ., Jacobsen, T.S., Gargas, E., and Buch, E, The Belt Project, Evaluation of the Physical, Chemical and Biological Measurements, The National Agency of Environmental Protection, Denmark, 1981.

2. Nielsen, G.Æ., Causes for and effects of eutrophication in the Kattegat and the Belt Sea., Proc. 22nd Nordic Symposium on Water Research, Iceland, Nordforsk, Miljövårdsserien publ. 1987:1, 1987 (in Danish).

3. Kullenberg, G, Water Exchange in the Danish Straits, In: ref. 4

4. Consensus report on Nitrogen and Phosphorous in the Water Environment, Danish Council of Research Planning and Policy, 1987 (In Danish).

Avoidance/Preference Reaction in Fish - a Method of Marine Environmental Management

INTRODUCTION

Toxicity testing forms part of hazard assessment of pollutants in the management of marine environmental quality on a routine basis. In recent years, however, the use of behavioral studies has been investigated, and it has become apparent that avoidance/preference tests may be valuable tools in environmental management of coastal marine waters.

AVOIDANCE/PREFERENCE REACTIONS

The chemosensory systems of fish and shrimps are extremely sensitive (22) to chemicals in their environment laboratory trials and field studies have demonstrated the ability of fish and shrimps to respond to the perception of certain toxic pollutants by moving away from the area of pertubation (avoidance; see Table 1). However, it is also well established that some toxic substances are not avoided and that others actually may attract fish (preference) (12, 22).

Avoidance reactions are often found at concentrations well below lethal levels (Table 2), sometimes at orders of magnitudes lower e.g. the avoidance threshold value of Cr (VI) is 0.0004 times that of the 96h LC_{50} concentration for rainbow trout (Salmo gairdneri) (1, 31).

TABLE 1. Examples of Pollutants Avoided by Fish and Shrimps.

Pollutants		References
Heavy metals:	Copper, zinc, mercury, chromium, lead	1,7,17,29,34
Pesticides:	DDT, Endrin, Dursban, 2,4-D, parathion, malathion, Sevin, dimilin-G1, toxapene, Dalapon, acrolein, fenitrothion, pesticide contaminated sediment.	7,8,9,10,20,25,28
Other pollutants:	Pulp mill effluents, oil contaminated sediment, ammonium, chlorine, Arochlor 1254, a-chloracetophenone, dichlornitrobenzene, thanite, sodium cyanide, cymene thiocyanate, ethyl alcohol, chloroform, formalin, para-cresol, ortho-cresol	2,11,16,18,19,23, 24,27,30

TABLE 2. Comparison of Lethal Levels and Avoidance Levels of Pollutants for Fish

Pollutant	Species	Lethal levels (mg/l)	Avoidance levels (mg/l)	Reference
Heavy metals:				
Cu	Salmo salar	0.125[1]	0.009[4]	31,34
Zn	Salmo gairdneri	3.000[2]	0.03-0.13[5]	29
Cr (VI)	Salmo gairdneri	69.000[1]	0.028[5]	1,31
Pesticides:				
Dursban	Gambusia affinis	4.000[3]	0.100[4]	10
Malation	Gambusia affinis	2.000[3]	0.050[4]	10
2,4D	Gambusia affinis	7.000[3]	1.000[4]	10
Endrin	Cyprinodon variegatus	0.003[3]	0.0001[4]	9
DDT	Cyprinodon variegatus	0.006[3]	0.005[4]	9
Xylene	Salmo gairdneri	13.500[1]	0.100[4]	7
Waste water from a pesticide factory	Anguilla anguilla	33.000[1,6]	0.1800[5]	4,6
	Plathichtys flesus	25.000[1,6]	0.3800[5]	4,6

[1] 96hLC$_{50}$, [2] 48h LC$_{50}$, [3] 24h LC$_{50}$, [4] lowest concentration tested, avoided by the fish, [5] estimated threshold level, [6] concentration in ml/l.

HAZARD ASSESSMENT AND AVOIDANCE/PREFERENCE STUDIES

Traditionally, hazard assessment of pollutants is based on comparisons of toxicity and bioaccumulation data with the expected pollutant levels in the receiving waters.

Under certain circumstances avoidance/preference studies may be of crucial importance in the overall assessment of the effects of pollutants and should be applied in addition to toxicity tests. If fish show preference reactions towards chemicals in the waste water, they may be attracted to lethal concentrations. Avoidance reactions at sublethal levels are advantageous as they provide a chance to escape lethal exposure. On the other hand avoidance of waste water may alter the distribution of fish in areas far from the pollution source causing disruption to fisheries. This may occur when a plume of waste water blocks the migratory routes of fish, such as occured at the Miramichi river in Canada. During spring and summer, Atlantic salmon (Salmo salar) migrate up the Miramichi to spawn. After the establishment of a drainage from a mine introducing copper and zinc to the river, it was observed that the immigrating fish in periods avoided the polluted water and returned to the sea. Upstream migration was blocked when the concentration of copper and zinc was greater than 0.35-0.43 times that of the incipient lethal level for juvenile salmons (34).

In Denmark, avoidance studies have recently been applied when setting requirements of the discharge from a pesticide plant (Cheminova) (4,5). The plant is situated at the mouth of Limfjorden, an enclosed lagoon in the northern part of Jutland to the North Sea through a pipeline. The waste water is mainly carried north by the prevailing northerly currents in the area (Fig. 1C).

The plant which was established in 1953 has caused severe environmental problems in the area, including:

- massive fish, lobster and bird kills
- tainting of fish
- high concentrations of pesticides and mercury in ducks, seals and fish (the mercury was leaked from burial and dump sites containing residues from the former production of organomercurial fungicides).

In 1965, waste water discharges were regulated and the toxicity of the discharge reduced considerably. In 1982, the plant was required to install better waste water treatment facilities. A comprehensive hazard assessment programme was initiated by Cheminova and local authorities. This comprised:

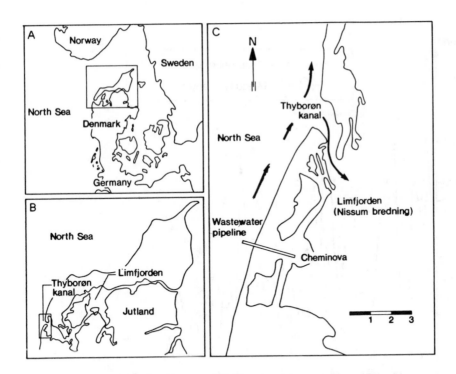

FIGURE 1. Location of Cheminova, Denmark. 1B Location of Limfjorden. 1C Location of Cheminova, the waste water pipeline and the entrance to Limfjorden (Thyboroen kanal). The prevailing currents in the area indicated.

- estimates of the dilution of the waste water in the North Sea
- investigations into the fate of waste water components
- comparison of the acute and chronic toxic effects of the present and future waste water to 17 marine species (protozoans, algae, crustaceans and fish)
- comparison of the avoidance inducing capacity of the present and future waste water.

Avoidance studies were included because the plant is situated near migratory routes for commercially important fish species. Stocks of eel, (*Anguilla anguilla*), plaice (*Pleuronectes platessa*) and cod (*Gadus morhua*) in Limfjorden are sustained by immigration from the North Sea through Thyboroen kanal (Fig. 1C). Formerly, Limfjorden was an important fishery area for eel and plaice. However, since the 1950s, catches of these species from Limfjorden have decreased drastically (21) (Fig. 2). One reason may be the increased eutrophication of the fjord causing seasonal oxygen depletion in certain areas. However, it has been suspected that the reduced catches may be due partly to reduced immigration caused by the waste water from Cheminova which may create a chemical barrier avoided by the fish at Thyboroen kanal (Fig. 1C). Fishery research has shown that immigration of plaice to the western part of Limfjorden (Nissum bredning) decreased from 1953 when Cheminova was established (Fig. 3) and pesticide contaminated sediment taken outside the plant in Nissum bredning was avoided by flatfish and shrimps (25).

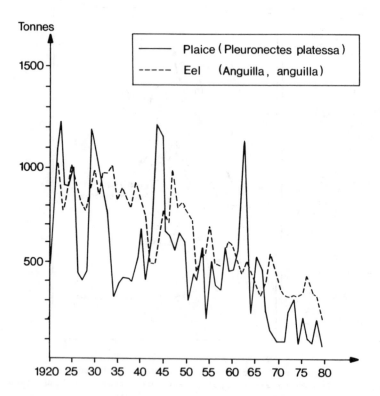

FIGURE 2. Total catches of plaice and eel from Limfjorden 1920-1979 (21).

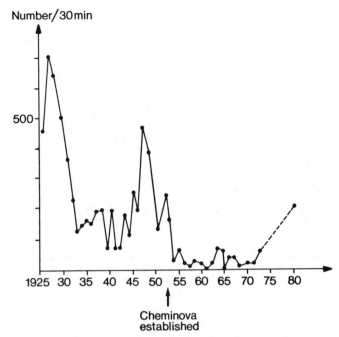

FIGURE 3. Relative density of plaice in the western part of Limfjorden (Nissum bredning). Standardized trawlings performed by research vessels 1925-1972 and in 1980. Number caught per half an hour of trawling (21).

Laboratory investigations into the behavior of juvenile eel and flounder (Platichthyes flesus) when exposed to present waste water and waste water from a pilot treatment plant (future waste water) were performed to see if the waste water contains chemicals in concentrations which may be avoided by fish   (4).

The experiments were performed in a tank, as shown in Fig. 4:

- diluted waste water and clean water enter the tank on each side of a partition wall (1).

- the water passes through a filter at the end of the partition wall and at the outlet (2).

- the design causes laminar water flow, producing an abrupt concentration gradient down the middle of the observation area (3). The design further allows alteration of the polluted and the clean water streams.

Five to ten fish were placed in the observation area for each test and their movements photographed every thirty seconds over a period of eighty minutes. During the first ten minutes of the experiment clean water was fed into both sides of the tank. Diluted waste water was then fed into one side of the tank for forty-five minutes, after which the streams were reversed. The degree of avoidance at a given dilution was expressed as the relative presence of fish in clean water for each period of forty minutes.

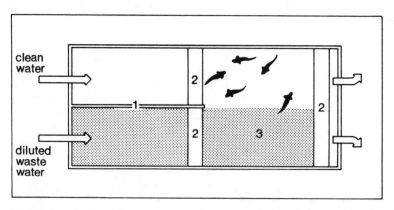

FIGURE 4. Laboratory set-up, principle. 1) Partition wall, 2) Filters, 3) Observation area.

Samples of the present waste water were tested against eel and flounder and samples of the future waste water were tested against eel, only. For each sample the degree of avoidance was determined for different dilutions and a dose-response curve estimated, according to a procedure described in ref. 32. Further, EC_{10}, EC_{50} and EC_{90} were estimated.

Figure 5 shows the estimated dose-response curve of eels for present and future waste water, respectively. Table 3 gives the estimated EC_{10}, EC_{50} and EC_{90} values of the present waste water. EC_{10} may be considered the approximate threshold avoidance level.

TABLE 3. Avoidance of Eel and Flounder of Present Waste Water. Estimated EC_{10}, EC_{50} and EC_{90}. 95% confidence interval in parentheses.

| Species | Effect concentrations (ml/1) | | |
	EC_{10}	EC_{50}	EC_{90}
Eel (Anguilla anguilla)	0.18 (0.08-0.41)	0.82 (0.57-1.17)	3.78 (1.60-8.95)
Flounder (Plathictys flesus)	0.38 (0.05-2.88)	0.81 (0.34-1.92)	1.75 (0.31-9.72)

Juvenile eel and flounders avoided the present waste water at concentrations >-0.18 and >-0.38 ml/1. The concentration of substances inducing avoidance is considerably lower in the future waste water in which juvenile eel did not avoid concentrations of up to 166 ml/1, the highest dilution tested.

In Table 4 the EC_{10} values are compared to estimated and measured waste water concentrations at different distances north of the outlet.

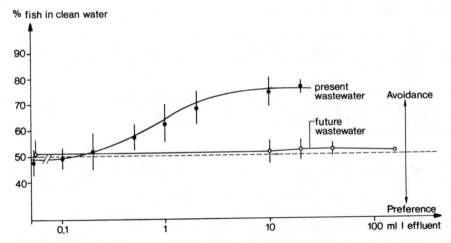

FIGURE 5. Eel avoidance of different dilutions of present and future waste water. Mean ± S.D. and the estimated dose-response curves are shown. Totally, 91 tests were performed.

At the outlet, and 0.5 km north waste water concentrations are well above the avoidance threshold values obtained for the present waste water, indicating that avoidance could be occurring in the natural environment. At a distance of 5.2 km from the outlet the discharge is diluted to concentrations comparable to the avoidance threshold. At Thyboroen kanal 7 km north of the outlet wastewater concentrations are well below the avoidance threshold indicating that avoidance is not likely to occur at present. Further the results indicate that avoidance problems will be eliminated when the new waste water treatment is installed.

TABLE 4. Comparison of Avoidance Threshold Values (EC_{10}) of Present and Future Waste Water and Waste Water Concentrations at Various Distances North of the Outlet.

Distance from outlet, km	Waste water concentration ml/l
0	2-10
0.5	0.6-1.1
5.2	0.2
7[1]	<0.1
- - - - - - -	
EC_{10}, present waste water	0.18-0.38
EC_{10}, future waste water	>166

[1] At Thyboroen kanal

299

LABORATORY VERSUS NATURAL ENVIRONMENT

Few studies have compared laboratory and field avoidance responses in fish (Table 5). In two studies avoidance-preference tests performed in the laboratory have predicted fairly well the concentrations of pollutants avoided by fish in the natural environment.

In the first study, the avoidance reactions of spotfin shiner (Notropis spilopterus) at a power plant in USA towards chlorine in the cooling water were generally similar between laboratory and field concentrations (0.21 mg/1 and 0.18 mg/1, respectively) (2).

In the second study avoidance of a blend of copper, chromium, arsenic and selenium by fathead minnows (Pimephales promelas) was observed in the laboratory and in the field (Adair Run at New River, Virginia, USA (13). In the spring the concentration avoided in the laboratory was similar to the concentration avoided in the field (0.071 mg/1 and 0.073 mg/1, respectively). However, in experiments performed in the summer the fish avoided considerably lower concentrations in the laboratory (0.034 mg/1) than in the field (0.073 mg/1). Differences in water hardness and turbidity were cited as possible causative factors; higher water hardness and turbidity may lessen the availability of the metals to fish due to complexation, precipitation and adsorption).

TABLE 5 - Comparison of Avoidance Levels in the Laboratory and in the Natural Environment.

Species	Chemicals	Avoidance level laboratory	Avoidance level natural environment	Reference
Notropis spilopterus	Chlorine	0.21 mg/1	0.18 mg/1	2
Pimephales promelas	Blend of copper, chromium, arsenic and selenium	0.071 mg/1 [1] 0.034 mg/1 [2]	0.073 mg/1	13
Salmo salar	Blend of copper and zinc	>0.02 of incipient lethal level	>0.35-0.45 of incipient lethal level	34

[1] dilution water taken in the spring from New River, Virginia
[2] dilution water taken in the summer from New River, Virginia.

Laboratory trials performed in connection with field investigations at Miramichi River into the avoidance of zinc and copper by mature salmon showed that juvenile salmon avoided zinc and copper at much lower concentrations in the laboratory than that observed in the field (approx. 20 times lower) (34). The difference may be due to the different sensitivity of juvenile and mature fish. However, it may be possible that the fish behave differently in the field. Mature salmon ascending upstream may thus be exposed to stimuli raising the avoidance threshold

These examples show that caution must be used in interpreting laboratory results, and appliying them to the field. However, if laboratory trials have demonstrated that avoidance reactions occur towards a pollutant, it provides strong evidence for similar reactions in the natural environment. Further, if field studies show a corresponding absence of fish from an area where this pollutant is discharged, evidence of avoidance in the field is strengthened.

CONCLUSION

Avoidance/preference tests may be a valuable, sensitive tool in the management of coastal water. However, they cannot generally replace toxicity tests when assessing the environmental effects of pollutants, but under certain circumstances and in certain areas they should be applied in addition. Avoidance/preference studies should be applied to areas where it is important that a zone of passage is maintained for migratory fish, e.g. juveniles migrating to nursery areas or mature fish migrating to spawning areas.

REFERENCES

1) Anestis, I. and Niufeld, R.J., Avoidance - Preference reactions of rainbow trout (Salmo gairdneri) after prolonged exposure to chromium (VI), Wat. Res., vol. 20, pp. 1233-1211, 1986.

2) Cherry, D.S. and Cairns, J., Biological monitoring, Part V - Preference and avoidance studies, Water. Res., vol. 16, pp. 263-301, 1982.

3) Cherry, O.S., Larrick, S.R., Giattina, J.D., Cairns, J. and van Hassel, J., Influence of Temperature Selection upon the Chlorine Avoidance of cold-water and warmwater fishes, Can. J. Fish. Aquat. Sci., vol. 39, pp. 162-173, 1982.

4) COWIconsult, Avoidance behavior of eel, flounder and common shrimp, Report to the Environmental Protection Agency. Ringkoebing County Council and Cheminova, 1986 (in Danish), pp. 47 + appendixes.

5) COWIconsult, Avoidance behavior of eel, Report to Cheminova, 1986 (in Danish), pp. 15.

6) COWIconsult, Acute toxicity of waste water from Cheminova to eel, flounder and common shrimp, Report to Ringkoebing County Council, 1986 (in Danish), pp. 12.

7) Folmar, L.C., Overt avoidance reaction of rainbow trout fry to nine herbicides, Bull. Environm. Contam. Toxicol., vol. 15, pp. 509 - 514, 1976.

8) Granett, J., Morang, S. and Hatch, R., Reduced movement of precocious male Atlantic Salmon parr into sublethal Dimilin-G1 and carrier concentrations. Bull. Environm. Contam. Toxicol. pp. 463-464, 1978.

9) Hansen, D.J., Avoidance of pesticides by untrained sheepshead minnows, Trans. Amer. Fish. Soc., vol. 3, pp. 426-429, 1969.

10) Hansen, D.J., Matthews, E., Nall, S.L. and Dumas, D.D., Avoidance of pesticides by untrained mosquitofish Gambusia affinis, Bull. Environm. contam. Toxicol. vol. 8, pp. 46-51, 1972.

11) Hansen, D.J., Schimmel, S.C. and Matthews, E., Avoidance of Aroclor 1254 by shrimp and fishes, Bull. Environm. Contam. Toxicol. vol. 12, pp. 253-256, 1974.

12) Hara, T.J. and Thompson, B.E., The reaction of whitefish Coregonus clupeaformis to the anionic detergent sodium lauryl sulphate and its effects on their olfactory responses, Water. Res., vol. 12, pp. 893-897, 1978.

13) Hartwell, S.I., Cherry, D.S. and Cairns, J., Field validation of avoidance of elevated metals by fathead minnows (Pimephales promelas) following in situ acclimation, Environment. Toxicol. and Chemistry, vol. 6, pp. 189-200, 1987.

14) Höglund, L.B., The reactions of fish in concentration gradients, Institute of Freshwater Research Drottningholm, Lund, Sweden, Report No 43, 1961.

15) Johnston, D.W. and Wildish, P.J, Avoidance of dredge spoil by herring (Clupea harengus harengus), Bull. Environm. Contam. Toxicol., vol. 26, pp. 307-314, 1987.

16) Jones, J.R.E., The reactions of Pygosteus pungitius to toxic solutions, J. exp. Biol., vol. 24, p. 110-122, 1947.

17) Jones, J.R.E., A further study of the reactions of fish to toxic solutions, J. Exp. Biol., vol. 25, p. 22-34, 1948.

18) Jones, J.R.E., The reactions of the minnow Phoxinus phoxinus to solutions of phenol, ortho-cresol and para-cresol, J. Exp. Biol., vol. 28, p. 261-270, 1950.

19) Jones, B.F., Warren, C.E., Bond, C.E. and Doudoroff, P., Avoidance reactions of salmonid fishes to pulp mill effluents, Sewage and Industrial Wastes, vol. 28, pp. 1403-1413, 1956.

20) Kynard, B., Avoidance behaviour of insecticide suspectible and resistant populations of mosquitofish to four insecticides, Trans. Amer. Fish Soc., vol. 3, pp. 557-561, 1974.

21) Limfjordskomiteen, Fishery research in Limfjorden 1980-81, Report from the Danish Institute of Fishery Research, 1982 (In Danish), pp. 144 + appendixes.

22) Lindahl, P.E. and Marcström, A., On the preference of roaches (Leuciscus rutilus) for trinitrophenol, studies with the fluviarium technique, Fish. Res. Bd. Canada, vol. 15, pp. 685-694, 1958.

23) Lubinski, K.S., Cickson, K.L. and Cairns, J., Effects of abrupt sublethal gradients of ammonium chloride on the activity level, turning and preference - avoidance behaviour of bluegills. In: Aquatic toxicology, ASTM STP 7078, (J.C. Eaton, PR. Parrish and A.C. Hendricks, Eds.) American Society for Testing and Materials, pp. 328-340, 1980.

24) McGreer, E.R. and Vigers, G.A., Development and validation of an in situ fish-preference avoidance technique for environmental monitoring of pulp mill effluents In: Aquatic Toxicology and Hazard Assessment: Sixth Symposium. ASTM STP 802, (W.E. Bishop, R.D. Cardwell and B.B. Heidolph Eds.) American Society for Testing and Materials. Philadelphia, pp. 519-529, 1983.

25) Møhlenberg, F. and Kiørboe, T., Burrowing and avoidance behaviour in marine organisms exposed to pesticide contaminated sediment, Mar. Pollut. Bull., vol. 14, pp. 57-60, 1983.

26) Olla, B.L., Pearson, W.H. and Studholme, A.L., Applicability of behavioral measures in environmental stress assessment, Rapp. P-v. Reun int. Explor. Mer., vol. 197, pp. 162-173, 1980.

27) Pinto, J.M., Pearson, W.H. and Anderson, J.W., Sediment preferences and oil contamination in the Pacific sand lance Ammodytes hexapterus, Marine Biology, vol. 83, pp. 193-204, 1984.

28) Scherer, E., Avoidance of fenitrothion by goldfish (Carassius auratus), Bull. Environm. Contam. Toxicol., vol. 13, pp. 492-496, 1975.

29) Sedgwick, R.W., The development of an apparatus to assess avoidance behaviour in fish, Annual study course of the Institutes of Fisheries Management, 1980.

30) Summerfelt, R.C. and Lewis, W.M., Repulsion of green sunfish by certain chemical, Water Poll. Control Fed. J., vol. 39, pp. 2030-2038, 1967.

31) USEPA, Quality criteria for water, U.S. Environmental Protection Agency, Washington D.C., 1976, pp. 256.

32) Vølund, Aa., Application of the four-parameter logistic model to bioassay: Comparison with slope ratio and parallel line models, Biometrics, vol. 34, 357-365, 1978.

33) Verschueren, K., Handbook of environmental data on organic chemicals, (Second edition), Van Nostrand Reinhold Company, 1983.

34) Warren, C.E., Biology and water pollution control, pp. 168-191, W.B. Saunders Company, 1971.

Ocean Waste Management

MICHAEL A. CHAMP
Cross-Disciplinary Research Division
Engineering National Science Foundation
Washington, D.C. 20550

IVER W. DUEDALL
Department of Oceanography and Ocean Engineering
Florida Institute of Technology
Melbourne, Florida 32901

ABSTRACT

From the earliest times, the ocean has been used for the disposal of wastes. The ocean (considered the ultimate sink with an unlimited assimilative capacity) provided a cheap outlet for waste disposal. Wastes either floated away, dissolved, or sank and thereby were out-of-sight, out-of-smell, and out-of-mind. However, in the mid-1970s, the ocean developed a protected status from the disposal of wastes due to public awareness of coastal pollution problems and "equality" practices in federal regulations. There is a need for comprehensive waste multimedia assessments (air, land, and water) that consider all media equally for the disposal of wastes and not the shunting of a waste to the media of least regulation. Ocean waste management strategies are in their infancy--in the hypothesis testing and evaluating stages. Scientific and policy issues for the use of ocean space for waste disposal have to be delineated so that they can be compared with those associated with other disposal media. To enhance the future utilization of the multimedia approach in the waste management decision process, a strategy for comprehensive and integrative waste management is needed, which utilizes the concepts of risk assessment and risk management, and which uses risk reduction to the total environment as the decision-making tool. Consideration is also given to the concept of assimilative capacity of a dumpsite as a waste management strategy.

INTRODUCTION

From the earliest times coastal civilizations have used the sea for waste disposal. Wastes from ships have always been thrown overboard. Fish processing wastes, seafood wastes, fish parts, culls, and inedible species have been discarded into the sea or at sea since the time humans began to fish. Also maritime nations have found it essential to maintain shipping lanes and harbors, which involves the dredging of deposited materials and subsequent disposal in adjacent coastal areas. The practice of ocean dumping of sewage sludge was initiated in 1887 in the United Kingdom with the dumping of London wastes in the Outer Thames Estuary. In the United States the first dumping of sewage sludge occurred in 1924 in the New York Bight.

The ocean provided the NIMBY (Not in My Back Yard) cheap outlet for waste disposal. With the advance of civilization and the

305

industrial revolution, not only have waste discharges increased
(with greater potential for oxygen depletion in the water column)
but the toxicity of the waste has increased by the addition of
heavy metals, synthetic organic substances, and persistent compounds. As nations became aware of the consequences of direct
municipal and industrial discharges to waterways, waste treatment
procedures and facilities were required. However, this cleaning
up of the discharge to waterways by removing or decreasing the
BOD (biochemical oxygen demand) levels, nutrients, and suspended
solids created a separate disposal problem for the removed materials, commonly referred to as sludges. The regulatory agencies
were now faced with the problem of what to do with the sludge.
They certainly could not permit the dumping or discharge of these
sludges back into the waters that the regulatory process was
trying to protect when the sludge was separated from the effluent.
Therefore, authorities, reacting to public pressures to maintain
clean coastal waters, began to consider offshore disposal by
pipelines, tankers, or barges.

The focus of this chapter is to (1) delineate global ocean waste
disposal practices (i.e., what is dumped where), (2) discuss
scientific and policy issues and public perceptions on the use of
ocean space for waste disposal, (3) review the ocean waste
disposal option from a waste management perspective, (4) define
and discuss the concept of assimilative capacity, and identify
physical, chemical, and biological information that is needed to
estimate the assimilative capacity of a body of seawater for a
given waste, and (5) discuss the concept of the multimedia approach utilizing risk assessment and risk management as the
decision-making tools and focusing on reducing the risk as a
waste management strategy.

1.   GLOBAL OVERVIEW OF OCEAN DUMPING PRACTICES

Ocean disposal of wastes has increased worldwide even though
stringent national and international regulations [since the mid-
1970s] have greatly reduced the dumping of hazardous or toxic
industrial wastes. Nevertheless, from a global perspective, the
desirability of using ocean space for waste disposal has increased
because of several factors: (1) the inherent stability of physical
and chemical environments in the ocean, makes it easier to predict
the fate and behavior of many types of wastes deposited there as
compared to on land, (2) our knowledge base is least advanced
particularly in the subsurface soil/ groundwater environment and
currently inadequate to protect public health and the environment,
and (3) ocean disposal is generally cheaper than land based
options or incineration alternatives. The following section will
discuss from a global perspective, the types and amounts of wastes
that have been disposed of in the ocean, and the location of
dumpsites.

1.1.   Categories of Wastes Ocean Dumped

The following categories of wastes and other matter have been
dumped in the ocean or disposed at sea from ships or platforms
(Champ and Park, 1982):

   Dredged materials
   Municipal wastes (sewage sludges)
   Industrial wastes (acids, alkali, organic and pharmaceutical)

Radioactive wastes
Chlorinated hydrocarbons (Ocean incineration)
Fish (seafood) wastes
Drilling fluids (drilling muds, cutting wastes, and
production waters)
Coal wastes (colliery wastes, fly ash, flue-gas desulfurization
sludges, and boiler bottom ash)
Ocean mining wastes (mining and processing)
Oil spill studies (research permits)

The international convention (global) for the regulation of ocean dumping is the "Convention on the Prevention of Marine Pollution by Dumping of Wastes and Other Matter," commonly referred to as the London Dumping Convention (LDC), for more information, see IMCO, 1972, IMO, 1982d, and Nauke, 1988. The Secretariat of LDC is the International Maritime Organization (IMO) [IMO prior to 1982 was named the Inter-Governmental Maritime Consultative Organization (IMCO)]. The IMO prepares an annual report providing global information on the number of dumping permits, quantities and types of wastes ocean dumped, and the exact location of dumpsites for that year. This report is drafted from the required information that Contracting Parties (Nations) to the London Dumping Convention are required to prepare and submit annually to the LDC Secretariat (IMO). The following section summarizes the annual reports for the years 1976-1982.

1.2. Permit Quantities

A review of the IMO (IMCO) annual reports (IMCO, 1979a-c, 1980a-b; IMO, 1984a-b) for the period of 1976-1982 finds about 2700 permits issued in this 7 y period reported to the LDC by Contracting Parties, with an increase each year of about 70 permits during the first 5 y (Champ and Park, 1981, 1988; Duedall et al., 1983; and revised in Duedall, 1985). It should be noted that these data do not contain information from Australia or Japan, which have recently become LDC Contracting Parties. Figure   1   (from Duedall, 1985) presents the annual ocean dumping permit quantities for municipal and industrial wastes, dredged materials, and a category for minor constituents. It is interesting to note that for the United States and the United Kingdom, where the volumes of sewage sludges dumped in the ocean have been increasing annually, that the percent solids in the sludges have been decreasing by almost 50% over this period due to shifts from primary to secondary treatment, which produces less solids (5-8% to 2-3% presently); therefore, the total contaminant loading to the dumpsites has not increased proportionally during this period (Norton and Champ, 1988).

The annual global tonnages covered by industrial permits have almost doubled during this period, even though in the United States, the EPA has significantly phased out a large number of U.S. industrial ocean dumping permits. Figures   2a,   2b, and 2c (from Duedall, 1985) presents data from nations that dump over 1% of the annual quantities of wastes (reported to LDC) dumped at sea by waste categories (sewage sludge, industrial wastes, and dredged materials), but these data are permit application quantities and not actual amounts dumped. We have found that the actual dumped quantities are around 80% of the reported permit quantities as deduced from the Oslo and Paris Conventions Annual Reports for the North Sea.

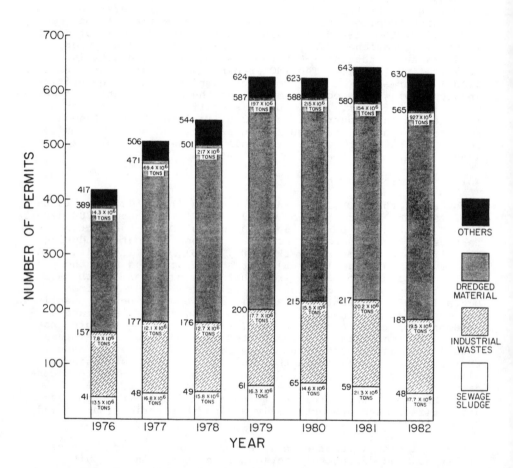

Figure 1.   Total number of permits issued and permit tonnages (metric tons) on a global basis for ocean dumped municipal and industrial wastes, dredged materials, and a category for minor constitutents from IMCO [IMO] annual reports for the years 1976 to 1982 as reported to the London Dumping Convention.   This figure is from Duedall (1985).

308

SEWAGE SLUDGE

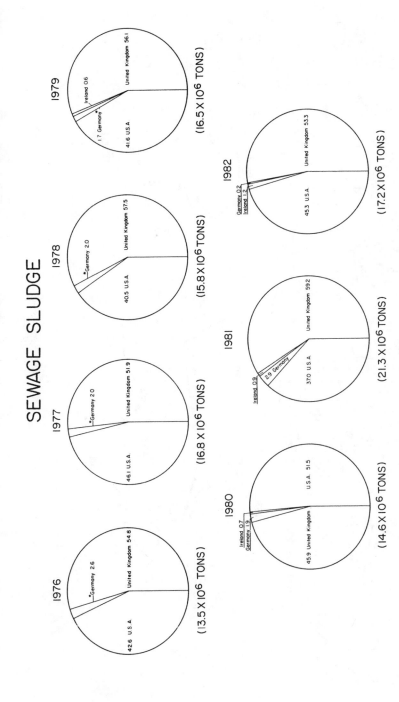

Figures 2a., 2b., and 2c. Percentages of the total estimated tonnage of ocean dumped wastes (by country for countries dumping over 1% of the total) from IMCO [IMO] data for the years 1976 to 1982 by waste categories (sewage sludge, industrial wastes and dredged materials). *Germany is the Federal Republic of Germany. The figures are from Duedall (1985).

309

INDUSTRIAL WASTES

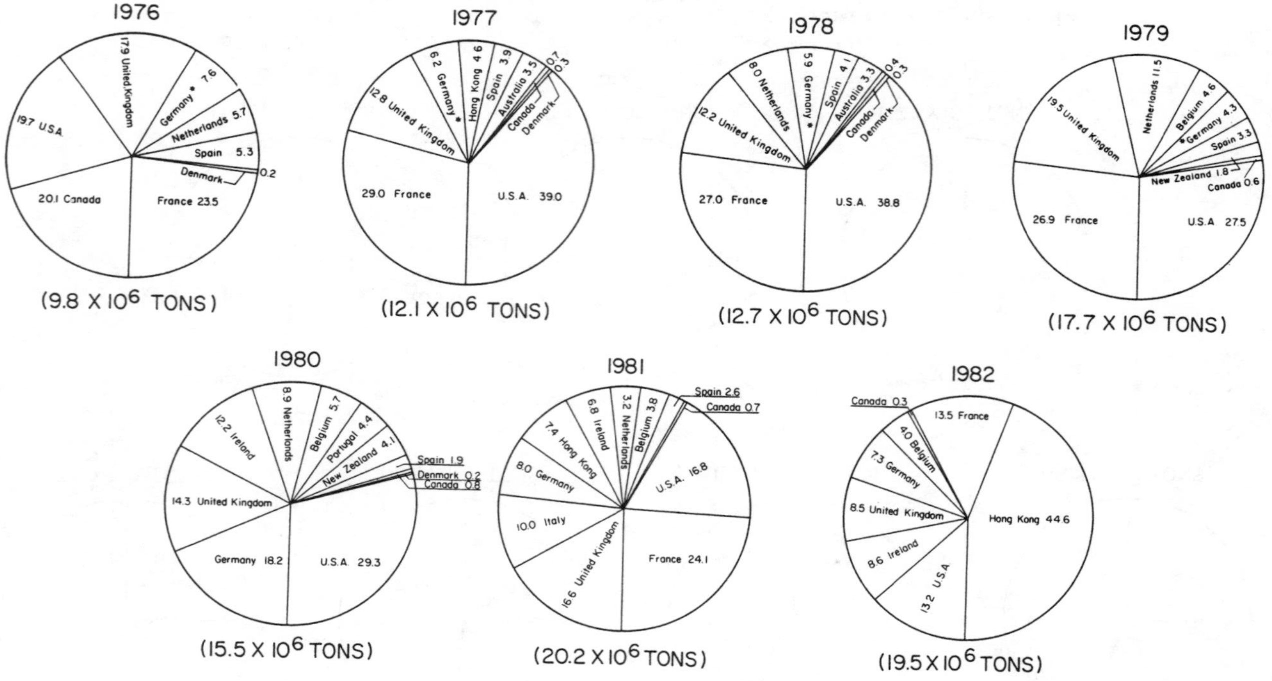

1976

Germany • 7.6
Netherlands 5.7
Spain 5.3
Denmark 0.2
France 23.5
17.9 United Kingdom
19.7 U.S.A
20.1 Canada

(9.8 X 10⁶ TONS)

1977

Denmark 0.7
Canada 0.2
Australia 3.5
Spain 3.9
Hong Kong 4.6
Germany • 6.2
12.8 United Kingdom
29.0 France
U.S.A. 39.0

(12.1 X 10⁶ TONS)

1978

Denmark 0.4
Canada 0.5
Australia 3.3
Spain 4.1
Germany • 5.9
8.0 Netherlands
12.2 United Kingdom
27.0 France
U.S.A. 38.8

(12.7 X 10⁶ TONS)

1979

Canada 0.6
New Zealand 1.8
Spain 3.3
Germany 4.3
Belgium 4.6
Netherlands 11.5
19.5 United Kingdom
26.9 France
U.S.A. 27.5

(17.7 X 10⁶ TONS)

1980

Spain 1.9
Denmark 0.2
Canada 0.1
Portugal 4.1
New Zealand 4.4
Belgium 5.7
8.9 Netherlands
12.2 Ireland
14.3 United Kingdom
Germany 18.2
U.S.A. 29.3

(15.5 X 10⁶ TONS)

1981

Spain 2.6
Canada 0.7
U.S.A. 16.8
Belgium 3.8
3.2 Netherlands
6.8 Ireland
7.4 Hong Kong
8.0 Germany
10.0 Italy
16.6 United Kingdom
France 24.1

(20.2 X 10⁶ TONS)

1982

Canada 0.3
13.5 France
4.0 Belgium
7.3 Germany
8.5 United Kingdom
9.8 Ireland
13.2 U.S.A.
Hong Kong 44.6

(19.5 X 10⁶ TONS)

Figure   2b

310

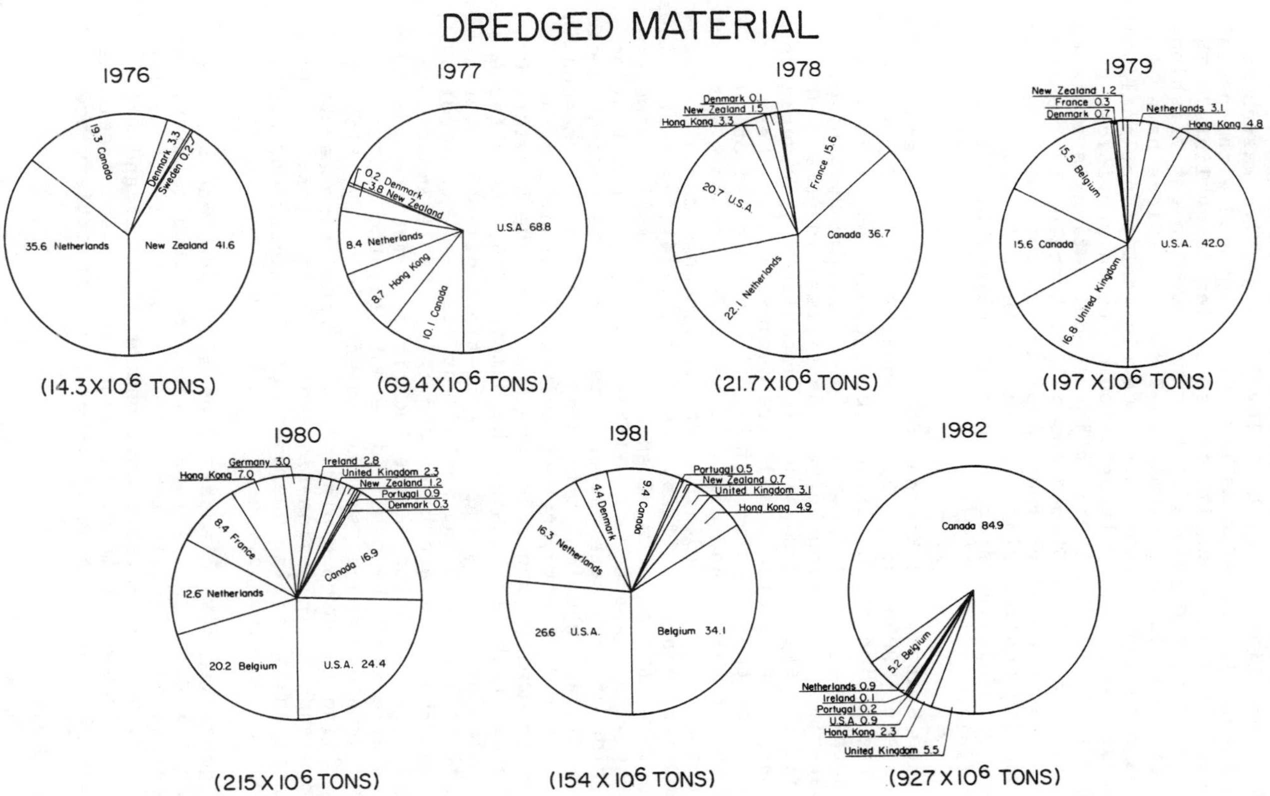

DREDGED MATERIAL

1976

New Zealand 41.6

35.6 Netherlands

19.3 Canada

Denmark 3.3
2.0
Sweden 0.2

(14.3 X 10⁶ TONS)

1977

U.S.A. 68.8

0.2 Denmark
1.3 New Zealand

8.4 Netherlands

8.7 Hong Kong

10.1 Canada

(69.4 X 10⁶ TONS)

1978

Canada 36.7

France 15.6

Denmark 0.1
New Zealand 1.5
Hong Kong 3.3

20.7 U.S.A

22 Netherlands

(21.7 X 10⁶ TONS)

1979

U.S.A. 42.0

Netherlands 3.1
Hong Kong 4.8

New Zealand 1.2
France 0.3
Denmark 0.7

16.8 United Kingdom

15.6 Canada

15.5 Belgium

(197 X 10⁶ TONS)

1980

U.S.A. 24.4

20.2 Belgium

12.6 Netherlands

8.4 France

Canada 16.3

Germany 3.0
Hong Kong 7.0

Ireland 2.8
United Kingdom 2.3
New Zealand 0.9
Portugal 0.9
Denmark 0.3

(215 X 10⁶ TONS)

1981

Belgium 34.1

U.S.A. 26.6

16.3 Netherlands

4.4 Denmark

9.4 Canada

Portugal 0.5
New Zealand 0.7
United Kingdom 3.1
Hong Kong 4.9

(154 X 10⁶ TONS)

1982

Canada 84.9

Belgium 5.5

Netherlands 0.9
Ireland 0.1
Portugal 0.2
U.S.A 0.9
Hong Kong 2.3
United Kingdom 5.5

(927 X 10⁶ TONS)

Figure 2c

311

1.3. LDC Guidelines and the U.S. Permitting Process

The basic strategy of the LDC is to prohibit the dumping of certain particularly hazardous substances listed in Annex I, the so-called "black list" (e.g. high level radioactive wastes are listed in Annex I). Substances that may be dumped with special care are contained in Annex II, the so-called "grey list." The Annex III presents provisions to be considered in establishing criteria for ocean dumping permits. These provisions have been prepared as guidance to support the development of national waste management strategies. Implementation of these provisions by Contracting Parties into their national legislation is a statutory requirement. The United States (as all Contracting Parties to the LDC) has implemented national legislation for the regulation of ocean dumping and formulated ocean dumping criteria and an ocean dumping permitting system under the guidelines of LDC (Wastler, 1981).

In the United States, the Marine Protection, Research and Sanctuaries Act (MPRSA) of 1972, (as amended 33 U.S.C. 1401 et seq.), is the national legislation (U.S. Congress, 1972a). It authorizes the EPA to grant permits for the ocean dumping of various materials. Section 102(a) identifies nine statutory factors that the EPA must consider in establishing criteria to evaluate permit applications and site designations. These are: (1) the need for the proposed dumping, (2) the effect of such dumping on human health and welfare, including economic, esthetic, and recreational values, (3) the effect of such dumping on fisheries resources, plankton, fish, shellfish, wildlife, shore lines and beaches, (4) the effects of such dumping on marine ecosystems, particularly with respect to: (a) the transfer, concentration, and dispersion of such material and its byproducts through biological, physical, and chemical processes, (b) potential changes in marine ecosystem diversity, productivity, and stability, and (c) species and community population dynamics, (5) the persistence and permanence of the effects of the dumping, (6) the effects of dumping particular volumes and concentrations of such materials, (7) appropriate locations and methods of disposal or recycling, including land-based alternatives and the probable impact of requiring use of such alternative locations or methods upon considerations affecting the public interest, (8) the effect of alternative uses of oceans, such as scientific study, fishing, and other living resource exploration, and nonliving resource exploration, and (9) in designating recommended sites, the [EPA] Administrator shall utilize wherever feasible locations beyond the edge of the continental shelf.

There are generally five classes of permits for ocean dumping: general, special, emergency, interim, and research permits. A general permit may be issued for the dumping of certain materials that will have minimal adverse environmental impact and will generally be disposed of in small quantities. Special permits may be issued for the dumping of materials that satisfy ocean

312

dumping criteria and have an expiration date of not longer than three years. Emergency permits may be issued for any of the materials listed in LDC Annex I, except as trace contaminants, after following the consultative processes of the LDC. This permit may be issued for the dumping of such materials, where it has been demonstrated that an emergency exists, which possess an unacceptable risk to human health, and where no other solution exists. Interim permits may be issued to dump materials that are not in compliance with ocean dumping criteria or for which an ocean dump site has not been designated. This type of permit is usually issued to allow the dumper time to add new pollution control technology to existing waste treatment or pretreatment facilities (e.g., source control or direct pretreatment) with the perspective that in a given time period the waste dumper would either comply with the ocean dumping criteria or dumping would be phased out. Research permits may be issued for a short-time period when it has been determined that the scientific merit of the proposed project outweighs the potential environmental or other damage that may result from the dumping. In the United States, permits for incineration of wastes at sea have only been issued as research permits.

1.4. Location of Dumpsites

In the IMO (IMCO) annual reports, the longitude and latitude for the location of each dumpsite is given. Figures 3a and 3b (from Duedall, 1985) present the location of ocean dumpsites as reported to the LDC for the years 1976 to 1979 and 1979 to 1982. Figure 4 presents the major municipal and industrial ocean dumpsites for the Northern Hemisphere.

2. HISTORICAL PERSPECTIVE OF U.S. OCEAN DISPOSAL POLICY

There was a piecemeal approach to the development of environ-mental and pollution legislation in the U.S., as each crises was detected, from water pollution to marine pollution to air pollution to endangered species. This lack of inte_gration developed from our focus on trying to control the point or source of introduction of the contaminant into the environ-ment. Regulatory agencies developed programs for different media (water, then air, then back to land), even programs for specific contaminants (pesticides, hazardous materials , radioactive materials, etc.), all attempting to control the problem at the source or its direct use or introduction into the environment. To make matters worse, a waste being released or dumped into one media could cut across all media. Also some environments (such as the ocean) were being perceived by public opinion as needing protection from waste disposal, and public interest became more focused on nearshore ocean dumpsites (Swanson and Devine, 1982; Swanson et al., 1985). Scientific uncertainities were great. Public misinformation still ranges greatly as has been evidenced by studies of public perception of ocean incineration in the mid 1980s (U.S. EPA, 1985a).

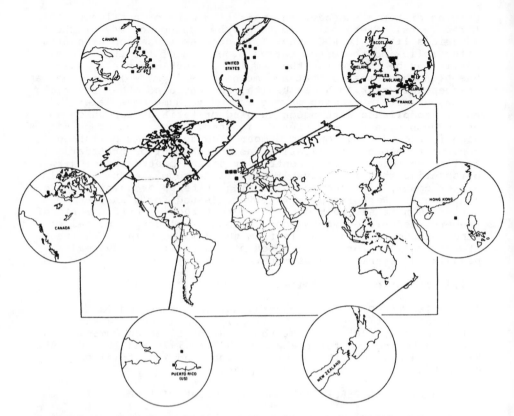

Figure 3a. Ocean dumpsites used in mid to late 1970's by Contracting Parties to the London Dumping Convention. Maps and locations
of dumpsites have been adapted from IMCO (1981a). This Figure is
from Duedall (1985).

2.1. Preferred or Protected Status

The factor that has most recently impeded the use of the ocean
for waste disposal was the development of the perspective that
the ocean should have a preferred or protected status from waste
disposal. This policy developed on the onslaught of increasing
waste discharges and illegal industrial dumping activities on
the high seas and when major environmental crises were identified
in several coastal regions of the world (the New York Bight, the

314

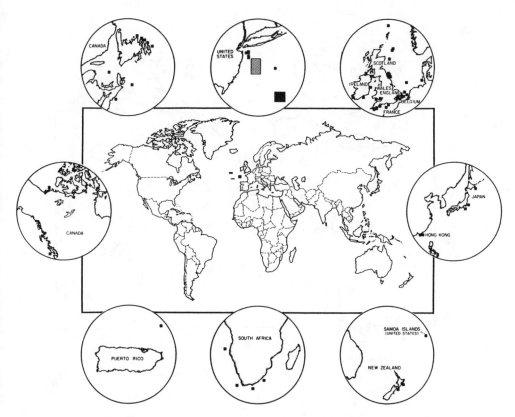

Figure 3b. Ocean dumpsites used in the early to mid 1980's by Contracting Parties to the London Dumping Convention. Maps and locations of dumpsites have been adapted from IMO (1984b). This Figure is from Duedall (1985).

Houston Ship Channel, Bay of Minamata, and the Mediterranean Sea). Pollution as defined was seen as the harbinger of these crises, and immediate measures were demanded of governments to protect the oceans (Champ, 1983). In the minds of many environmentalists, scientists, engineers, members of the public, elected officials, and decision makers, there was a need for the ocean to have preferred or protected status until more information was available. The feeling was that the ocean should only be used for waste disposal as an alternative of last resort.

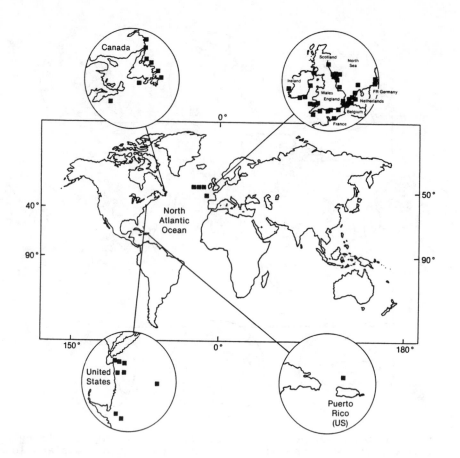

Figure 4. The major industrial and municipal and ocean
dumpsites for the Northern Hemisphere.

With the benefit of the last 15 y experience, the preferred or
protected status policy was very appropriate. It provided time
for extensive research to be conducted by federal, state, academic
and private institutions and regulatory authorities on the fate,
behavior, and effects of contaminants disposed of in the ocean.
Also, the regulatory process aided the information collection
process because the discharger or dumper was required to monitor
chemical and biological effects in the field and laboratory and
to report them to the regulatory agency in a specified time
period as part of the process of being granted an ocean dumping
permit (Champ and Panem, 1985).

However, our current waste management practices have been described as "fuelish" in an example recently cited by Spilhaus (1985) which is an excellent lesson for the future. Mine rubble from gold mines in Johannesburg, South Africa, was piled in mountainous crushed rock heaps at the turn of the century. When the cyanide process for extracting gold was invented, it was found that these discarded wastes contained enough residual gold to make reprocessing profitable. These residues were again heaped in piles only to be reprocessed in the 1950s when it was learned that they contained enough uranium to be economically worth extracting. Spilhaus's point is that the throw-away-discard mentality may solve short-term needs, but society's long-term needs are not being considered. What is a waste today may not be so tomorrow. Research and development into waste utilization is a very important long-range goal for society.

2.2. Scientific Perspectives

A spectrum of scientific perspectives on the use of the ocean for waste disposal can be identified. The debunking of the preferred status of the ocean for waste disposal has been voiced by Osterberg (1981; 1982a-c; 1983; 1985a-b, 1986a-b). He believes that "Planet Water" is a better name than Planet Earth since 71% of the Earth's surface is covered with seawater. Over 99% of all liquid water on earth is saltwater with 1% fresh or drinkable. This small amount of freshwater and land supports some 4.9 billion people. Osterberg holds that the laws that protect the oceans are forcing the most toxic wastes to go on land, placing humans in jeopardy by contaminating the limited volume of drinkable water. He notes that, when considering sources of food, the oceans only provide about 1% (Roels, 1982). The key question is: "what is more critical to mankind--the 99.4% of the water polluted with salt, which provides so few of our needs, or the tiny 0.6% of fresh liquid water that contributes nearly all of our food, fiber, and shelter?" (Osterberg, 1983).

Other perspectives have been presented by Goldberg (1981) and Kamlet (1981a). Goldberg (1981) makes the argument that the ocean can be used for waste disposal based on the assimilative capacity of a body of seawater for wastes. This capacity he defines as "the amount of a given material that can be contained within it without producing an unacceptable impact on living organisms or nonliving resources." He would determine this amount by the use of titration endpoints, in which the polluting substances in the discharged material becomes evident within the body of water at an endpoint. An example of one of these endpoints would be the U.S. Food and Drug Administration's (FDA) safe-maximum level of contaminants in marine resources consumed as food by humans. This approach could be flawed if the endpoint protected only human health and not the health of the organisms in the ecosystem. Kamlet's (1981a) rebuttal to Goldberg's position was an argument for keeping the preferred status policy of the ocean for waste disposal and presented four critical perspectives for the disposal of any waste on land, in the air, or at sea: (1) The ocean, as a commonly owned (or unowned) resource, is not protected by market-place and political forces; consequently, ocean disposal should not be permitted for persistent, toxic materials unless disposal in other media has at least marginally greater environmental impacts, (2) The ocean, as the prototypical dispersal medium, is an inappropriate place to dispose of persistent, toxic materials: the land, which if properly managed is the exemplary containment medium, is, in general, a sounder choice for the management of such wastes, (3) No waste management strategy can be totally free of risk to health or the environment; management decisions should be based on multimedium comparisons and risk minimization, and (4) In view of the rudimentary ability of marine science to detect, much less correct, problems associated with waste disposal, we cannot prudently rely on a permissive approach based on crude assimilative capacity models in the hope that after-the-fact monitoring and a decades-long response time will ensure that health and the environment are protected.

Kamlet fears for the use of the "ocean as a fall-back waste heap for any material deemed too toxic or too controversial to dump on the land." He argues that the ocean "is nobody's backyard and fish do not vote." He believes that if ocean disposal decisions are made on the basis of the socio-political acceptability of landbased alternatives, the ocean will always be the disposal medium of choice. He calls it the "tragedy of the commons" where the ocean is the "global commons" described by Hardin (1968). Kamlet (1981a) questions a basic premise: "is it better to put a persistent toxic material in a dispersal medium or in a containment medium?" For persistent synthetic chemicals, such as PCBs, Kepone, DDT, and the like, he believes that they should be isolated and contained to the fullest possible extent if they cannot be destroyed by high temperature incineration.

The issues identified during the 1970s reflected both scientific and policy concerns: (1) there were great scientific (laboratory, and field) uncertainties about the short- and long-term effects of ocean-dumped wastes, and the potential impact on human health, (2) any large addition of chemicals to the ocean (such as from ocean dumping) was predicted to disrupt the stable chemical and osmotic balance that exists between marine organisms and seawater, (3) due to the complexity and interdependence in marine food webs, they were considered very delicate and easy to disrupt, (4) as the oceans were a "free commodity" they could be subject to abuse, (5) there was limited ability to take remedial action after the wastes were dumped in the ocean, whereas contamination on land was thought to remain confined and available to future disposal options, and (6) the costs of using the ocean were considered inversely related to incentives for the waste disposer to develop pretreatment, recycling, or land-based alternatives.

Nations during the 1970s in attempting to control marine pollution, were faced with implementing global or regional conventions and drafting their own national legislation and regulations. At the same time, they were evaluating the results of marine waste disposal activities or options through millions of dollars of research and monitoring.

Policy and decision makers will be always faced with an increasing volume of wastes, a lack of acceptance for either landbased (NIMBY and potential groundwater contamination) or ocean alternatives, and implementing waste management strategies in which wastes are disposed of in the manner and medium (air, land, or water) that minimizes the risk to human health and the environment, and all at a price that people are willing to pay.

Unfortunately, policy that supports a preferred ocean status could force the development of an environmentally undesirable land disposal option, which could led to the contamination of a drinking water aquifer. Also, it is quite possible that the costs (to develop and operate) and the benefits of a land-based alternative may never balance against the environmental and public health risks of ocean disposal (U.S. NRC, 1984).

However, it should be noted that land containment options have three important waste disposal characteristics: (1) monitoring can be on a small scale with emphasis on release from the containment system, (2) the persistence of long-term toxicity can be followed by time interval bioassays, and (3) if the containment system is leaking, the waste can be moved and repackaged, providing better "fail safe" mechanisms and better isolation from the environment (Champ and Panem, 1985).

The critical aspect is to determine the degree of risk from an option, predicting the degree of uncertainty from what is known and or unknown. The driving force of the multimedia (air, land, or water) assessment approach is to provide comparable information on the environmental, social, and economical implications of all possible disposal alternatives. Due to the political unpopularity of land-based or ocean alternatives, there is need for a common denominator to equally weigh these options against each other so that the public and environmental good is best served.

In the U.S., legislation and regulations for the protection of the ocean from pollutants has been directed toward prevention of pollutant inflow into the coastal waters with emphasis on a single media approach. This, of course, is expensive and basically unachievable.

Recognition of the ineffectiveness of the single media approach to pollution control has been long in coming. Waste disposal is not a single media problem--all wastes must be disposed of onto land or into air or the oceans--thus, waste disposal must be considered from a multimedia (land, air, and water) perspective. In consideration of waste disposal from a multimedia perspective, it is assumed that even though disposal occurs into one of the media, the wastes, in one form or another, generally are transported between these media, and in many cases, actually end up in the oceans. A compounding factor to make this dilemma worse is the fact that in the United States, environmental legislation has developed in a disjointed process because each piece of legislation that developed in the late 60s and early 70s was in a step-by-step response to individual crises as they were identified. For example, in the U.S., disposal of wastes from ships and barges at sea is regulated under the Ocean Dumping Act, while pipe discharges (outfalls) into the nation's waterways and oceans fall under the Clean Water Act. The incineration of wastes is regulated under the Clean Air Act when on land, and under the Ocean

Dumping Act when at sea (U.S. Congress, 1967; 1972a, b). It was also concluded that waste disposal regulated by this medium-by-medium approach (land, air, or water) would be responsible for shifting the risk posed by individual classes of wastes to the medium of least regulation rather than to the medium of least risk (U.S. NACOA, 1981).

2.3. Inherent Problems - Policy Formulation

The implementation of a national strategy for ocean waste management should be a straight-forward exercise of following the guidelines presented in Annexes I and II of the LDC. However, the rapidly rendered harmless and trace contaminants provisions (exceptions) and the special care provisions, when combined with the requirements of estimating dumpsite assimilative capacity, develop an inertia in the permitting process because of the uncertainties in specific information available. In the United States, ocean dumping criteria (U.S. EPA, 1973a, b,; 1977) focused on LDC Annex I and II stipulations and toxicity tests (i.e., 96 hr. bioassays) and on limiting permissible concentrations in the mixing zone behind the barge. The U.S. District Court for the District of New York ruled in the City of New York v. the Environmental Protection Agency that the EPA's conclusive presumption that sewage sludge that is deemed harmful to the marine environment (pursuant to EPA regulations, i.e., 96-h bioassay tests), and therefore will degrade the environment, was arbitrary and capricious and inconsistent with the intent of the 1972 Ocean Dumping Act and 1977 Amendments (U.S. District Court, 1981). The decision in this case left the 1981 deadline intact for the phasing out of ocean dumping of harmful sewage sludge. However, the decision allows for the continued ocean dumping of sewage sludge until there is an evaluation of whether this sewage sludge will cause unreasonable degradation in the context of all relevant environmental, social, and economic factors as referenced in Section 102(a) of the Ocean Dumping Act. It should be noted that Kamlet in reviewing this case, identified several legal questions of which the major one is that the courts decision could be in direct violation of the legal requirements of LDC Annex I, which prohibits the ocean dumping of certain toxic substances, regardless of the need for the dumping or consideration given to the balancing of all relevant social, economic, and environmental factors.

The EPA in 1984 revised its regulatory approach to ocean dumping in an attempt to reflect the intentions of Congress (as mandated by the Ocean Dumping Act); the U.S. District Courts recent interpretations; and the need to preserve environmentally, economically, and socially acceptable alternatives for waste disposal, regardless of the medium. The EPA's new ocean dumping policy can be summarized as: (1) as an overall principle, protect the oceans from significant adverse effects of waste disposal, and particularly assure that it is not used for "cheap" waste disposal as a matter of short-run economic considerations alone, (2) in any specific case, allow ocean dumping of a waste only if the applicant can show that no practicable alternative(s) are available that have less impact on the total environment and that the EPA will apply a rule of reason in determining practicability,

and (3) for the long-run, actively encourage environmentally
beneficial approaches such as waste minimization, recycling, or
reuse (U.S. EPA, 1984a).

Sebrek (1983), in his excellent review of the gap between environmental science and policy making, identified seven factors which
add to the problem: (1) international and national legislative
priorities reflect the degree of popular media attention, rather
than the scientific priorities, (2) monitoring of marine pollution
is still inadequate to identify cause and effect relationships in
the environment, (3) independent scientific input into diplomatic
conferences which adopt international rules and standards is
insufficient, (4) when international conventions are being
adopted, environmental and scientific values have to compete with
economic, social, and political consideration, (5) there is poor
enforcement or "policing" of international rules or regulations,
because the administrative and technical infrastructures to support them is either lacking or inadequate, (6) when agreement
cannot be reached on a critical aspect of a regulation, the environmental issues are often addressed by nonbinding resolutions
and are waived by the assumption that the problem will be correct-
ed by the action (e.g., legislation, etc.), and (7) environmental
cost-benefit analyses are seldom carried out or independently
audited as part of essential preparations for diplomatic
conferences which adopt relevant rules.

3.   THE OCEAN WASTE DISPOSAL OPTION

The ocean waste disposal option can be defined as the disposal
of wastes with an offshore perspective with the aim to have
subsequent mixing occur offshore and not inshore into estuaries
or bays and in which oceanic processes dominate mixing with
subsequent dispersion into the maximum water volume available.
The ocean disposal option can be further defined as the disposal
of wastes at sea by direct discharge from ocean outfalls or
structures, or by dumping from barge or vessel, or incineration
aboard ship or platform at sea, or by subseabed implacement in
bottom sediments.

The general advantages of using ocean space for waste disposal
are: (1) the costs for land-based waste disposal alternatives
are increasing, (2) municipalities are having difficulty in managing the disposal of large volumes of sewage sludges using land-
based alternatives, due to public dislike for locating NIMBYs
(incinerators, composing facilities, and landfills) in urban
areas, and the saturation of local sludge markets, and (3) there
is a growing public attention to the environmental and public
health hazards, such as groundwater contamination from land-based
waste management alternatives. The variables in waste disposal
by ocean dumping that can be managed are: (1) waste quality, (2)
waste quantity, (3) dumpsite location, and (4) discharge engineering technology. There appear to be only four ocean disposal
strategies for waste management that can be developed: (1) containment (drums or capping), (2) isolation (deep-sea burial), (3)
dispersion, and (4) ocean incineration.

The capping method is the covering of contaminated material, which has been placed in a depression (such as a burrow pit from marine sand and gravel extraction), with clean sand. This provides a barrier over the contaminated material in order to isolate it from biological and chemical activity, which would release contaminants into the water column. The strengths of the containment strategy are that it restricts the environmental impact to a small area, and there may be potential recovery of the waste with future relocation and management. The weaknesses of the containment approach are that it is difficult to monitor release or escape of the contaminant and to estimate the potential for large-scale release from catastrophic (episodic) events.

The concerns of isolation (e.g. deepsea burial, trenches) are the retrievability of the waste from great depths, the monitoring at these depths, the long-term effects (hundreds of generations), and potential displacement by natural catastrophic events.

Dispersion strategies have two important strengths: (1) appropriateness for naturally occurring compounds and (2) small initial impact zone, if diluted rapidly. However, dispersion also has weaknesses: (1) potential contamination of a larger area and (2) it may not be appropriate for synthetic or persistent compounds. The major concerns for utilization of a dispersive ocean disposal strategy are the unknown speciation of contaminants into more toxic forms and the bioavailability of contaminants.

3.1. Oceanic Processes and Factors to be Considered in Ocean Waste Management

When a waste is disposed or dumped into the ocean, a series of physical, chemical, and biological processes begin to alter the waste's distribution, chemical properties and its subsequent environmental effects. As we better understand these processes, we can better predict the consequences of using ocean space for waste disposal. Also, in managing ocean waste disposal, we can use criteria for the selection of dumpsites that maximize the dispersal of wastes, thereby achieving the greatest dilution and subsequent minimum exposure concentrations to marine resources. Another step to add to the permit evaluation process is to require the estimation of dumpsite assimilative capacity for a given waste and accumulative wastes as part of the review process.

3.1.1. Physical, Chemical, and Biological Processes

To develop a risk assessment protocol for the disposal of wastes at sea, there are unique informational needs, which are specific to physical, chemical, and biological ocean processes that play a role in dispersion, dilution, fate, and behavior of wastes and subsequent environmental effects. The assessment needs the following information: (1) waste characterization (Wastler and Offutt, 1975) with respect to prediction of the wastes reaction in seawater and waste disposal methods (engineering technology) with respect to technology that maximizes dispersion or containment; (2) dumpsite selection criteria (GESAMP, 1982), because some sites are more dispersive than others and may not accumulate

wastes (Champ et al., 1984, Norton and Champ 1988); and (3) due
to the uncertainties in an ocean risk assessment, a monitoring
program is needed to verify the assessment predictions (IAEA,
1983).

Dilution is a central concept in managing waste discharges, but
dilution is not a self-purification mechanism since it does not
lead to a reduction of the constituent within environment.
However, the importance of dilution mechanisms comes from the
reduction of toxicity due to a decrease in waste toxicant concentration to biota during the initial exposure period. A waste
discharged from a moving vessel at (6 knots) obtains an initial
dilution of up to a factor of 10^4 or 25-20 times better than
initial mixing from a typical, well designed oceanic sewage
outfall (Csanady, 1981). Subsequent dispersion and dilution at
the dumpsite or discharge site depends on the physical properties
of the waste, such as specific gravity, suspended solids, and
particle size, and physical oceanographic processes of a dumpsite,
which can determine the postdumping waste exposure concentrations
and time periods to biota. Appropriate dumpsite selection is the
key in the preassimilative capacity matrix. Careful site selection can maximize the spatial boundries of a dumpsite for waste
dispersion because it decreases the waste concentration per unit
volume of seawater. In sites where waste materials are rapidly
dispersed by water movement there is minimum accumulation, providing a greater degree of assimilative capacity. At highly
dispersive dumpsites, long-term adverse effects have not been
detected (O'Connor et al., 1983; Norton and Champ, 1988). A
semiquantitative model (Champ et al., 1984, and Devine et al.,
1986) has been developed from findings of Norton and Champ (1988)
for assessing the dispersion at a site being considered for dumping. The model uses dumpsite physical characteristics such as
water depth, tidal currents, wave induced currents, distance from
the coast, degree of open water exposure, and seasonal storms to
determine local (dumpsite) and regional dispersiveness of sewage
sludge at eight major sewage sludge dumpsites worldwide.

After wastes have been dumped, the following dilution and
dispersion processes have been determined influential to the
fate and behavior of wastes dumped at ocean dumpsites: settling
rates; flocculation and particle dynamics; vertical mixing rates
within and below the mixed layer; horizontal dispersion rates,
mass fluxes to the seafloor, and bioturbation (magnitude, depth,
and rates). Biological processes that include bioavailability,
bioaccumulation, biomagnification, uptake processes, pathways
depuration kinetics, and detoxification processes also determine
the subsequent environmental effects and impacts.

3.1.2.   Dumpsite Selection Criteria

Dumpsite selection criteria to be considered in the issuance of
permits are specified in Annex III of the LDC (see Nauke, 1988).
In 1975, GESAMP (Joint Group of Experts on the Scientific Aspects
of Marine Pollution) published its first version of scientific
criteria for the selection of dumpsites. It was updated in 1982
to cover new information and added a section on ocean incineration.

The second report drifted off from being a report on site selection criteria to being a discussion on waste disposal and is
highly recommended reading. Both of these reports advocate the
dumping of material permitted under the LDC to avoid, or minimize,
undesirable effects by: (1) ensuring maximum initial dilution,
through an appropriate means of disposal, (2) selecting areas
where dispersive processes (transport and mixing) are active, and
(3) avoiding particularly sensitive areas. The reports also
recommend that dumpsites should be selected to minimize interference with other present and potential uses of the sea, such as
fishing, aquaculture, mining and drilling for minerals (oil and
gas), recreation, transportation, and national defense (GESAMP,
1982).

A simple, yet specific, list of dumpsite selection criteria has
been developed by the U.S. EPA (1977): (1) geographical position,
depth of water, bottom topography, and distance from the coast,
(2) location in relation to breeding, spawning, nursery, feeding,
or passage areas of living resources in adult or juvenile phases,
(3) location in relation to beaches and other amenity areas, (4)
types and quantities of wastes proposed to be disposed of, and
proposed methods of release, including methods of packing the
waste, if any, (5) feasibility of surveillance and monitoring, (6)
dispersal, horizontal transport and vertical mixing characteristics of the area, including prevailing current direction and
velocity, if any, (7) existence and effects of current and previous discharges and dumping in the area (including cumulative
effects), (8) interference with shipping, fishing, recreation,
mineral extraction, desalination, fish and shellfish culture,
areas of special scientific importance and other legitimate uses
of the ocean, (9) the existing water quality and ecology of the
site as determined by available data or by trend assessment or
baseline surveys, (10) potentiality for the development or recruitment of nuisance species in the disposal site, and (11)
existence at or in close proximity to the site of any significant
natural or cultural features of historical importance. The GESAMP
(1982) report also discusses the physical, chemical, and biological processes that should be considered in conducting an assessment to predict the effects and impacts of a given waste at a
given dumpsite.

3.2. Etymology of the Concept of Assimilative Capacity

The concept of assimilative capacity probably originated from
engineering waste-load allocation studies. Cairns (1976) defined
assimilative capacity in biological terms "as the ability of an
ecosystem to receive waste discharges without significant alteration of the structure and/or function of the indigenous community."
Cairns also felt that "exceeding the assimilative capacity of an
ecosystem will produce biologically significant impact." The
farsightedness of Cairns is also revealed in the same paper when
he suggested that there is an "element of risk" from the introduction of a material into an aquatic ecosystem and that "most
wastes at proper concentrations may be introduced into aquatic
ecosystems without significant damage, and further, that many
wastes may be transformed into less harmful and perhaps even
useful materials." Cairns also charged that "instead of being

324

distracted by the white hat/black hat (good guy/bad guy) argument over whether assimilative capacity does or does not exist, that the research community might well address the task of developing a protocol that will enable decision-makers to estimate the risk of discharging a particular waste into a particular water ecosystem." To give further direction to regulatory programs, Cairns also recommended "an ecological monitoring program consisting of biological, chemical, and physical data gathering which could enable ecologists and others to check the validity of the predictions and to generate the necessary feedback of information to make effective corrections and alterations in the models developed." This information would also tend to reduce ecological catastrophies and provide an earlier warning than has been possible in the past when dead fish and massive algal blooms were the first symptoms of ecosystem disfunction (Cairns, 1976).

Stebbing (1981) noted that Cairns (1977) has reworded his definition for assimilative capacity as "the ability of a receiving system or ecosystem to cope with certain concentrations or levels of waste discharges without suffering any significant deleterious effects." He also pointed out that "it seems to the detractors of the concept of assimilative capacity that the introduction of chemicals and fibers in the form of a leaf is not regarded as a threat to the biological integrity of the receiving system (i.e., the maintenance of the structure and function of the aquatic community characteristic of that locale) but that the introduction of the same chemicals and fibers via an industrial municipal discharge pipe would be deleterious."

Bascom (1974) also suggested that the ocean is the plausible place for humans to dispose of some of their wastes and that, if the process were thoughtfully controlled, it would do no damage to marine life. He also recommended that mankind should "rigorously prohibit the disposal in the ocean of all synthetic and radioactive materials, halogenated hydrocarbons (such as DDT and PCBs), and other synthetic organic materials that are toxic and against which marine organisms have no natural defenses."

Webster (1979) defines assimilation with a biological emphasis as "the incorporation or conversion of nutrients into protoplasm that in animals follows digestion and absorption and in higher plants involves both photosynthesis and root absorption." The critical assumption on which the concept of assimilative capacity of the ocean is based is that the marine environment can tolerate some wastes, because various physical, chemical, and biological processes will degrade these wastes. Assimilative capacity is an ecosystem concept that relates to the loading of waste materials that can be accommodated without changing the ecological balance. To measure natural assimilative capacity, one must determine the rate of biodegradation of a given waste material in a body of water that does not alter the distribution or abundance of organisms (Champ, 1984). Early awareness that sewage sludge dumpsites had different assimilative capacities was noted in comparing the accumulation or dispersion properties of dumpsites for wastes. Figure 5 presents a two-dimensional illustration of the assimilative capacity of major sewage sludge dumpsites from different parts of the world. The illustration has been constructed using as accumulating factors the buildup of heavy metals or organic materials in sediments and alteration of the benthic communities. The dispersing factors are those physical processes involved in advection, diffusion, particle settling, tidal currents, and dispersion by oceanic currents with further dilution by natural processes, such as turbulence and storms. No scales are given in Fig. 5, because the placement of a dumpsite has been arbitrary from interpretation of the data available from the sites. Also, following further discussion and review of the data with scientists from the United Kingdom, we have switched the Firth of Clyde dumpsite with the New York Bight dumpsite from how it was originally published in Champ and Park (1981).

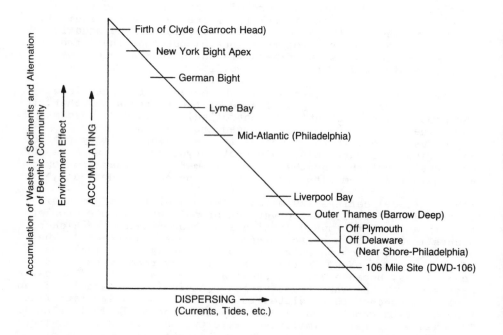

Figure 5. A two-dimensional illustration of the assimilative capacity of the major sewage sludge dumpsites from different parts of the world, plotting accumulation and dispersive properties (from Champ, 1984).

There are at least four steps involved in determining assimilative capacity: (1) representative biological indicators must be selected, (2) contaminants for which assimilative capacity is to be determined must be identified, (3) all sources of selected contaminants must be identified, including pathways, transformations, and sinks, and (4) a threshold of what constitutes an "unacceptable impact" (see Fig. 6. from Champ and Hebard, 1981) or a limit of "unreasonable degradation" must be set as discussed by O'Connor and Swanson (1982).

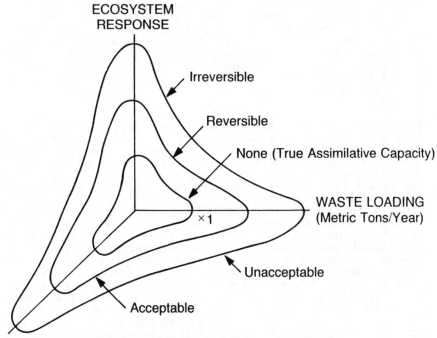

Figure 6. Waste loading and assimilative capacity (from Champ and Hebard 1981).

4. The Concept of the Risk Assessment--Risk Management Approach

The task of reducing the total risk in the whole environment from exposure to toxic chemicals is perhaps the most complex management initiative ever undertaken. The variety and degree of scientific uncertainty in the information available, and the need to protect a variety of environmental values, as well as human health, all work to complicate the decision-making process. Many times regulatory actions unwittingly transfer contaminants from one environmental medium to another via pollution control technology or by other uncoordinated regulation of the same industry or of the same substance by different regulatory offices of agencies.

Since 1945, over 65,000 different chemicals have been listed as being in commercial production in the United States alone [updating the data of Maugh (1978)]. Approximately 1200 new chemicals are being added to this list every year. The toxicological potential of many of these chemicals have not been characterized. A few thousand have demonstrated some toxic effects and are used or released to the environment in such a magnitude that there is concern. Management of toxic chemicals is a critical aspect of waste management. The contamination of clean fine sediments in a harbor with highly toxic chemicals, such as PCBs in the New York Harbor can immediately create an inverse ratio between the available options for disposal and costs of disposal for maintenance dredging.

The risk assessment and risk management approach uses risk reduction as a decision-making tool. This regulatory approach is quite different from developing discharge guidelines or regulations based on available technology and cost, because, by nature, the risk approach is an integrating management concept. Risk assessment has been distinguished and separated from risk management by the National Academy of Sciences, National Research Council (U.S. NRC, 1983). Risk assessment is the scientific study to identify the problem(i.e., how toxic, to whom, what-when-where, and why); while risk management is deciding what to do about the identified problem and is a decision-making process aimed at control strategies and cost considerations. In the risk management process for waste disposal, the key issues that need public debate are the "values" used by the policy and decision makers, acceptability of the costs of control measures, of alternatives and environmental effects, the acceptability of environmental and public health risks and benefits, and the measure of confidence -the degree(s) of uncertainty in the information used in the decision process. The key to risk management in waste disposal is not only to balance risk reduction against effects and control costs, but to balance risk and reduce the total risk in the whole environment. These approaches should make regulatory judgements more consistent and more reflective of the state of scientific understanding.

In essence, the next sections discuss the concept of the risk assessment and risk management process. The only approach that appears to meet the requirements of considering and balancing environmental, economic, public health, and social aspects and can be applied across media is the application of risk assessment and risk management techniques in the decision-making process.

4.1. Risk Assessment

There is no standard protocol for conducting risk assessments. Users need to understand from the onset, it is an analytical tool which must be tailor-made to the needs of each use. Risk assessment focuses on two measurable factors: hazard and exposure For a chemical to cause a risk, it must be toxic and hazardous and be present in the ocean at levels above background, if it occurs there naturally. The principles of risk assessment and risk management have been outlined by the U.S. National Academy of Sciences, National Research Council (U.S. NRC, 1983). These principles have been further delineated by the EPA (U.S. EPA, 1984b).Essentially, the principles of risk assessment involve the analysis of an environmental contaminant (whether chemical, physical, or biological) from exposure through demonstration of toxic effects. The process has been divided into four discrete components (see Fig. 7):

Hazard Identification: The determination of whether a particular chemical is or is not causally linked to particular health effects.

Dose-Response Assessment: The determination of the relation between the magnitude of exposure and the probability of occurrence of the health effects in question.

Exposure Assessment: The determination of the extent of human exposure before or after application of regulatory controls.

Risk Characterization: The description of the nature and often the magnitude of human risk, including attendant uncertainty and given assumptions (U.S. NRC, 1983).

An ocean disposal risk assessment has to integrate physical, chemical, and biological processes, and identify target organisms, critical pathways, and food webs [to predict the actual levels of exposure to the organisms or to a population and the subsequent effects]. The term hazard evaluation (Dortland, 1981; Cairns et al., 1985; Rose and Ward, 1981) has been also commonly referred to as hazard assessment (Gentile et al., 1988). Risk assessment is a process that provides the necessary data and interpretive framework for estimating the probability of harm to mankind or to the environment.

A generic risk assessment protocol for ocean disposal is presented in Fig. 8. It requires the integration of exposure assessment (toxicity), with environmental fate and behavior data of wastes (following dumping), and subsequent environmental (biological) effects. Exposure assessment consists of estimating the duration and intensity of waste exposure for potentially affected focuses on estimating the biological response of these populations or communities in terms of toxicity or bioaccumulation. The risk assessment protocol is structured to be sequentially tiered,

ELEMENTS OF RISK ASSESSMENT AND RISK MANAGEMENT

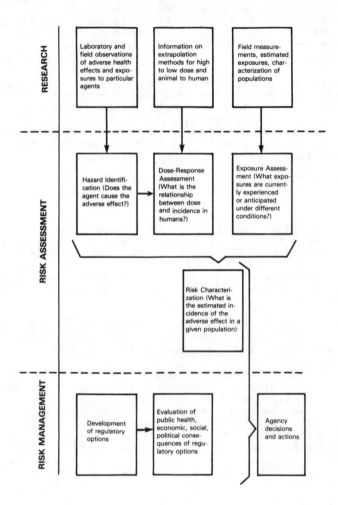

Figure 7. An overview of the risk assessment and risk management process, (from U.S. EPA, 1984a).

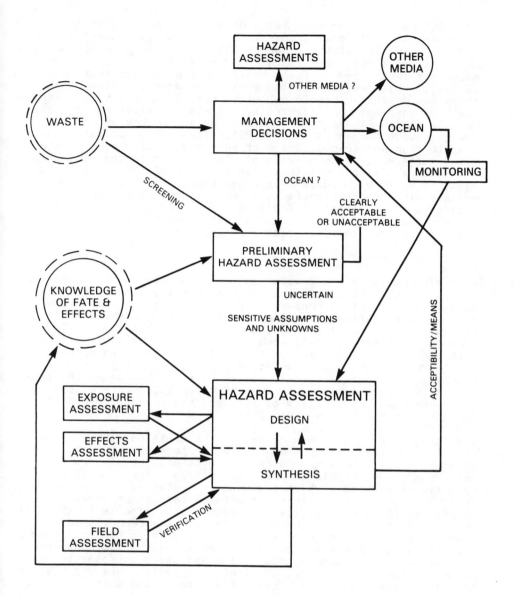

Figure 8. A generic risk assessment protocol for ocean waste
disposal, in which physical, chemical, and biological processes,
target organisms and critical pathways, food webs must be integrated to predict the actual levels of exposure to organisms or
to a population and the subsequent effects (from Cross and Boesch,
1983).

331

with information from each level of analysis being evaluated to determine if additional information is necessary to arrive at a disposal decision with a prescribed level of confidence. Gentile and Schimmel (1984) have recommended that in order to interpret data from laboratory-toxicity tests, that a complimentary database is needed that predicts environmental exposure conditions, and these two databases should be integrated within the hazard assessment process and be evaluated by a field verification program. Gentile et al. (1988) have conducted an hazard assessment for sewage sludge dumped at the 106-Mile site. They estimated the distribution and concentration of PCBs and total copper in the mixed surface layer downstream of the dumpsite (under minimum and maximum mixing conditions) for a period of 100 daily dumps. Next, using these predicted exposure contour levels with appropriate bioconcentration factors (BCFs), they calculated the bioaccumulation potential for PCBs and copper in fisheries in these contours. These values did not exceed the FDA safe limit levels in commercial fisheries for human consumption.

4.2. Risk Management

Risk management is the development of regulatory options and the evaluation of public and environmental health, economical, social, and political consequences of each option. Risk management has been defined to embody the following general principles: "It is an agency decision-making process that entails consideration of political, social, economic, and engineering information with risk-related information to develop, analyze and compare regulatory options and to select the appropriate regulatory response to a potential chronic health hazard. The selection process necessarily requires the use of value judgements on such issues as the acceptability of risk and the reasonableness of the costs of controls" (U.S. EPA, 1984b).

The final step of risk management is the decision-making process of selecting the best option and then implementing it. The risk management process is made difficult by the uncertainties that stem from the extrapolation of effects from hazard and dose-response assessments, which are usually conducted in the laboratory in single species bioassays with perhaps three organisms (algae, zooplankton, and fish) to predict the effects in the field (White and Champ, 1983, and White, 1985). Laboratory studies with mice and rats are also used to predict effects in humans. Another area of uncertainty that should be identified and discussed is that of the values used by the policy and decision maker in the process. Several tools are available to predict this uncertainty: benefit-cost analysis, risk-benefit analysis, and cost-effectiveness analysis.

The U.S. EPA is in the process of further defining risk management principles. For example, cost-effectiveness analysis is at present the most frequently used risk management tool at the EPA. It involves the determination of the least-cost path to achieve a given goal, such as the achievement of a protective standard. One of the ways it has been used recently at the EPA to compare different ways of obtaining some specified degree of risk reduction. The use of inter-media analysis has been

advocated to control those risks first that cause the greatest
harm in the appropriate medium (U.S. EPA, 1984b). It should be
emphasized that controlling one medium, ultimately and sometimes
unwittingly, involves the transfer of harmful substances to
another medium. These intermedia transfers have only recently
been recognized as a problem and as an environmental trade-off.
Therefore, it must be emphasized that in order to ensure an
equal balancing of options in a multi-media assessment, the
principles of risk assessment and risk management must be used
in the decision-making process.

Given this, how does one structure the risk assessment - risk
management process for sewage sludge? Sewage sludge poses both
a potential risk and a benefit. However, municipalities have
recognized for some time that sewage sludge is becoming the most
expensive residual to manage in the waste-water treatment
process. Sludge, like other wastes, can be land-filled, sanitary-
dumped or treated on land (composting), thermally destroyed via
incineration, or ocean dumped. Whatever option is chosen, the
decision maker needs to weigh the risks, costs, and benefits
equally. National policies can provide special status to certain
options, and, therefore, restrict this equal balancing approach.
For example, in the United States, disposal in landfills is
becoming discouraged as an unproductive use of land and a
potential threat to groundwater contamination. In addition, the
oceans since the early 70s have been given a preferred status
against ocean disposal. However, in the long run for public
good, what is needed is equal balancing approach, and the tools of
risk assessment and risk management should provide this equality
to both the national and local decision-making process. However,
we cannot emphasize enough that these tools are not ends in themselves, or fully developed, but simply procedures to identify and
define risks within options.

There is a need for the risk assessment and risk management
process because of the emotional reactions surrounding a NIMBY
and the general lack of a balancing mechanism to convince policy
and decision makers and the public, that the option selected by
policy and decision makers is the best one and that it has the
stronger scientific, technical, and economic base. Therefore, in
terms of public good, the public will receive a higher degree of
human health and environmental protection while conserving scarce
local tax dollars. For more discussion and an example of a multi-
disciplinary intermedia risk assessment and risk management
approach for the disposal of sewage sludge, see Champ et al.,
1988.

4.3.   Information Requirements for Waste Management
        Decision-Making

A working group on Assessment Technologies at the International
Ocean Disposal Special Symposium Ocean Waste Management (IODS,
1983) developed a generic flow chart to represent the various
information-collecting activities as they relate to the decision-
making process required for the ocean disposal option for waste
disposal (see Fig.   8., from Cross and Boesch, 1983). These
include theoretical and experimental methods designed to predict

effects, or prospective methods, and observational approaches that lead to definitions of effects that have occurred, or retrospective methods. The term hazard assessment in this figure is analogous to the term risk assessment and it is used to refer to the synthesis activity wherein knowledge of the environmental fate of contaminants (exposure assessment) is coupled with the knowledge of the effects of the contaminant (effects assessment) to make a prediction of the effects likely to occur in a given situation. The only way to validate these predictions is by field verification (research and monitoring).

In 1982, the International Atomic Energy Agency (IAEA) convened a Technical Committee to provide additional guidance to the LDC's Annex III for Contracting Parties on the preparation and evaluation of environmental assessments, relevant to the issuance of a special permit for dumping of radioactive wastes under Annex II of the Convention. The Committees report (IAEA, 1983), which delineates the information requirements for an environmental assessment for ocean disposal of radioactive wastes, can be condensed and modified into a generic type of protocol that could be used for a multimedia assessment for any type of waste being considered for ocean disposal. However, in the process of identifying and assessing factors, the degree(s) of risk and uncertainty of each factor must be evaluated.

Multimedia management strategies are aimed at developing waste management alternatives that optimize the following parameters: (1) engineering feasibility and reliability, (2) health issues, (3) environmental effects, (4) social acceptance, and (5) costs. In this effort, there is a need to assess each option in detail for comparison. The entire process can be seen as optimization. One of the difficult management decisions is to determine the amount of effort that should be devoted to comparing options and alternatives. The process is usually simplified by focusing on those components that differ markedly from one system to another (IAEA, 1983).

5. Future Considerations

Industrialization cannot survive with the present rates of resource utilization and waste production. In the United States every person uses approximately 20,000 kg of raw materials per year (Champ et al., 1984/1985). These materials are used to build roads, buildings, materials, products and goods, and are used in services. The volume of wastes produced has increased exponentially with the rate of resource utilization. The EPA estimates that the present volume of industrial wastes produced in the United States each year is about 3×10^8 metric tons. In the United States, waste management (treatment, processing, hauling, and dumping) is the third largest business (by dollar). The stability of industrialized societies and their standard of living cannot be sustained with these rates of increase. Ling (1979) stated it very eloquently: "There is only one land mass, one atmosphere and a finite supply of water for us to share. To survive, let alone maintain dignified life, requires that we make the best- and least-use of these resources. If we do not sustain them, they will not sustain us." Within the next 20 years, western societies will be forced to accept waste reduction to maintain their high standard of living. Because net purchasing power is declining and is not being offset by gross income increases due to advances in technology or productivity. A consumption plateau is being created by the high cost of waste management, or simply stated, a substantial proportion of the U.S. gross national product is being diverted from production and profits to waste. Ironically, the key to the future is not waste treatment but to not produce a waste in the first place, there by producing more profits.

5.1. Global Requirements for Management of Hazardous Wastes

The World Commission on Environmental Development (WCED) initiated in 1985 a study to review global aspects of hazardous waste manage-ment and to identify policy options. The following section has been prepared as a brief summary from notes and background papers prepared for their first meeting in Oslo, Norway, June 24-28, 1985.

The worldwide production of hazardous wastes that were produced in 1984 has been estimated to be in the range of 325 to 375 million metric tons with the industralized nations producing 90% of the total. The sources were industry, military, agriculture and even households. It was also estimated that the annual growth in hazardous on a global basis ranges between 2 to 4%. An alarming finding was that on the average, a shipment of hazardous wastes crosses a border (national/federal/state) more than once every 5 minutes.

Rapid industralization in developing countries is perhaps bringing about more severe problems than in developed countries because of the absence of a well developed environmental regulatory infra-structure and the financial capability to deal with hazardous wastes. A major concern of the Conference was the occurance of a market in developing countries that engaged in "trade" in hazard-ous waste disposal, establishing disposal sites in return for hard currency. In these cases, industry is not looking for cheap labor but for a cheap environment for waste disposal. As a country or region participates in this "trade" it may be compromising its basis for future development. The subsequent economic burdens such as the costs for clean-up (which in the U.S. have exceeded those of prevention by a factor of 10 times), and care, not to mention victum compensation, could become enormous. As the NIMBY attitude prevails, there will be tremendous pressures in develop-ing countries to become "traders" of hazardous wastes. An example of such activities, is that China is currently being paid to receive radioactive wastes from Germany for storage in the vast deserts of China. It was anticipated that this trade may be sustained for some period of time, given comparative levels of development, awareness and information flows. However, such trade in the long-term can not be sustainable neither economically or politically. The exporting countries are running the risk of being charged with taking advantage of unequal relations, of neo-colonialism or worse. The entire process simply removes the pressure to find environmentally and economically acceptable ways of managing wastes within the industries, communities and coun-tries generating it.

There is a need for international guidelines governing the transport or shipment of hazardous wastes across borders, and for the development of an organization to monitor international shipments of wastes. Most countries have had little or no knowledge of consignments of hazardous wastes entering or leaving their borders. National and international rules and regulations exist to monitor and control the shipment of more than one million shipments of dangerous goods.

It seems very logical that hazardous wastes are some category of dangerous goods and should fall under that same rules and regulations. Goods having an economic value are less likely to become damaged in route or lost in shipping, at least not intentionally. Nevertheless the rules and protocols concerning the transport of dangerous goods would serve as an excellent framework in formulating policies for the transport of hazardous wastes.

The strengthening of international cooperation in this area is also seen as key international issue. On December 6, 1984, the Council of Ministers of the European Communities adopted a Directive on the Supervision and Control of Transfrontier Shipments of Hazardous Waste within European Communities, which entered into force on October 1, 1985. On February 1, 1985, the OECD Council decided that Member countries "shall control the transfrontier movements of hazardous waste and, for this purpose, shall ensure that the competent authorities of the countries concerned are provided with adequate and timely information concerning such movements." The United Nations Environment Program (UNEP) has drawn up draft guidelines for the environmentally sound management of hazardous wastes. They include a provision for transfrontier movements, safety, packaging, labelling, licensing, documentation and international cooperations to minimize export from developed countries to developing countries (UNEP, 1984). The LDC is also considering its policies with regard to a LDC Contracting Party shipping a waste to a non-LDC Contracting Party for ocean disposal.

Several key policy principles can be identified with regard to transfrontier shipments of hazardous wastes: (1) prior notification, (2) prior consent, (3) full disclosure, (4) timeliness, and (5) cradle to grave monitoring. Policy questions include: (1) liability from the cradle to grave, and past the grave, (2) transferability of title or liability from generator to a second party for transport and/or a third party for disposal, (3) insurance requirements, and (4) questions of reasonable and equitable means to assess and cover the risks involved.

5.2. Future Goals in Waste Management

A key to comprehensive waste management is the short-term is to emphasize control at the source and to assess all waste disposal options from a risk perspective. The global objectives for future waste management goals should be to reduce the volume and toxicity of wastes by: (a) source reduction, (b) recycling, (c) detoxification, and (d) development of low waste products and technology. Mankind, in an attempt to prevent environmental problems due to wastes, has shifted disposal policies from one medium to another as contamination has occurred. First we dumped or discharged wastes on land, then into rivers or lakes, and eventually into estuaries and coastal waters. Today we are moving offshore looking for more dilution by using the ocean. At each of these steps, saturation of the receiving medium led to the use of the next convenient option. In the future, it may even be feasable for rockets to transport wastes to outer space.

Application of the concept of assimilative capacity for ocean disposal of municipal and industrial wastes is in its infancy, and may in the future provide workable solutions, if we make major advances in our knowledge of physical, chemical, and biological oceanographic processes. Nevertheless, it offers a predictive tool, and research and monitoring into cause and effect relationships will increase our understanding of the capacity of the sea to receive wastes. Figure 9 presents a rational, pragmatic, and scientific approach to the use of a medium for waste disposal. However, this approach is still limited and should be approached in an experimental manner. Past problems with waste disposal stem from the fact that in many cases the use of marine and/or terrestrial sites for such purposes was carried out with a degree of finality and with insufficient heed to the application of scientific knowledge and principles.

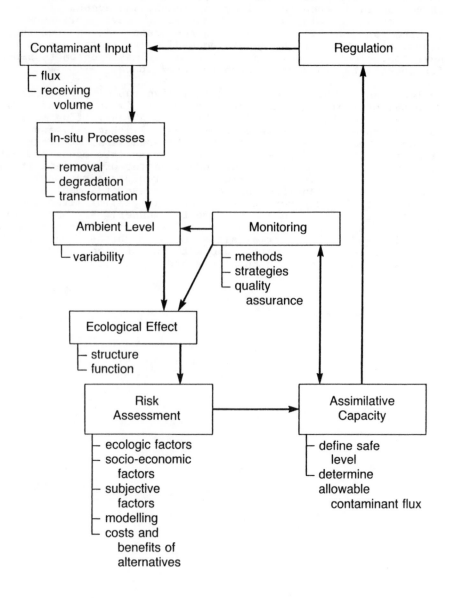

Figure 9. A scheme for a rational, pragmatic and scientific approach to the use of a medium for waste disposal.

ACKNOWLEDGMENTS

We thank the following for their assistance in reviewing earlier
drafts of this manuscript: Thomas P. O'Connor, Douglas A. Wolfe,
and Paul Kilho Park. We also appreciate very much the technical
editing and sugestions on earlier drafts by Joyce G. Nuttall. The
views expressed in this chapter are those of the authors and may
not reflect the offical views of their respective organizations,
or the United States Government.

REFERENCES

Bascom, W. 1974. The disposal of wastes in the ocean. Scientific
    American, 231(2), 16-25.

Cairns, J. Jr. 1976. Estimating the assimilative capacity of
    water ecosystems. In: The Biological Significance of
    Environmental Impacts, R. K. Sharma, J. D. Buffington, J.
    T. McFadden (Eds.), National Technical Information Service,
    Springfield, Virginia, pp. 173-189.

Cairns, J. Jr. 1977. Quantification of biological integrity.
    Proceedings of a Symposium: The Integrity of Water. U.S.
    EPA, Office of Water & Hazardous Materials, U.S. Government
    Printing Office, Washington, D.C., Stock No. 055-001-01068-
    1, pp. 171-187.

Cairns, J. Jr., K. L. Dickson, and A. W. Maki. 1978. Estimating
    the hazard of chemical substances to aquatic life. ASTM STP
    657, American Soc. for Testing and Materials, Philadelphia,
    Pennsylvania.

Champ, M.A., and P.K. Park. 1981. Ocean dumping of sewage sludge:
    A global review. Sea Technology. 22(2):18-24.

Champ, M.A., and J.F. Hebard. 1981. Legislative history of U.S.
    national marine pollution strategies. In: The Management
    of Oceanic Resources--The Way Ahead. Proceedings of an
    International Conference. Organized by the British
    Committee for ECOR. London. Pub. by the University of
    Southern California, Los Angeles. pp. 389-409.

Champ, M.A., P.K. Park. 1982. Global Marine Pollution
    Bibliography: Ocean Dumping of Municipal and Industrial
    Wastes. IFI/Plenum. New York. 399p.

Champ, M.A. 1983. Etymology and use of the term "pollution."
    Procedings of the Conference on Pollution in the North
    Atlantic Ocean. J.W. Farrington, J.H. Vandermeulen, and
    D.G. Cook (Eds.). Canadian Jour. of Fisheries and Aquatic
    Sciences. Vol. 40. Suppl. 2. pp. 5-8.

Champ, M.A. 1984. A global overview of ocean dumping, with
    discussion of the assimilative capacity concept for sewage
    sludge. In: The Law of the Sea and Ocean Industry: New
    Opportunities and Restraints, D.M. Johnston and N.G. Letalik
    (Eds.). Law of the Sea Institute. University of Hawaii.

Champ, M.A., M.G. Norton and M.F. Devine. 1984. A semi-quantitative model for the assessment of dispersion at near shore ocean dumpsites. International Council for the Exploration of the Sea. Symposium on Contaminant Fluxes Through the Coastal Zone. 14-16 May 1984. Nantes, France. Paper No. 59. 27p.

Champ, M.A., and S. Panem. 1985. The utilization of ocean space for waste disposal: Changing perspectives. In: Proceedings of the International Symposium on Ocean Space Utilization. June 1985, Tokyo. Springer Verlag. Tokyo. Vol. 2. pp. 297-306.

Champ, M.A., W.P. Dillon, and D.G. Howell. 1984/85. Non-living EEZ Resources: Minerals, Oil and Gas. Oceanus. 27(4):28-34.

Champ, M.A., and Ned A. Ostenso. 1984/85. Future uses and research needs in the EEZ. Oceanus. 27(4):62-69.

Champ, M.A., and M.A. Conti, and P.K. Park. 1988. Multimedia Risk Assessment and Ocean Waste Management. Chapter I. In: Marine Waste Management: Science and Policy, M.A. Champ, and P.K. Park (Eds.). Vol. 3. Krieger Publishing Co. Malabar, Florida.

Champ, M.A., and P.K. Park. (Eds.). 1988. Marine Waste Management: Science and Policy. In: Oceanic Processes in Marine Pollution. Vol. 3. Krieger Publishing Co. Malabar, Florida.

Cross, F.A., and D.F. Boesch. 1983. Assessment technologies. Workshop Summary. In: Background Papers. IODS Special Symposium on Waste Management in the Ocean: Policy and Strategies. M.A. Champ (Chairman). Pub. by The Center for Academic Publications. Florida Institute of Technology. Melbourne, Florida. Paper No. 38. 11p.

Csanady, G.T. 1981. An analysis of dumpsite diffusion ex periments. In: Ocean Dumping of Industrial Wastes, B.H. Ketchum, D.R. Kester, and P.K. Park. (Eds.). Plenum Press. New York. pp. 109-129.

Devine, M.F., M.G. Norton, and M.A. Champ. 1986. Estimating particulate dispersiveness and accumulation at nearshore ocean dumpsites. Mar. Poll. Bull. 17(10):447-452.

Dortland, R.J. 1981. Problems in evaluating the hazards as-sociated with dumping chemical waste in the marine environ-ment. In: Disposal of Chemical Wastes in the Marine En vironment: Implications of the International Dumping Conven-tions. E.H. Hueck- van der Plas (Ed.). Chemosphere. Special Issue. 10(6):677-692.

Duedall, I.W., B.H. Ketchum, P.K. Park, and D.R. Kester (Eds.). 1983. Industrial and Sewage Wastes in the Ocean. Vol. 1 Wastes in the Ocean Series. John Wiley and Sons. N.Y. 431p.

Duedall, I.W. 1985. International Conventions, and Overview. Final Report prepared for U.S. Congress, Office of Technology Assessment. Report No. 85/3-13. Florida Institute of Technology. Melbourne, Florida. 107p.

Farrington, J.W., J.M. Capuzzo, T.M. Leschine, and M.A. Champ. 1982/83 (Winter). Ocean dumping. Oceanus. Vol. 25(4):39-50.

Gentile, J.H., and S.C. Schimmel. 1984. Strategies for utilizing laboratory toxicological information in regulatory decisions. In: Concepts in Marine Pollution Measurements, H.H. White (Ed.). University of Maryland Sea Grant College. College Park, Maryland. pp. 57-80.

Gentile, J.H., W.A. Walker, and D.C. Miller. 1988. A hazard assessment research strategy for ocean disposal. In: Marine Waste Management: Science and Policy. M.A. Champ, and P.K. Park (Eds.). Chapter 16. Krieger Publishing Co. Malabar, Florida.

GESAMP (Joint Group of Experts on the Scientific Aspects of Marine Pollution). 1975. Scientific criteria for the selection of waste disposal sites at sea. Reports and Studies No. 3. IMCO/FAO/UNESCO/WMO/WHO/IAEA/UN/UNEP. Pub. by IMCO. London. 21p.

GESAMP (Joint Group of Experts on the Scientific Aspects of Marine Pollution). 1982. Scientific criteria for the selection of waste disposal sites at sea. Reports and Studies. No. 16. IMCO/FAO/UNESCO/WMO/WHO/IAEA/UN/UNEP. Pub. by IMCO. London. 60p.

Goldberg, E.D. 1981. The oceans as waste space: The argument. Oceanus. Vol. 24(1)2-9.

Hardin, G. 1968. The tragedy of the commons. Science. Vol. 162. pp. 1243-1248.

IAEA (International Atomic Energy Agency). 1983. Environmental assessment methodologies for sea dumping of radioactive wastes. IAEA Tec. Doc. 296. Vienna. 55p.

IMCO (The Inter-Governmental Maritime Consultative Organization). 1972. Convention on the Prevention of Marine Pollution by Dumping of Wastes and Other Matter. London. 36p. + 1979 Amendments. (Pub. No. 76. 14 E).

IMCO (The Inter-Governmental Maritime Consultative Organization). 1979a. Report of permits issued for dumping in 1976. IMCO Doc. LDC 2/Circ. 31. Inter-Governmental Maritime Consultative Organization. London. 25p.

IMCO (The Inter-Governmental Maritime Consultative Organization). 1979b. Report of permits issued for dumping in 1977. IMCO Doc. LDC 2/Circ. 33. Inter-Governmental Maritime Consultative Organization. London. 23p.

IMCO (The Inter-Governmental Maritime Consultative Organization).
1979c. Report of permits issued for dumping in 1978. IMCO
Doc. LDC 2/Circ. 47. (12 February 1980). Inter
Governmental Maritime Consultative Organization. London.
24p.

IMCO (The Inter-Governmental Maritime Consultative Organization).
1980a. Report of permits issued for dumping in 1979. IMCO
Doc. LDC 2/Circ. 47. Inter-Governmental Maritime
Consultative Ogranization. London. 24p.

IMCO (The Inter-Governmental Maritime Consultative Organization).
1980b. Summary report of permits issued in 1980. IMCO Doc.
LDC 7/INF. 12 (28 January 1983). 29p. Organization.
London. 19p. + 9 Annexes.

IMO (International Maritime Organization). 1982. The
Provisions of the London Dumping Convention, 1972 and
Decisions Made by the Ccnsultative Meetings of Contracting
Parties, 1975-1981. IMO Doc. LDC 7/INF. 3 (16 December
1982). International Maritime Organization. London. 47p.
+ 13 Annexes

IMO (International Maritime Organization). 1984a. Draft report
of permits issued in 1981. IMO Doc. LDC SG.8/INF.2 (20
December 1984). International Maritime Organization.
London. 33p.

IMO (International Maritime Organization). 1984b. Draft report
of permits issued in 1982. IMO Doc. LDC SG. 8/INF. 3 (20
December 1984). International Maritime Organization.
London. 33p.

IODS (International Ocean Disposal Symposium). 1983. Ocean
Waste Management: Policy and Strategies. Background Papers.
M.A. Champ (Chairman). Pub. by the Center for Academic
Publications. Florida Institute of Technology. Melbourne,
Florida. 40 Chapters.

Kamlet, K.S. 1981a. The oceans as waste space: The rebuttal.
Oceanus. Vol. 24(1):10-17.

Kamlet, K.S. 1981b. Court decision makes future of sewage
sludge ocean dumping uncertain. Coastal Ocean Pollution
Assessment News. 1:4647.

Ling, J.T. 1979. Forward. In: Pollution Pays. M.G. Royston,
Pergamon Press. Toronto, pp. ix.

Maugh, T.M. 1978. Chemicals: How many are there? Science.
199:162.

Nauke, M. 1988. The London Dumping Convention. In: Marine
Waste Management: Science and Policy. M.A. Champ, and P.K.
Park (Eds.). Vol. 3. Marine Pollution Processes.
Krieger Press. Malabar, Florida.

Norton, M.G., and M.A. Champ. 1988. The influence of site
    specific characteristics on the effects of sewage sludge
    dumping. In: Wastes in the Oceans, I.W. Duedall et al.
    (Eds.). Vol. 4. Krieger Publishing Co. Malabar, Florida.

O'Connor, J., and R.L. Swanson. 1982. Unreasonable dedgradation
    of the marine environment. In: Marine Pollution Papers.
    Oceans '82. Reprinted by NOAA Office of Marine Pollution
    Assessment. Rockville, Maryland. pp. 1125-1132.

O'Connor, T.P., A. Okubo, M.A. Champ, and P.K. Park. 1983.
    Projected consequences of dumping sewage sludge at a deep
    ocean site near New York Bight. In: Proceedings of the
    Conference on Pollution in the North Atlantic Ocean. J.W.
    Farrington, J.H. Vandermeulen, and D.G. Cook (Eds).
    Canadian Jour. of Fisheries and Aquatic Sciences. Vol. 40.
    Suppl. 2. p. 228-241.

Osterberg, C. 1981. The inviolate ocean. Jour. of Soil and
    Water Conservation. 36(6):311.

Osterberg, C. 1982a. The Ocean--Nature's Trash Basket. In:
    Waste Management - 82. R. Post (Ed.). Vol. 2. Low Level
    Wastes. University of Arizona. Tucson. pp. 407-414.

Osterberg, C. 1982b. Why not in the ocean? IAEA Bulletin.
    International Atomic Energy Agency. Vienna. Vol. 24(2):3-
    34.

Osterberg, C. 1982c. Waves of the future. Cornell Executive.
    Cornell University. Ithaca, New York. 8(2):35-38. p. B3.

Osterberg, C. 1983. Whither the waste on a water planet? In:
    Ocean Waste Management: Policy and Strageties. Pub. in the
    Background Papers for the International Ocean Disposal
    Symposium. Pub. by the Center for Academic Publications.
    Florida Institute of Technology. Melbourne, Florida.
    Paper No. 13. 9p.

Osterberg, C. 1985a. Waste disposal: Where should it be?
    or sea? The Siren. UNEP. No. 28. May. Geneva,
    Switzerland. pp. 9-16.

Osterberg, C. 1985b. Nuclear power wastes and the ocean.
    Chapter 4. In: Wastes in the Ocean, I.W. Duedall, (Eds.).
    Vol. 4. Wastes in the Ocean. John Wiley and Sons. New
    York. pp. 127-162.

Osterberg, C. 1986a. Factors influencing the land/freshwater
    verse the sea option. In: Basis for the role of the ocean
    as a Waste Disposal Option. The D. Reidel Publishing Co.
    Dortrecht, Holand. pp. 39-53.

Osterberg, C. 1986b. Ocean dumping policy and old nuclear
    submarines. Marine Policy Review. University of Delaware.
    March, 1986. Vol. 8(5):1-5.

Roels, O. 1982. Mariculture fertilized by upwelling. Sea Technology. Vol. 23(8):63-67.

Rose, C.D., and T.J. Ward. 1981. Principles of aquatic hazard evaluation as applied to ocean-disposed wastes. ASTM TP 737. American Society for Testing and Materials. Philadelphia, Pennsylvania. pp. 138-158.

Sebek, V. 1983. Bridging the gap between environmontal science and policy-making: Why public policy often fails to reflect current scientific knowledge. Amb,io. 12(2):118-120.

Spilhaus, A. 1985. Keynote Address of Dr. Athelstan Spilhaus. Pub. in the Proceedings of the First Annual Environmental Law Symposium. Jour. of Law and the Environment. University of Southern California Law Center. Vol. 1. pp. 57-64.

Stebbing A.R.D. 1981. Assimilative capacity. Mar. Pollut. Bull. 12(11):362-363.

Swanson, R.L., and M.F. Devine. 1982. Ocean dumping policy: The pendulum swings again. Environment. 24(5):14-20.

Swanson, R.L., M.A. Champ, T.P. O'Connor, P.K. Park, J.S. O'Connor, G.F. Mayer, H.M. Stanford, E. Erdheim, and J.L. Verber. 1986. Sewage sludge dumping in the New York Bight apex: A com parison with other proposed ocean dumsites. In: Near-Shore Waste Disposal. B.H. Ketchum, J.M. Capuzzo, W.V. Burt, I.W. Duedall, P.K. Park, and D.R. Kester (Eds.). Vol. 6. Wastes in the Ocean Series. John Wiley and Sons. New York. pp. 461-488.

U.S. Congress. 1967. Air Quality Act of 1967 (The Clean Air Act). 42 U.S.C. 7401 et seq. as amended.

U.S. Congress. 1972a. The Marine Protection, Research and Sancturies Act of 1972. (The Ocean Dumping Act). 86 Stat. 1052. 33 U.S.C. 1401 et seq. as amended.

U.S. Congress. 1972b. FWPCA Amendements of 1972. (The Clean Water Act). 86 Stat. 816. 33 U.S.C. 1251 et seq. as amended.

U.S. District Court, Southern District of New York. 1981. 80 Civ. 1677. A.D. Sofaer.

U.S. EPA (Environmental Protection Agency). 1973a. Ocean dumping interm regulations governing transportation for dumping and disposal of material into ocean waters. U.S. Federal Register. Title 40. Subchapter H - Ocean Dumping (Rules and Regulations). April 5. Vol. 38. pp. 8726-7830.

U.S. EPA. 1973b. Ocean dumping criteria. Title 40. Subchapter H Ocean Dumping (Rules and Regulations). May 16. U.S. Federal Register. Vol. 38. pp. 12872-12877.

U.S. EPA (Environmental Protection Agency). 1977. Ocean dumping, final revision of regulations and criteria. January 11. U.S. Federal Register. Vol. 42. pp. 2462-2490.

U.S. EPA. 1984a. Final designation of site. May 4. U.S. Federal Register. Vol. 49. pp. 19005-19012.

U.S. EPA. 1984b. Risk assessment and risk management: Framework for decision-making. EPA 600/9-85-002. Washington, D.C. 35p.

U.S. EPA. 1985. Assessment of incineration as a treatment method for liquid organic hazardous wastes. Background Report No. V. Public concerns regarding land-based and ocean-based incineration. Office of Policy, Planning and Evaluation. Washington, D.C. 42p.

U.S. NACOA (National Advisory Committee on Oceans and Atmosphere). 1981. The role of the ocean in a waste management strategy. A Special Report to the President and the Congress. U.S. Gov. Print. Office. Washington, D.C. 103p. + Appendicies.

U.S. NRC (National Research Council). 1983. Risk assessment in the federal government: Managing the process. Commission on Life Sciences. National Academy Press. Washington, D.C. 191p.

U.S. NRC. 1984. Disposal of industrial and domestic wastes: Land and sea alternatives. Board on Ocean Science and Policy. National Academy Press. Washington, D.C. 210p.

Wastler, T.A., C.K. Offutt. 1975. Chemical needs for the regulation of ocean disposal. In: Marine Chemistry in the Coastal Environment. T.M. Church (Ed.). American Chemical Society. Washington, D.C. pp.416-423.

Wastler, T.A. 1981. Ocean dumping permit program under the London Dumping Convention in the United States. In: Disposal of Chemical Wastes in the Marine Environment: Implications of the International Dumping Conventions, E.H. Hueck-van der Plas (Ed.). Chemosphere Special Issue. 10(6):659-699.

Webster's New Collegiate Dictionary. 1979. G&C Merriam Co. Springfield, Massachusetts. 1531p.

White, H.H., and M.A. Champ. 1983. The great bioassay hoax, and alternatives. In: Hazardous and Industrial Solid Waste Testing. R.A. Conway and W.P. Gulledge (Eds.). ASTM STP 805. American Society for Testing and Materials. pp. 299-312.

White, H.H. (Ed.). 1985. Concepts in Marine Pollution Measurements. University of Maryland Sea Grant. Publication 84-03. College Park. 743p.

World Commission on Environment and Development (WECD). 1985. Global aspects of hazardous waste menagement: Policy options, WCED/85/17, Oslo, Norway, 21pp.

345

■ International Programs for Marine Pollution Studies and the Role of the Oceans as a Waste Disposal Option: An Overview

GUNNAR KULLENBERG
Intergovernmental Oceanographic Commission
7, Place de Fontenoy
75700 Paris, France

INTRODUCTION

International marine pollution research and monitoring programmes on a regional and global basis include:

- programmes generated through direct agreements between neighbouring countries;
- programmes generated under the umbrella of an international organization as a regional project;
- programmes governed by regional intergovernmental conventions;
- programmes of global character within the framework of an intergovernmental organization.

The scope of these programmes can cover both basic scientific studies and monitoring and surveillance. The objectives of monitoring programmes are: the protection of human health; detection of long-term trends of contamination; quantification of biological effects; provision over the region of an indication of the state of health of the marine environment.

The concept of health of the oceans has emerged through the realization of the need to summarize the state of pollution of the world's oceans with respect to different uses of the ocean and its resources. Different uses will require different qualities of marine environment, including for fishing, aquaculture, recreation, exploitation of non-living resources and shipping, and hence different states of the health of the oceans in order to ensure a safe use.

The legal framework is important in deciding the character and scientific content of the programmes, as well as the financial and institutional basis. Regional conventions define priorities: of the region on the basis of existing information. Coastal zone management is often emphasized and methods for monitoring conditions in the coastal zone receive high priority.

THE GIPME PROGRAMME

The Intergovernmental Oceanographic Commission (IOC) adopted the Global Investigation of Pollution in the Marine Environment (GIPME) programme in response to Recommendation 90 of the United Nations Conference on the Human Environment (Stockholm, 5-16 June 1972). The overall objective of GIPME is to provide a scientifically sound basis for the assessment and regulation of marine contamination and pollution, including sensibly planned and implemented international, national and regional monitoring programmes. The programme covers all matters related to marine pollution research and associated monitoring

347

activities required for the assessment of marine pollution. A specific objective of the programme is to provide information on contamination in the marine environment which will lead to such an assessment of pollution. Unesco (1976); Unesco/IOC (1984); IOC (1986); Kullenberg et al (1986).

The development and implementation of the programme is based on a scientifically defensible approach that comprises four major sequential stages:

1. Mass balance determinations, which imply calculation of the balance of a given contaminant in a given segment or phase of the marine environment (e.g., water, particulate matter, sediments and biota, together with inflows and outflows) to ascertain whether the amount of the contaminant is increasing, decreasing or in a steady state. This first step requires that comparable data are available so that the necessary elements in the equation can be calculated. It thus includes the development and inter-comparison of the necessary analytical methodologies.

2. Contamination assessment, which is essentially the result of the mass balance calculation showing the level of contamination of the segment or phase.

3. Pollution assessment, which also requires an evaluation of the biological effects of the contaminants upon the ecosystem, or components thereof, in the given segment or phase of the marine environment.

4. Regulatory action by appropriate bodies.

The scientific and methodological development required to complete these steps are carried out under the guidance of three Expert Groups: on Methods, Standards and Intercalibrations (GEMSI), jointly sponsored by IOC and the United Nations Environment Programme (UNEP); on Effects (Biological) of Pollutants (GEEP), jointly sponsored by IOC, UNEP and the International Maritime Organization (IMO); on Standards and Reference Materials, jointly sponsored by IOC, International Atomic Energy Agency (IAEA) and UNEP. Regional components of the Marine Pollution Monitoring System are monitoring activities at the international level covering a regional sea programme and national efforts of concerned Member States. The major proviso governing the suitability of any data made available through the Programme lies in the assurance of the quality and comparability of such data.

In developing GIPME and initiating its implementation, priority has been assigned to studies of trace metals, petroleum hydrocarbons and organochlorines, with the developmental stages for monitoring trace metals progressing more rapidly than those for organochlorines.

From an ideal point of view, all phases (i.e., water, biota and sediment) of the marine environment should be analyzed. Only in this way can a comprehensive mass-balance be obtained. However, recognizing that this may not be possible in all cases, the analysis of any phase provides an input to the data base, as long as the methods employed are shown to produce reliable data. Nevertheless, to make a valid assumption of an equilibrium in the distribution of contaminants between the various phases requires knowledge of biogeochemical processes.

Several organisms have been proposed and used as indicators of contaminant distribution. Mussels, notably Mytilus edilus, other bivalves, barnacles, crabs, limpets, macroalgae and fish have frequently been suggested. The monitoring of biota is often conducted for reasons other than for purposes of describing the distribution of contaminants specifically (e.g., for assessing the hazards to human health through direct consumption of marine products). It is therefore not surprising that a potentially larger data collection system exists for this type of monitoring activity. The concept of "Musselwatch" was initially proposed as a rational monitoring programme for contaminants introduced into the marine environment. Bivalves are known to concentrate all of the contaminant groups addressed by the programme, with the exception of nutrients. The "Musselwatch" approach naturally has disadvantages, as well as advantages, since no single sentinel can be expected to respond to a variety of inputs of environmental contaminants.

The concept of pollution implies that observed levels of contamination can be interpreted in terms of biological consequences. It is, therefore, necessary both to link the objects of chemical analysis (e.g., water, sediment and fauna) with biological components of potential interest, and to arrive at agreed and standard methods for evaluating biological impact.

Water column surveys present a number of inherent difficulties, such as the possibility of high natural variability in the nearshore environment, low contaminant levels and contamination problems concerned with the collection, storage and analyses of samples. The applicability of monitoring this phase in the high dynamic nearshore environment should be carefully considered in the light of these points, prior to embarking on an expensive monitoring activity. For example, in using the aqueous phase for determining river inputs to the coastal zone, aside from requiring analytical capabilities for making the measurements, an understanding of the possible modification of the contaminant being measured is also required as it progresses from the river to the sea cross the land/sea boundary (e.g., through the estuarine regime).

Methodology being employed for determining the levels of a contaminant is not universal. Indeed, it is clearly unrealistic to believe that all participants will possess equal analytical capabilities or facilities. Nevertheless, any method that is being used has been demonstrated to produce comparable data with other methods. Such demonstrations result from intercalibration activities, inter-laboratory comparisons and internationally co-ordinated exercises at all stages. All methods previously accepted by the Groups of Experts on Methods, Standards and Intercalibration, or on Effects of Pollutants, as appropriate for the intended measurements, are proposed as the methods of choice in the GIPME Programme. In all cases, complete descriptions of methods are requested with all reported data, as well as statistical aspects of the data (e.g., details of replication, variability of results, blank values and descriptions of standards or reference materials used). It has been necessary, in some cases, to employ less discriminatory, more easily used methods of analysis to set up a framework of laboratories for purposes of training, intercalibration and familiarization with sampling and logistics requirements and then plan to move towards wider employment of more specific, sensitive discriminatory techniques of measurements. For example, using ultra-violet fluorescence estimates of dissolved and dispersed petroleum in seawater, as was done in the pilot project on Marine Pollution Monitoring of Petroleum Hydrocarbons.

The objectives and components of the programme have been established on scientifically credible principles. Data gathering, reporting and exchange requires stringent control of the quality of the information retrieved, which in turn dictates the development and testing of standard methodology, its widespread adoption and intercomparison of methods and refinement of original methods or hypotheses through feedback.

Reliable standards and reference materials are required to enable laboratories to intercalibrate and perform in-house quality controls. Experience has demonstrated that the availability of such materials may be the single most severe problem facing both developed and developing laboratories coupled

with, but not always restricted to, access to the state-of-the-art instrumentation
Inexperience in performing the particular analyses is also a major contributing
factor of failure to meet standards of intercomparison.

THE REGIONAL COMPONENT OF THE GIPME MARINE POLLUTION MONITORING PROGRAMME
(MARPOLMON) FOR THE CARIBBEAN AND ADJACENT REGIONS (CARIPOL)

Since 1979, IOC's Regional Subcommission for the Caribbean and Adjacent Regions
(IOCARIBE) has conducted a project of monitoring for petroleum pollution in
the area known as the Wider Caribbean, which includes the Gulf of Mexico, the
Straits of Florida and the Eastern approaches to the Caribbean Sea. The
programme is one of the regional marine pollution research and monitoring
efforts conductedwithin the IOC/GIPME framework, and is known as CARIPOL.

Methodology used successfully in the IOC-World Meteorological Organization (WMO)
Marine Pollution Monitoring Pilot project (Petroleum) (MAPMOPP) (Levy, et al,
1981) was adapted for use in the CARIPOL project. This included monitoring of
  the following three parameters;

Tar on beaches. Tar is collected from the water line to backbeach along one
meter transects, weighed and reported as grams-tar/metre of beach front.
Floating tar. A one metre wide neuston net is towed from a vessel and outside
the vessel wake for a known time and vessel speed. The tar collected is
weighed and reported as mg-tar/square metre of sea surface.
Dissolved/dispersed petroleum hydrocarbons (DDPH). A one gallon sample is
collected from a carefully cleaned, small mouth bottle suspended on a one metre
tether from a surface float. The sample is extracted twice with 50 ml. aliquots
of nanograde hexane and the concentration of petroleum type hydrocarbons in
the hexane phase is estimated using an ultraviolet spectrofluorescence technique
with chrysene as the primary standard.

The only difference between these parameters and those used in MAPMOPP is the
use of hexane as the primary extractant for DDPH samples.

Training workshops were conducted in Costa Rica in 1980, during which personnel
from governments throughout the region received training in the CARIPOL methods
described above. On the basis of results from these workshops a methods manual
for analysis and data submission was written in both English and Spanish and
distributed to participants (CARIPOL, 1980). A data archiving system was
established by co-operation of IOCARIBE with the US National Oceanic and
Atmospheric Administration (NOAA) National Oceanographic Data Center (USNODC)
in Washington, DC, and the NOAA Atlantic Oceanographic and Meteorological
Laboratory (AOML) in Miami, Fl. Training of new participants was accomplished
by establishing CARIPOL training centres at the University of Costa Rica in
San José, Costa Rica, the University of Puerto Rico in Mayaguez, Puerto Rico,
and at AOML. Propsective participants were sent to these centres and scientists
from these centres visited interested regional governments (Atwood et al 1987).

Since 1979, the CARIPOL Petroleum Pollution Monitoring Project has generated
well over 8,000 observations which are archived in the CARIPOL data base.
Results were presented at a symposium conducted with UNEP/CAP and IOC funding
at the University of Puerto Rico (Mayaguez) in December 1985 (CARIPOL 1987).

THE GLOBAL ENVIRONMENT MONITORING SYSTEM (GEMS)

The role given to GEMS is to co-ordinate international monitoring activities
conducted throughout the world, particularly within the UN system, and stimulate
the initiation of new activities or the expansion of on-going ones.

The monitoring activities that UNEP currently supports, fall into five major programmes, which have built-in provisions for training and for technical assistance to ensure the participation of countries that may be inadequately provided with personnel and equipment: climate-related monitoring; monitoring of long-range transport of pollutants; health-related monitoring; ocean monitoring; and terrestrial renewable-resources monitoring (Gwynne 1982).

Ocean monitoring is considered from two aspects - the monitoring of open oceans, and the monitoring of regional seas and coastal waters. The monitoring of pollutants in open oceans has proved both expensive and technically difficult. It is a field, however, in which remote sensing is obviously destined to play a major role but, at the moment, the available technology is still far from adequate. This situation is, however, improving rapidly, and the GEMS PAC is keeping in touch with latest developments, with a view to the future establishment of open ocean pollutant monitoring programmes.

To date, there has been only one open ocean pilot activity under direct GEMS PAC responsibility. This was carried out in co-operation with Intergovernmental Oceanographic Commission (IOC) of Unesco and WMO, and was aimed at monitoring, within the framework of the Integrated Global Ocean Station System (IGOSS), marine pollution by petroleum hydrocarbons along major shipping lines.

MARINE POLLUTION (PETROLEUM)MONITORING PILOT PROJECT (MAPMOPP)

In 1970, the Scientific Committee on Problems of the Environment (SCOPE) of the International Council of Scientific Unions (ICSU) was requested to make recommendations on the design, parameters and technical organization for a global marine environmental monitoring system which, for practical reasons, would make maximum use of existing and planned national, regional and international programmes. The ultimate objectives of such a system were to provide a measure of the distribution of polluting substances in the world oceans, to provide information on long-term trends in significant characteristics of the marine environment, and to give warning when certain pollutants exceed maximum permissible concentrations in water or marine biota. In 1971, the IOC and WMO governing bodies decided that the Integrated Global Ocean Station System (IGOSS) provided a suitable framework for the co-ordination of marine pollution monitoring activities in respect of physical and some chemical parameters. The joint IOC-WMO Planning Group for IGOSS (IPLAN), initiated work toward the design and implementation of a Marine Pollution Monitoring Programme and recommended that as a first step, a pilot project should be undertaken which was endorsed by the IOC and WMO governing bodies.

It should be emphasized that this exercise was considered to be a pilot project and that its primary purpose was not to provide a detailed picture of oil pollution on a global scale, but rather to demonstrate the technical and organizational problems that would have to be overcome in any future co-ordinated monitoring programme.

Four components of MAPMOPP were identified: monitoring of oil slicks; monitoring of floating particulate petroleum residues; monitoring of dissolved and dispersed petroleum residues; and monitoring of tar on beaches.

The areas to be monitored were determined, in part, on the basis of existing national and regional programmes, but consideration was also given to the main oil transportation routes.

Data for the four components of MAPMOPP were collected by 26 participating IOC and WMO Member States and forwarded by their National Co-ordinators to the

IGOSS Responsible National Oceanographic Data Centre in Washington or Tokyo.
A summary of results is given in Levy et al., 1981.

ICES MONITORING PROGRAMMES ON CONTAMINANTS IN MARINE POLLUTION

The International Council for the Exploration of the Sea (ICES) monitoring
activities to date, have concentrated mainly on studies of contaminants in fish
and shellfish. The two main reasons for this focus are (Pawlak 1986).

- fish and shellfish accumulate contaminants in their tissues, thus presenting
  a possible risk to man or other higher organisms consuming them, and
- owing to this bioaccumulation, the concentrations of contaminants are
  generally high enough to permit relatively easy determination, thus providing
  a better intercomparability among laboratories participating in an international programme.

Under the Co-operative ICES Monitoring Studies Programme, concentrations of
contaminants in fish and shellfish are monitored in the North Atlantic area,
excluding the Baltic Sea (except for certain programmes, such as the Baseline
Study described below). The programme first began in the North Sea in 1974 and,
over the next two years, extended to other areas of the North Atlantic.
Countries participating in this programme include Canada, Belgium, Denmark, the
Federal Republic of Germany, France, Ireland, the Netherlands, Norway,
Portugal, Sweden, the United Kingdom and the United States. Initially, the
programme was based on the national monitoring schemes of the participating
countries, but a review of results conducted in 1980-1981 indicated that new
guidelines were needed for the programme to improve the ability to interpret
the data. Equally important to this programme is the maintenance of the
quality of the analytical data produced by the participating laboratories;
the intercomparability of these data are checked periodically through participation in intercalibration exercises and maintained via the appropriate use
of reference materials and good laboratory practice.

The revision of the original monitoring programme took effect in 1982 with the
implementation of new guidelines for sampling, sample preparation and reporting
of data, which were specified separately for each of the three objectives of
the revised programme, entitled to the Co-operative ICES Monitoring Studies
Programme. The three separate objectives for monitoring contaminants in the
marine environment using fish and shellfish as indicator species are:

- the provision of a continuing assurance of the quality of marine foodstuffs
  with respect to human health;
- the provision over a wide geographical area of an indication of the health
  of the marine environment in the entire ICES North Atlantic area;
- the provision of an analysis of trends over time in pollutant concentrations
  in selected areas, especially in relation to the assessment of the efficacy
  of control measures.

The contaminants analyzed in this programme include the following trace metals:
mercury, copper, zinc, cadmium, and lead. The comparability of data among
participating laboratories on these metals is generally good, except for lead
at low levels. Some laboratories also analyze for arsenic, chromium, and
nickel, but the level of comparability of data for the latter two metals has
not been fully established.

Data are also reported for the determination of a number of organochlorine
residues, including the DDT group, PCBs, hexachlorobenzene (HCB), dieldrin,
chlordane, and the α-, β-, γ- isomers of hexachlorocyclohexane (HCH).

352

Intercomparison exercises, however, have shown that the level of comparability of data on these substances between laboratories is not satisfactory.

The guidelines to this monitoring programme specify that to meet Objective 1 (assessment of contaminant levels in fish and shellfish with respect to human health), samples should be collected every second year, starting in 1982. A sample should consist of 25 fish or larger crustaceans, such as lobsters or crabs, 50 mussels or other molluscs, or 100 small crustaceans, such as shrimps.

The second objective (to provide an indication of contaminant dsitribution over a wide geographical area) is essentially a baseline study. This type of study is to be carried out every five years, beginning in 1985. In this type of study, fish should be selected to be representative of an area, i.e., they should not be recent immigrants to or on passage through an area. Each sample should consist of 25 fish or 50 mussels. Samples should be collected from as many areas as possible throughout the ICES field of interest.

In 1985, the Baseline Study of Contaminants in Fish and Shellfish covered many coastal areas, primarily in the Northeast Atlantic, and most areas of the Baltic Sea. Species of interest in the context of this type of study are those with a broad geographical distribution. The species samples in the Baltic component of this study include cod (Gadus morhua), herring (Clupea harengus), dab (Limanda limanda), flounder (Platichthys flesus), plaice(Pleuronectes platessa), mussels (Mytilus edulis), Mesidotea entomon and Macoma balthica. With the exception of the latter two organisms, the species that have been sampled in the North Atlantic are the same as those for the Baltic Sea. In this Baseline Study, particular emphasis has been given to the collection of mussels in as many coastal areas around the North Atlantic, including the Baltic Sea, as possible, so that the study has a strong Mussel Watch component.

The third objective of the Co-operative ICES Monitoring Studies Programme is the monitoring of trends in the concentrations of contaminants over time. To meet this objective, sampling should be conducted annually at the same time of the year, from the same areas and from the same stock of organisms. Species which have been identified as of particular interest in the ICES context are cod, plaice, flounder, mussels, hake (Merluccius merluccius) and mackerel (Scomber scombrus).

The sampling areas chosen for this objective are generally those with the greatest problems of contamination. Data from a minimum of three years are required to conduct the statistical analyses for temporal trends.

JOINT MONITORING PROGRAMME OF THE OSLO AND PARIS COMMISSIONS

The Joint Monitoring Programme of the Oslo and Paris Commissions, focuses on marine organisms and sea water.

Monitoring of marine organisms (fish, molluscs and crustacea) is carred out for the following purposes:

1. the assessment of possible hazards to human health;
2. the assessment of harm to living resources and marine life (ecosystems);
3. the assessment of the existing level of marine pollution;
4. the assessment of the effectivenss of measures taken for the reduction of marine pollution in the framework of the Conventions;

Monitoring of sea water is carried out for the following purposes:

1. the assessment of the existing level of marine pollution;
2. the assessment of the effectiveness of measures taken for the reduction of
marine pollution in the framework of the Conventions.

These monitoring activities take place in geographical locations chosen by
the Contracting Parties themselves. In some areas organisms only are sampled;
in others only one (or more) of the purposes are achieved. Analyses of organisms
are performed for cadmium, mercury and PCBs, and of sea water for cadmium and
mercury.

From 1986 the Joint Monitoring Programme covers most of the Convention Area,
the areas being situated mostly along the coastline although there are some
relatively open water sites in the Irish Sea, North Sea and Skagerrak.

With regard to compartments monitored, most of the countries now measure
seawater and biota; the compartment of sediments is not yet sufficiently
covered on a routine basis.

THE BALTIC MONITORING PROGRAMME (BMP)

The Baltic Monitoring Programme of the Baltic Marine Environment Protection
Commission (Helsinki Commission), is the joint monitoring programme prepared
and adopted for use by the Contracting Parties to the Helsinki Convention. All
seven Baltic Sea states are participating in the programme which began in 1979.

The objective of the programme is to provide data for periodic assessments of
the state of the Baltic marine environment, by repeated measurements of selected
determinands, including pollutant concentrations, which will follow changes in
the marine environment over a period of time.

The programme is intended for implementation in several stages. The first stage,
1979-1983, had an experimental character. The programme was adjusted after the
first five year period. The second stage will last until the end of 1988, but
a group of experts has already started the work to revise the BMP for the third
stage, taking into account the experience gained during the first and second
stages of the programme , Helsinki Commission (1983), Melvasalo (1986).

Frequency of sampling. For physical, chemical and pelagic biological determinands
the frequency of sampling should be four seasons (February-March, if possible;
May-June, August-September and November-December).

At the representative stations, the minimum frequency of sampling shall be six
time per year. By co-ordination of timetables a higher frequency at the
representative stations should be aimed at. At the other stations higher
frequency of sampling should be encouraged.

For zoobenthos, the frequency of sampling should be twice a year (spring and
autumn) at the monitoring stations and once a year (spring) at the shallow
area zoobenthos stations.

For harmful substances in selected species, the frequency of sampling should
be once a year.

Physical and chemical determinands.   Temperature; salinity; density structure;
oxygen; hydrogen sulphide; pH; alkalinity; nutrients (phosphate, total phosphorus,
ammonia, nitrate, nitrite, total nitrogen); silicate; heavy metals (Hg, Cd, Zn,
Cu and Pb) in sea water and sediments; petroleum hydrocarbons (PCH's) in sea
water at 1 m. depth at representative stations; chlorinated hydrocarbons (DDT

and PCB groups) in sea water and sediments.

Levels of harmful substances in selected species. Species to be sampled:
Macoma baltica (only metals); Clupea harengus, herring; Gadus morrhua, cod;
Mytilus edulis, blue mussel; Platichthys flesus, flounder; Mesidothea entomon;
Crangon crangon, common shrimp.

It is recommended that in addition to this, the Contracting Parties include
in their programmes sampling of at least one species of marine birds.

Substances to be analyzed: hazardous substances (DDT, DDD, DDE, PCBs, PHC's
and Toxaphene) and Noxious substances (Hg, Cd, Zn, Pb, Cu).

Biological parameters to be measured. Phytoplankton primary production;
phytoplankton chlorophyl a; phaeopigment; phytoplankton; mesozooplankton;
protozooplankton; soft bottom macrozoobenthos; micro-organisms.

REGIONAL SEA PROGRAMME OF UNEP: THE LONG-TERM PROGRAMME FOR POLLUTION
MONITORING AND RESEARCH IN THE MEDITERRANEAN (MED POL) - PHASE II - AN EXAMPLE

In the light of the results of the Stockholm Conference on the Human Environment,
the United Nations General Assembly decided, in 1972, to establish the United
Nations Environment Programme (UNEP) to "serve as a focal point for environmental action and co-ordination within the United Nations system". Subsequently,
the Governing Council of UNEP chose "Oceans" as one of the priority areas in
which it would focus efforts to fulfil its catalytic and co-ordinating role
and the Regional Seas Programme was initiated by UNEP in 1974.

At present, the Regional Seas Programme includes ten regions and has over 120
coastal States participating in it.

The Mediterranean Action Plan was the first one developed in the framework of
the Regional Seas Programme. It was adopted in early 1975 in Barcelona and
since then has made remarkable progress.

In the framework of the Mediterranean Action Plan, the pilot phase of the
Co-ordinated Mediterranean Pollution Monitoring and Research Programme (MED
POL - PHASE 1.) was initiated in 1974 and formally approved in 1975 as the
scientific/technical component of the Mediterranean Action Plan.

Based on the recommendations made at various expert and intergovernmental
meetings, the Long-term (10 years) Programme for Pollution Monitoring and
Research (MED POL - PHASE II) was formulated and formally approved in 1981 .

The general long-term objective of MED POL - PHASE II is to further the goals
of the Barcelona Convention by assisting the Parties to prevent, abate and
combat pollution of the Mediterranean Sea and to protect and enhance the marine
environment of the Area. The specific objectives are designed to provide, on
a continuous basis, the Parties to the Barcelona Convention and its related
protocols with (UNEP 1983).

-   information required for the implementation of the Convention and the
    protocols;
-   indicators and evaluation of the effectiveness of the pollution prevention
    measures taken under the Convention and the protocols;
-   scientific information which may lead to eventual revisions and amendments
    of the relevant provisions of the Convention and the protocols and for the
    formulation of additional protocols;

355

- information which could be used in formulating environmentally sound national, bilateral and multilateral management decisions essential to the continuous socio-economic development of the Mediterranean region on a sustainable basis;
- periodic assessment of the state of pollution of the Mediterranean Sea.

These objectives will be achieved through the evaluation of the information on the sources, amounts, levels, trends, pathways and effects of pollutants in the Mediterranean which will be collected, analyzed and reported   on a systematic basis, using commonly agreed methods, and taking into account data available from other sources.

Monitoring has been organized on several levels:

- monitoring of sources of pollution providing information on the type and amount of pollutants released directly into the environment;
- monitoring of neashore areas, including estuaries, under the direct influence of pollutants from idenifiable primary (outfalls, discharge and coastal dumping points) or secondary (rivers) sources;
- monitoring of the transport of pollutants to the Mediterranean through the atmosphere, providing additional information on the pollution load reaching the Mediterranean Sea.

Sampling and analytical techniques used in monitoring have been based on recommended reference methods. Other methods could also be used, including remote sensing, subject to a satisfactory intercomparison (Civili and Jeftic 1985).

THE ROLE OF THE OCEANS AS A WASTE DISPOSAL OPTION

Human activities produce wastes. Part of these wastes can be recycled or used after certain cleaning processes, which often give rise to a remainder which must be disposed of. Other parts of the wastes cannot be recycled with presently available technology, but have to be disposed of directly.

In order to select the most appropriate waste disposal option, it is necessary to consider the total environment - the earth, freshwater, atmosphere and ocean, as common resources. This consideration should include the amount and availability of the resource, the interactions between the different parts of the environment and their importance to the support of life and human society on earth, as well as their relative availability. In relation to the oceans, elements of such consideration were presented at a recent symposium (Kullenberg, editor, 1986).

The ocean covers about 70% of the earth, with a volume of approximately $1370.10^6 km^3$ sea water, or roughly 99% of all liquid water on earth. Only 1% liquid water is fresh water. The ocean is inhabited by living organisms to the greatest depths so the whole volume of the sea constitutes the ocean biosphere. Assuming that the soil/land biosphere is 10 metres deep, the volume of the terrestrial biosphere is $0.01 \times 154.10^6 km^3$ or $1.54.10^6 km^3$. The ratio of the oceanic to the terrestrial biospheres is approximately 900 (Østerberg 1986). The total production on earth is approximately 155.10^9 tons carbon as net primary production per year, with about 100.10^9 tons on land and 55.10^9 tons in the ocean. The ocean produces 1% of man's food, or 3% of his protein, or 10% of his animal protein. The food producing chains for man are 1-2 steps on land and up to 5 in the ocean. The soil on earth is a very valuable resource, and the oceans constitutee a viable option for disposal of certain types of waste.

Examples of common wastes include sewage sludge formed after cleaning processes. The European sewage sludge production is presently approximately $7 \cdot 10^6$ tons dry solid per year, expected to rise to $9.5 \cdot 10^6$ by 1990. The US production is presently $4.5 \cdot 10^6$ tons dry solid per year. Raw sewage sludge contains $2-7\%$ solids. The sewage sludge can be disposed of on land or in the sea. In Europe, the ocean disposal is about 7% of the production and in the US about 18% (Wood, 1986). The global input of nitrogen compounds to the world ocean is about 10.10^6 tons annually and of phosphorous compounds 2.10^6 tons, while the oceanic reserves are $191.800.10^6$ tons, respectively, assuming a production of 10 gram N/person per day and 2 gram P/person per day (Topping, 1976).

Several metals are indispensable to present human activities and the production is considerable. The waste amounts are also considerable.

Dredging is an activity known to most seafarers, keeping harbours and channels open. Much dredge spoils are disposed at sea, often in such a way that the material is carried back to its dredging origin, to become dredged once more.

Much persistent material produced and used on land ultimately reaches the ocean. More than 25% of all DDTs produced by 1970 have reached the sea.

It is difficult to obtain a definition of waste which is generally acceptable. It is equally difficult to identify the many circumstances under which a hazardous and potentially toxic material would have a deleterious effect on the biological system.

The ICSU Committee on Toxic Wastes adopted a slightly modified definition used by the European Economic Community, as:

"Waste is defined as any substance or object which the holder disposes of or is required to dispose of, pursuant to the provision of national law in force. Toxic and dangerous means any substance containing or contaminated by the substances or materials of such a nature, in such quantities or in such concentrations as to constitute a risk to health or to the environment if disposed of inappropriately. (It is recognized that a substance can only display toxicity by interaction with an organism)".

Through improving techniques, changing processes and instituting management and control the amounts of wastes can be considerably reduced. Examples are provided by the pulp and paper industries and by various chemical industries. However, human wastes will always be produced, moving of material will be necessary and accidents or breakdowns of systems can occur. In addition, the need for production increases gives rise to indirect releases to the environment of substances that may have deleterious effects. Examples are the DDT usage and the use of fertilizers in many countries. Much of the fertilizing material is wasted away before entering the soil and is cast to the sea where it can influence the nearshore and coastal environment.

A waste disposal operation of any considerable size, under present day conditions needs to be subject to careful planning, management and control. Consideration must also be given to the various disposal options and the method of disposal - on land, in the sea or the atmosphere; containment or dispersal, with inter- mediate possibilities. Control and management is also required in relation to transportation of wastes and other potentially hazardous materials.

The ocean environment and its resources is, to some extent, a common heritage and, hence, subject to a common control and associated global conventions. In order to use and protect the ocean environment and its resources in an optimal

357

way, it is necessary to take into account the characteristics of the ocean, the processes governing the natural conditions there, and the basis for the operation and functioning of the ocean environment as a major life supportiong system of nature. Monitoring systems are required to watch the development on both regional and global levels, so as to provide a database for a periodic evaluation of the state of health of the oceans (e.g., Kullenberg, 1986).

REFERENCES

Atwood, D.K., F.J. Burton, J.E. Corredor, G.R. Harvey, A.J. Mata-Jimenez, A. Vazquz-Botello, and D.A. Wade (1987). Results of the CARIPOL petroleum pollution monitoring project in the Wider Caribbean.Mar. Pollut. Bull. 18, 540-548.

CARIPOL (1980). CARIPOL Manual for Petroleum Pollution Monitoring. Atlantic Oceanographic and Meteorological Laboratory, 4301 Rickenbacker Causeway, Miami, FL 33149.

CARIPOL (1987). Proceedings of a Symposium on the Results of the CARIPOL Petroleum Pollution Monitoring Project. Caribbean Journal of Sci., 23.

Civili, F.S. and L. Jeftić (1985). The Co-ordinated Programme for Research and Monitoring of Pollution in the Mediterranean - Third International Symposium on Environmental Pollution and its Impact of Life in the Mediterranean Region. Istanbul.

Gwynne, M.D. (1982). the Global Environment Monitoring System (GEMS) of UNEP, Environmental Conservation, 9, 35-41.

Helsinki Commission (1983). Guidelines for the Baltic Monitoring Programme for the Second Stage, Baltic Sea Environment Proceedings No. 12, Helsinki Commission.

T. Melvasalo (1986). Baltic Monitoring Programme of the Baltic Marine Environment Protection Commission. Helsinki Commission.

IOC (1986). Report of the IOC Scientific Committee for the Global Investigation of Pollution in the Marine Environment., Sixth Session, Paris.

Kullenberg, G. (1986). The Health of the Oceans and the Need for its Monitoring. Environmental Monitoring and Assessment, 7, 47-58.

Kullenberg, G., editor (1986). The Role of the Oceans as a Waste Disposal Option. NATO ASI Series C, Mathematical and physical sciences, Vol. 172, D. Reidel Publishing Company, Dordrecht, Holland, 725 pp.

Kullenberg, G. et al. (1986). The IOC Programme on Marine Pollution, Mar. Pollut. Bull., 17, 341-352.

Levy, E., M.M. Ehrhardt, D. Kohnke, E. Sobtchenko, T. Suzuoki and A. Tokuhiro (1981). Global Oil Pollution, Results of MAPMOPP, the IGOSS Pilot Project on Marine Pollution (Petroleum) Monitoring. IOC Unesco.

Østerberg, C. (1986). Basic factors affecting the land/freshwater versus the sea option for waste disposal. In Kullenberg G., editor (1986), pp. 39-54.

Pawlak, J. (1986). ICES Monitoring Programmes on Contaminants in the Marine Environment, ICES.

UNEP (1983). Long-Term Programme for Pollution Monitoring and Research in the Mediterranean (MED POL) - Phase II, UNEP Regional Seas Reports and Studies No. 28.

Topping, G. (1976). <u>Sewage and the Sea.</u> In Marine Pollution, R. Johnston, editor, pp. 303-351. Academic Press, London, New York, San Francisco.

Unesco (1976). <u>A Comprehensive Plan for the Global Investigation of</u> <u>Pollution in the Marine Environment and Baseline Study Guidelines</u>. IOC Technical Series, No.14, 42 pp.

Unesco/IOC (1984). <u>A framework for the implementation of the Comprehensive</u> <u>Plan for the Global Investigation of Pollution in the Marine Environment</u>. IOC Technical Series No.25. 28 pp.

Wood, P.C. (1986). <u>Sewage sludge disposal options</u>. In Kullenberg G, editor (1986), pp. 111-124.

Index

Acids:
    chlorophenoxy, 105, 108, 123
    humic and fulvic, 108, 110, 113, 242, 243
    fatty, 235, 238, 244, 246, 248
Action Plans, 290, 347, 356
Absorbents, 230, 233, 241–243, 252–253
Absorption mechanisms, 99, 107, 242, 243, 252, 253
Aerosols (see atmospheric pollution), 234, 235, 255, 256, 258
Aldrin, 244, 246, 248
Aliphatic hydrocarbons, 117, 118, 168, 185, 189, 190, 196, 197, 231, 233, 235–238, 240, 244, 246, 248, 250, 267, 268
    biodegradation, 117
Alkylbenzenes, 20, 112, 119, 121, 240
Amberlite, 167, 242
Analytical:
    methods, 179, 188, 223, 241, 251, 253–265, 349
    techniques, 118, 201–217, 231, 233, 267
    sensitivity, 202, 204, 206, 210–217
Anodic stripping voltrametry, 215, 217, 265
Antarctica, 21, 22, 154
Arabian Gulf, 177, 181
Aromatic hydrocarbons, 106, 107, 142, 157, 158, 169, 206, 213, 225, 233–236, 240, 250, 260, 261, 267, 268, 350
    biodegradation, 118–121
    detoxication, 134, 137
    photooxidation, 112, 113
Arsenic, 244, 257, 264

Asbestos, 157
Asphaltenes, 111, 113
Assimilative capacity, 10, 18, 283, 317, 324–327, 338
Atlantic ocean, 314–316, 352
Atmospheric:
    exchanges, 9, 10, 16, 17, 28, 107, 155, 160, 161
    pollutants, 229–231, 233, 235, 285
    sampling, 230, 231
Atomic spectroscopy:
    absorption, 214–216, 231, 260, 262–264
    emission, 214
    fluorescence, 216
Atrazine, 113

Bacteria, 114–119, 121, 122, 124, 161, 224
Baltic sea, 23, 98, 165, 166, 286, 288, 354
Barcelona Convention, 355
Benzo(a)pyrene, 235, 236, 261
Beryllium, 257, 264
Bioconcentration, 99, 159, 162–164, 294
Biodegradation, 6, 115, 117–122
Biogeochemical cycles, 5, 15, 98
Bioindicators (see mussel watch), 162–164, 352
Biological effects, 13, 131–149
    reactions, 98, 99, 114
    responses, 131, 133, 137, 140, 143–145, 147, 293–301
Birds, 13, 19, 23, 160
Bivalves (see mussels), 162–165, 185–187, 196–199, 352, 353

Cadmium, 23, 191, 193, 197, 244, 246,
    248, 250, 262, 265
California, 162, 163, 165, 166
Capitella Capitata, 163
Carbamates, 104, 105
Cariaco Trench, 98
Caribean Sea, 350
Caripol, 350
Chemical:
    markers, 19
    reactions, 97, 98, 101
China, 335
Chlordane, 162, 233
Chlorinated compounds, 17, 21, 104, 105,
    110, 122, 123, 156, 157, 162, 225,
    231, 233, 236, 240,244, 246, 248,
    250, 255, 261, 293, 348, 352, 355
Chlorfluoromethanes, 155, 231
Chromatography (see gas- and liquid-)
Chromium, 244, 246, 248, 250, 262
Chromophores, 108
Clean Water Act, 18
Coastal zones, 8, 18, 21, 23, 108, 153,
    155, 162, 185, 279-285, 347
Cobalt, 244, 246, 262
Conventions:
    Barcelona, 355
    London, 307, 314, 315
    Oslo, 353
    Paris, 353
Copper, 103, 194, 196, 197, 244, 246,
    248, 250, 262, 265
Coprostanol, 20
Cycles (see pollutants)
Cytochrome P-450, 134, 135

Danish waters, 286-292, 295
DDE, 15, 16, 105, 164, 165, 233, 236,
    244, 246, 248, 250
DDT 1, 16, 20, 24, 103, 105, 154, 163,
    236, 242, 244, 246, 248, 250
Degradation (see biodegradation, photooxi-
    dation)
Detectors, chromatographic, 204, 206, 207,
    210, 212
Detergents, 20, 119, 125, 259
Detoxication, 123, 134, 135
Dieldrin, 158, 233, 236
Diffusive fluxes, 36-38, 41-58, 76-82
Dioxins and dibenzofurans, 158, 159
Dispersion flow, 61, 65-76, 78, 83, 84
Dissolved matter, 4, 5, 98, 167-169, 237,
    241-245

Ecotoxicology (see biological effects), 131,
    293-298, 329, 332
Electroanalytical techniques, 217
Environmental:
    management (see monitoring), 10,
        279-285, 291, 292, 305,
        332-338
    quality objectives, 283, 284
    stress, 132, 137, 140, 143, 145, 148,
        149
Enzyme activity, 134, 135, 137
EPA, 312, 320, 324, 332
Escherichia coli, 163
Estuarine processes, 9, 82, 83, 108, 113
Eutrophication, 286-292
Extraction (see sample handling), 241, 252

FAO, 21
Fish, 23, 153, 154, 159, 178, 185, 187,
    196-199, 287, 293-301, 352, 353
Fluid motion, 35, 41-44, 61-76
Fluorescence (see photochemical processes),
    100, 101
Fluxes:
    diffusive, 36-38
    of particulate matter, 22, 76
    turbulent, 39-41
Fossil fuels (see petroleum), 20
Fugacity, 161
Fulvic acids, (see acids)

Gas chromatography, 188, 203-205, 231,
    233, 259, 261, 264
    infrared spectroscopy, 212
    mass spectrometry, 207-210
GEEP, 348
GEMS, 350
GEMSI, 348
GIPME, 8, 347

Helsinki Commission, 354
Herbicides, 103-106, 110
Hexachlorobenzene, 225, 233, 244, 246,
    248, 267, 268, 352
Hexachlorocyclohexanes, 16, 25, 155, 157,
    160, 225, 236, 244, 246, 248, 267,
    268, 352
Humic acids (see acids)

Hydrocarbons (*see* aliphatic, aromatic)
  biodegradation, 117–121
  chlorinated, 17, 21, 104, 105
  petroleum, 17, 110, 116, 142, 143,
     166, 180, 181, 253–255, 260,
     350
Hydrography, 288–290
Hyphenated techniques, 210–212

IAEA, 334
ICES, 352
IGOSS, 351
IMO, 307, 313
Inductively coupled plasma, 216
Infrared spectroscopy, 212
Insecticides, 106
Intercalibration exercises, 266, 349
Interfaces:
  air-sea, 237
  analytical, 206, 207, 210, 212
International programmes, 347–356
IOC, 347
Iron, 244, 246, 250, 263

Kenya, 160
Ketones, 112
Kuwait, 178

Laboratory studies, 98, 102, 112, 115, 146,
    297, 298, 300, 330, 332
Lead, 5, 7, 17, 18, 20, 23, 123, 166, 191,
    192, 197, 198, 231, 244, 246, 248,
    250, 263, 265
Lindane (*see* hexachlorocyclohexanes)
Liquid chromatography, 205–207, 233, 259
  -mass spectrometry, 210, 211
London Convention, 307, 314, 315
Los Angeles, 19, 24, 162
Luminescence spectroscopy, 213
Lysosomes, 137–139

Management (*see* risk, waste)
  coastal, 279–301
  ocean waste, 305–338
Manganese, 244, 246, 248, 250, 263
Marine:
  biota (*see* Fish), 185, 196–198, 249,
    254–257, 287, 352–355
  birds, 13, 19, 23, 162
  compartments, 2, 9, 10, 97, 114, 284

environment:
  management (*see* assimilative capac-
    ity), 279, 301
  physicochemical features, 2, 97, 114
  pollution, 11–14
  transformation of pollutants, 5,
    97–125
  transport processes, 5, 35, 177
  mammals, 13
  photochemistry (*see* photochemical pro-
    cesses), 100
  productivity, 3, 20
Mass balance, 8, 17, 348
Mass spectrometry, 207, 211
mediterranean Sea, 17, 19, 23, 189, 191,
    196, 198, 355
Mercury, 23, 154, 155, 197, 198, 231,
    244, 248, 250, 256, 257, 261
  methyl, 123, 231, 257, 260, 264
Methoxychlor, 104
Methyl iodide, 103, 107
Metals, 8, 9, 21, 23, 103, 106, 124, 135,
    158, 185, 191, 196–198, 225, 244,
    246, 248, 250, 293, 294, 348, 352,
    355
  analytical techniques, 188, 215, 231,
    256, 258, 260, 267, 268
  binding proteins, 135, 136
  organic interactions, 113, 135
  metabolism, 123, 136, 163
  methylation, 123, 124
Microbial metabolism (*see* biodegradation),
    115, 117–121, 123, 124
Microlayer:
  seawater, 237, 238
Microorganisms (*see* bacteria)
Models:
  environmental quality, 284
  transport, 44–82, 178
Molibdenum, 244
Monitoring, 153, 177, 186, 225, 284, 288,
    338, 347, 348–356
  programmes, 288, 291, 292, 347–356
Mussel watch, 23, 155, 164, 349, 353
Mussels (*see* bivalves), 19, 129, 134, 137,
    138, 140–148, 162–166, 176, 250,
    348

Netherlands coastal waters, 24
New Bedford, 24
New York Bight, 18, 153, 305, 328
Nickel, 244, 246, 248, 250, 263
Nitrites, 103, 107

Nitroarenes, 107, 157
Nitrogen oxides, 107
Nitrophenols, 125, 157
Nonylphenols, 20, 125
North Sea, 308, 354
Norway, 141, 143
Nutrients, 289, 290

Ocean dumping (see risk assessment),
    306–311, 356, 358
  Act, 319
  criteria, 312–318, 323, 324
  effects, 312, 318, 329
  policy, 313–317, 320, 321
  sites, 313–317
Oil spills, 4–6, 111, 117, 138, 177–181,
    350, 351
Oman Gulf, 178
Organic carbon matter, 3, 98, 166, 167
Organisms (see marine biota)
Organochlorine compounds (see chlorinated
    compounds)
Organometallic compounds, 231, 260
Organophosphorous compounds, 206, 211,
    240, 260
Oslo Commission, 353
Oxidation (see photochemical processes),
    99, 111, 113
Oxygen:
  consumption, 140, 141, 286, 289
  singlet, 110
Oxygenases, 115, 135
Oxygenated compounds, 110–113,
    117–121
Oysters (see bivalves), 178

Pacific Ocean, 16, 160
Paris Commission, 353
Particulate matter, 4, 5, 76–82, 108,
    167–169, 237, 245–247
PCBs, 21, 24, 103–105, 122, 123, 154,
    157, 162, 165, 236, 237, 242, 244,
    248, 250, 328
Peruvian Coast, 98
Pesticides, 24, 103, 110, 157, 211, 293,
    294
Petroleum hydrocarbons (see oil spills), 17,
    19, 24, 103, 116, 177–181, 189,
    225, 350, 351
Phenols, 20, 103–105, 118–123, 125, 157,
    236
Phosphates, 240

Photochemical processes, 99–113, 116,
    121
Phthalates, 158, 233, 240
Plastics, 25
Pollutants (see metals, hydrocarbons, chlo-
    rinated compounds, etc.)
  absorption (see absorbents)
  biological transformation, 114–125
  chemical transformation, 97–113, 226
  cycles, 15, 98, 123
  effects (see biological responses)
  organic, 24, 25, 293
  partitioning, 159
  priority, 155–158, 223, 281, 312
  properties, 159–169
  solubility, 159
  transport, 35–91, 160, 161, 181
Pollution:
  concepts, 1, 11, 12, 131, 347
  effects (see biological effects), 13,
      131–149, 312, 318
  prevention, 281, 282
  priority problems, 154
  research programmes, 347–406
Polyurethane foam, 167, 232, 242
Processes, 99, 322, 326
  coastal, 288, 289
  estuarine, 82, 83, 108, 113
  microbial, 6, 115–125
  oceanic, 322, 323
  photochemical, 6, 99–113, 116, 121
  transport, 35–91
Productivity, 3, 20, 289, 356
Proteins, 135, 140, 141

Radioactive wastes, 1, 20, 22, 335
Reaction:
  mechanisms, 103, 104, 107, 111–113
  rates, 105, 117–123
Redox potential, 98
Research programmes, 347–406
  approaches, 102, 112, 146
  strategies, 98
Residence times, 8, 18
Responses, biological, 133–149, 293–301
Risk assessment, 329–332, 348
  management, 332, 333

Sample handling, 251–253
Sampling, 223, 226, 354
  atmosphere, 230–236

Sampling (*Cont.*):
  organisms, 249
  sediment, 165, 247
  variability, 227
  volatiles, 239
  water, 166, 179, 188, 237
San Francisco Bay, 168
Saudi Arabia, 178–181
SCOPE, 15, 351
Scope for growth, 140–144
Sediments, 3, 114, 123, 159, 165, 178,
    180, 247, 248, 253, 255–258, 326
Selenium, 248, 264
Shrimp, 187, 196, 197
Solar radiation, 100, 102
Solubility, 159
Sources:
  continuous, 53–76
  instantaneous, 46–53
Spectroscopy (*see* atomic), 179, 206,
    212–214
Standards, emission, 280, 281
Statistical analysis, 146, 177, 227–229,
    269
Sterols, 20
Strategies:
  environment management, 279–285,
    321–329, 334,
  monitoring, 291, 292
  research, 98, 348
Sulfur compounds, 111, 231
Suspended matter, 76, 98, 108, 159, 167
Synthetic compounds, 24
Syria, 185–198

Tars, 178, 350
Tenax, 231, 233, 242
Thames estuary, 305
Tidal dispersion, 84, 87
Tin, 123
  alkyl, 260
Titanium, 246
Taxophene, 10, 16, 24, 155, 160
Toxicity (*see* biological effects and ecotox-
    icology)

Transport:
  atmospheric, 16, 17, 160, 161
  diffusive, 36
  estuarine, 82
  models, 44, 178
  pollutant, 35, 177, 181, 226, 336
  turbulent, 39, 76
Triazines, 104, 105
Trifuralin, 106
Turbulent motion, 39–41, 44–76
  of suspended matter, 76–82

UNEP, 336, 348, 335
Ureas, 105
UV-Spectroscopy, 179, 206, 213, 214, 260,
    261, 350

Vibrio sp., 114, 117, 122
Volatile compounds, 157, 160, 232, 233, 239

Wadden Sea, 13
Walvis Bay, 98
Wastes, 20, 103, 187, 290, 294, 301,
    306–311, 335
  concept, 357
  disposal, 309–321, 338, 356
  emission standards, 280, 281
  management, 305–338
Water column, 2, 100, 114, 116, 237, 288,
    289
  hydrocarbons, 5, 178–181, 189, 190,
    253, 254
  metals, 5, 191–196, 244, 245, 246,
    256–258, 261
  monitoring, 177–181, 189–196, 349
  organic pollutants, 238, 240–242, 244,
    255, 259
  phases (*see* dissolved and particulate
    matter), 237, 241, 245, 484
  properties, 244, 354
  sampling, 166, 167, 179, 188, 237

Zinc, 195–197, 244, 246, 250, 264, 265